MACMILLAN
DICTIONARY
OF

PHYSICS

MACMILLAN DICTIONARY OF
PHYSICS

M. P. LORD

MACMILLAN
REFERENCE
BOOKS

First published 1986 by
THE MACMILLAN PRESS LTD
London and Basingstoke

Associated companies in Auckland, Delhi, Dublin, Gabornoe, Hamburg, Harare, Hong Kong, Johannesburg, Kuala Lumpur, Lagos, Manzini, Melbourne, Mexico City, Nairobi, New York, Singapore, Tokyo

British Library Cataloguing in Publication Data

Lord, M. P.
  Macmillan dictionary of physics.
  1. Physics—Dictionaries
  I. Title
530′.03′21   QC5

ISBN 0-333-39066-0
ISBN 0-333-42377-1 Pbk

Photoset by Styleset Limited, Warminster, Wiltshire and printed in Great Britain

Contents

Preface

This Dictionary is an enlarged and updated version of the Physics content of a Dictionary of Physical Science edited by Dr John Daintith, Pan Books 1978.

It is hoped that the book, which contains over 4000 entries and about 200 line drawings, will be of use to Physics students at first year undergraduate and lower levels, and also to amateur enthusiasts and lay persons. The aim is to state essential principles of the subject clearly and concisely and to cover topics, having a Physics content, from other sciences, including medical science.

I wish to thank my husband, Professor Arthur M. James for many interesting discussions and much forbearance during the compilation.

I express my appreciation to the staff of Macmillan Reference Books for their help and encouragement and am indebted also to all the referees, especially Valerie Illingworth who, in addition, edited the manuscript.

Notes on Use

Many, but not all, of the head words defined in the dictionary are shown in small capitals when they occur in another entry; so, if a word, not appearing in small capitals causes difficulty, it could be helpful to try looking it up.

For ease of reference any term whose description appears within an entry is usually italicized. For example, the entry for **drift tube** is *See* LINEAR ACCELERATOR. In the entry **linear accelerator,** *drift tube* is defined.

See also in an entry refers the reader to an entry which has some relevance to the one being read.

Some units appear as headwords and also, in the tables at the front of the dictionary.

Table 1. Base and supplementary SI units

Physical Quantity	Unit Name (base)	Unit Symbol
amount of substance	mole	mol
current	ampere	A
length	metre	m
luminous intensity	candela	cd
mass	kilogramme	kg
thermodynamic temperature	kelvin	K
time	second	s
	(supplementary)	
plane angle	radian	rad
solid angle	steradian	sr

Table 2. Derived SI units with special names

Quantity Name	Symbol	Unit	Unit Name	Unit Symbol
capacitance	C	$A^2 \ s^4 \ kg^{-1} \ m^{-2}$	farad	F
charge	Q	$A \ s$	coulomb	C
conductance	G	$A^2 \ s^3 \ kg^{-1} \ m^{-2}$	siemens	S
energy	E	$kg \ s^{-2}$	joule	J
force	F	$kg \ m \ s^{-2}$	newton	N
frequency	ν, f	s^{-1}	hertz	Hz
illuminance	E_v, E	$cd \ sr \ m^{-2}$	lux	lx
inductance, self	L	$kg \ m^2 \ s^{-2} \ A^{-2}$	henry	H
inductance, mutual	M	$kg \ m^2 \ s^{-2} \ A^{-2}$	henry	H
luminous flux	Φ_v, Φ	$cd \ sr$	lumen	lm
magnetic flux	Ψ	$kg \ m^2 \ s^{-2} \ A^{-1}$	weber	Wb
magnetic induction	B	$kg \ s^{-2} \ A^{-1}$	tesla	T
potential difference	V	$kg \ m^2 \ s^{-3} \ A^{-1}$	volt	V
power	P	$kg \ m^2 \ s^{-3}$	watt	W
pressure	p	$kg \ m^{-1} \ s^{-2}$	pascal	Pa
resistance	R, r	$kg \ m^2 \ s^{-3} \ A^{-2}$	ohm	Ω

Table 3. Prefixes used with SI units

Factor	Prefix	Symbol
10^{-1}	deci-	d
10^{-2}	centi-	c
10^{-3}	milli-	m
10^{-6}	micro-	μ
10^{-9}	nano-	n
10^{-12}	pico-	p
10^{-15}	femto-	f
10^{-18}	atto-	a
10	deca-	da
10^2	hecto-	h
10^3	kilo-	k
10^6	mega-	M
10^9	giga-	G
10^{12}	tera-	T
10^{15}	peta-	P
10^{18}	exa-	E

Table 4. Symbols and SI Units for some physical quantities

Name of quantity	Symbol	SI unit
absorptance	α	1
acceleration	a	m s^{-2}
angular frequency	ω	Hz $\equiv$ s^{-1}
angular momentum	L	J s
angular velocity	ω	rad s^{-1}
area	A, S	m^2
atomic number, proton number	Z	1
Bohr magneton	μ_B	A m^2
bulk modulus	K	N m^{-2}
charge, volume density of	ρ	C m^{-3}
compressibility	κ, k	N^{-1} m^2
cross section	σ	m^2
Curie temperature	θ_C, T_C	K
decay constant	λ	s^{-1}
degeneracy (multiplicity) of an energy level	g	1
density	ρ	kg m^{-3}
Dirac constant	$\hbar$	J s

Table 4 (*contd*)

Name of quantity	Symbol	SI Unit
efficiency	η	1
electric displacement	D	$C\ m^{-2}$
electric field strength	E	$V\ m^{-1}$
electric flux	ψ	C
electric polarization	P	$C\ m^{-2}$
electric susceptibility	χ_e	1
electromotive force	E	V
emissivity	ε	1
enthalpy	H	J
entropy	S	$J\ K^{-1}$
Gibbs function	G	J
half life	$T_{1/2}, t_{1/2}$	s
Hamiltonian function	H	J
heat capacity: at constant pressure	C_P	$J\ K^{-1}$
heat capacity: at constant volume	C_V	$J\ K^{-1}$
Helmholtz function	A, F	J
impedance, electric	Z	Ω
internal energy	U, E	J
irradiance	E_e, E	$J\ m^{-2}$
Joule-Kelvin coefficient	μ	$N^{-1}\ m^2\ K$
kinematic viscosity	ν	$m^2\ s^{-1}$
kinetic energy	T, E_k, K	J
Lagrangian function	L	J
linear absorption coefficient	α	m^{-1}
linear attenuation (extinction) coefficient	μ	m^{-1}
linear expansivity	α	K^{-1}
luminance	L_v, L	$cd\ m^{-2}$
luminous emittance	M_v, M	$lm\ m^{-2}$
magnetic field strength	H	$A\ m^{-1}$
magnetic moment	m	$A\ m^2$
magnetic quantum number	M, m_i	1
magnetic susceptibility	χ, χ_m	1
magnetization	M	$A\ m^{-1}$
magnetomotive force	F_m	A

Table 4 (*contd*)

Name of quantity	Symbol	SI Unit
mass number, nucleon number	A	1
mean free path	λ, l	m
mean life	τ	s
moment of inertia	l	kg m^2
momentum	p	N s
Néel temperature	θ_N, T_N	K
neutron number	N	1
nuclear magneton	μ_N	A m^2
nuclear radius	r	m
nuclear spin quantum number	I, J	1
osmotic pressure	Π	N m^{-2}
packing fraction	f	1
period	T	s
permeability, absolute	μ	H m^{-1}
permittivity, absolute	ε	F m^{-1}
Planck function	Y	J K^{-1}
polarizability	α, γ	C m^2 V^{-1}
potential energy	E_p, V, Φ	J
principal quantum number	n, n_i	1
quantity of heat	q, Q	J
radiance	L_e, L	W m^{-2} sr^{-1}
radiant excitance	M_e, M	W m^{-2}
radiant flux, radiant power	Φ_e, Φ	W
radiant intensity	I_e, I	W sr^{-1}
radioactivity	A	Bq*
reactance	X	Ω
reflection factor	ρ	1
refractive index	n	1
relative atomic mass	A_r	1
relative density	d	1
relative permeability	μ_r	1
relative permittivity (dielectric constant)	ε_r	1
relaxation time	τ	s
resistivity	ρ	Ω m
Reynolds number	(Re)	1

*See RADIATION UNITS

Table 4 (*contd*)

Name of quantity	Symbol	SI Unit
shear modulus	G	$N\ m^{-2}$
specific heat capacity: at constant pressure	c_p	$J\ kg^{-1}\ K^{-1}$
specific heat capacity: at constant volume	c_v	$J\ kg^{-1}\ K^{-1}$
specific optical rotatory power	α_m	$rad\ m^2\ kg^{-1}$
specific volume	v	$m^3\ kg^{-1}$
speed	u	$m\ s^{-1}$
spin quantum number	S, s	1
surface tension	γ, σ	$N\ m^{-1}$
susceptance	B	S
thermal conductivity	λ	$W\ m^{-1}\ K^{-1}$
thermal diffusivity	α	$m^2\ s^{-1}$
torque	T	$N\ m$
transmittance	τ	1
viscosity, coefficient of	η	$kg\ m^{-1}\ s^{-1}$
volume	V, v	m^3
wavelength	λ	m
wave number	σ	m^{-1}
weight	G	N
work	w, W	J
work function	Φ	J
Young's modulus	E	$N\ m^{-2}$

Table 5. Fundamental constants in SI units

Constant	Symbol	Value
acceleration due to gravity (standard value)	g	$9.806\,65$ m s^{-2}
Avogadro constant	N_A or L	$6.022\,169 \times 10^{23}$ mol^{-1}
Boltzmann constant	k	$1.380\,622 \times 10^{-23}$ J K^{-1}
charge on electron	e	$-1.602\,1917 \times 10^{-19}$ C
charge on proton	$-e$	$1.602\,1917 \times 10^{-19}$ C
electric constant	ε_0	$8.854\,1853 \times 10^{-12}$ F m^{-1}
electronic radius	r	$2.817\,939 \times 10^{-15}$ m
Faraday constant	F	$9.648\,670 \times 10^4$ C mol^{-1}
fine structure constant	α	$7.297\,351 \times 10^{-3}$
gravitational constant	G	6.6732×10^{-11} N m^2 kg^{-2}
Loschmidt's number	N	$2.687\,19 \times 10^{25}$ m^{-3}
magnetic constant	μ_0	$1.256\,64 \times 10^{-6}$ H m^{-1}
Planck constant	h	$6.626\,196 \times 10^{-34}$ J s
rest mass of electron	m_e	9.1095×10^{-31} kg
rest mass of neutron	m_n	$1.674\,92 \times 10^{-27}$ kg
rest mass of proton	m_p	$1.672\,62 \times 10^{-27}$ kg
Rydberg constant	R	$1.097\,373\,12 \times 10^7$ m^{-1}
speed of light	c	$2.997\,924\,58 \times 10^8$ m s^{-1}
Stefan's constant	σ	5.6697×10^{-8} W m^{-2} K^{-4}
universal gas constant	R	$8.314\,35$ J K^{-1} mol^{-1}

Table 6. Conversion factors for some other units

A

Length	m	cm	in	ft	yd
1 metre	1	100	39.3701	3.280 84	1.093 61
1 centimetre	0.01	1	0.393 701	0.032 808 4	0.010 936 1
1 inch	0.0254	2.54	1	0.083 333 3	0.027 777 8
1 foot	0.3048	30.48	12	1	0.333 333
1 yard	0.9144	91.44	36	3	1

	km	mile	n. mile
1 kilometre	1	0.621 371	0.539 957
1 mile	1.609 34	1	0.868 976
1 nautical mile	1.852 00	1.150 78	1

1 light year = $9.460\,70 \times 10^{15}$ metres = $5.878\,48 \times 10^{12}$ miles.
1 astronomical unit = 1.496×10^{11} metres.
1 parsec = 3.0857×10^{16} metres = 3.2616 light years.
1 fathom = 6 feet.

B

Volume

1 minim	1/9600 pint	
1 fluid drachm	60 minims	
1 fluid ounce	8 fluid drachms	
1 gill	5 fluid ounces	
1 pint	4 gills	0.568 26 dm³
1 quart	2 pints	
1 gallon	4 quarts	4.546 09 dm³

1 gallon	0.160 544 cubic feet	
1 cubic foot	6.228 82 gallons	0.028 316 8 m³

1 cubic decimetre (1000 cc) is 1 *litre*, although this name is not used
in precision measurement.
1 gallon (US) is equivalent to 0.83268 gallon (UK)

C

Velocity	m s⁻¹	km h⁻¹	mile h⁻¹	ft s⁻¹
1 metre per second	1	3.6	2.236 94	3.280 84
1 kilometre per hour	0.277 778	1	0.621 371	0.911 346
1 mile per hour	0.447 04	1.609 344	1	1.466 67
1 foot per second	0.3048	1.097 28	0.681 817	1

1 knot = 1 nautical mile per hour = 0.514 444 metre per second.

Table 6 (*contd*)

D

Mass	kg	g	lb	long ton
1 kilogramme	1	1000	2.204 62	$9.842\,07 \times 10^{-4}$
1 gramme	10^{-3}	1	$2.204\,62 \times 10^{-3}$	$9.842\,07 \times 10^{-7}$
1 pound	0.453 592	453.592	1	$4.464\,29 \times 10^{-4}$
1 long ton	1016.047	$1.016\,047 \times 10^{6}$	2240	1

Apothecaries' Units of Mass

1 grain	1/7000 pound (avoirdupois)
1 scruple	20 grains
1 drachm	3 scruples
1 ounce (apoth)	8 drachms
1 pound (apoth)	12 ounces (apoth)

The grain (1/7000 pound avoirdupois) has the same value in the avoirdupois, troy and apothercaries' systems.

Avoirdupois Units of Mass

1 grain	1/7000 pound	0.064 799 g
1 dram	1/256 pound	
1 ounce	16 drams	
1 pound	16 ounces	0.453 592 kg
1 stone	14 pounds	
1 quarter	2 stones	
1 hundredweight	4 quarters	50.802 kg
1 ton	20 hundredweights	1016.047 kg

1 gramme	0.0353 ounces
1 kilogramme	2.204 62 pounds

The avoirdupois hundredweight and ton are sometimes called the *long hundred-weight* and *long ton* to distinguish them from the US measures, the *short hundred-weight* (100 pounds) and *short ton* (2000 pounds).

Troy Units of Mass

1 grain	1/7000 pound (avoirdupois)
1 carat	4 grains
1 pennyweight	6 carats
1 ounce (tr)	20 pennyweights
1 pound (tr)	12 ounces (tr)
1 hundredweight (tr)	100 pounds (tr)
1 ton (tr)	20 hundredweights (tr)

Table 6 (*contd*)

E

Force	N	kg	dyne	poundal	lb
1 newton	1	0.101 972	10^5	7.233 00	0.224 809
1 kilogramme force	9.806 65	1	$9.806\ 65 \times 10^5$	70.9316	2.204 62
1 dyne	10^{-5}	$1.019\ 72 \times 10^{-6}$	1	$7.233\ 00 \times 10^{-5}$	$2.248\ 09 \times 10^{-6}$
1 poundal	0.138 255	$1.409\ 81 \times 10^{-2}$	$1.382\ 55 \times 10^4$	1	0.031 081
1 pound force	4.448 22	0.453 592	$4.448\ 23 \times 10^5$	32.174	1

F

Pressure	N/m²	kg/cm²	lb/in²	atm
1 newton per square metre	1	$1.019\ 72 \times 10^{-5}$	$1.450\ 38 \times 10^{-4}$	$9.869\ 23 \times 10^{-6}$
1 kilogramme per square centimetre	980.655×10^2	1	14.2234	0.967 841
1 pound per square inch	$6.894\ 76 \times 10^3$	0.070 306 8	1	0.068 046
1 atmosphere	$1.013\ 25 \times 10^5$	1.033 23	14.6959	1

1 newton per square metre = 10 dynes per square centimetre.
1 bar = 10^5 newtons per square metre = 0.986 923 atmosphere.
1 torr = 133.322 newtons per square metre = 1/760 atmosphere.
1 atmosphere = 760 mmHg = 29.92 in Hg = 33.90 ft water (all at 0° C).

G

Work and Energy	J	cal$_{IT}$	kW hr	btu$_{IT}$
1 joule	1	0.238 846	$2.777\ 78 \times 10^{-7}$	$9.478\ 13 \times 10^{-4}$
1 calorie (IT)	4.1868	1	$1.163\ 00 \times 10^{-6}$	$3.968\ 31 \times 10^{-3}$
1 kilowatt hour	3.6×10^6	$8.598\ 45 \times 10^5$	1	3412.14
1 British thermal unit (IT)	1055.06	251.997	$2.930\ 71 \times 10^{-4}$	1

1 joule = 1 newton metre = 1 watt second = 10^7 erg = 0.737 561 ft lb.
1 electronvolt = $1.602\ 10 \times 10^{-19}$ joule.

Table 7. The Greek Alphabet

Letters		Name
Capital	Small	
A	α	alpha
B	β	beta
Γ	γ	gamma
Δ	δ	delta
E	ε	epsilon
Z	ζ	zeta
H	η	eta
Θ	θ	theta
I	ι	iota
K	κ	kappa
Λ	λ	lambda
M	μ	mu
N	ν	nu
Ξ	ξ	xi
O	ο	omicron
Π	π	pi
P	ρ	rho
Σ	σ	sigma
T	τ	tau
Υ	υ	upsilon
Φ	φ	phi
X	χ	chi
Ψ	ψ	psi
Ω	ω	omega

A

ab. A prefix used to denote electromagnetic CGS UNITS.

Abbé condenser. A simple two-lens system used as a compound MICROSCOPE condenser.

A bomb. *See* NUCLEAR WEAPON.

aberration. (1) A defect in an optical system producing distortion of the image. The various types of aberration are ASTIGMATISM, CHROMATIC ABERRATION, COMA, CURVATURE OF FIELD, optical DISTORTION and SPHERICAL ABBERATION.
(2) The apparent displacement in a star's position resulting from the Earth's orbital motion.

abrasive. A substance used for wearing away a solid surface as in polishing or cleaning. Common examples are emery and pumice.

absicissa. The *x* co-ordinate of a point on a two-dimensional Cartesian graph, i.e. the distance of a point from the *y* axis, measured along the *x* axis from the origin. *Compare* ORDINATE.

absolute coefficient of expansion. *See* COEFFICIENT OF EXPANSION.

absolute current measurement. *See* AMPÈRE BALANCE.

absolute permeability. *See* PERMEABILITY (def. 1).

absolute permittivity. *See* PERMITTIVITY (def. 1).

absolute pressure. The actual pressure at a given point due to all causes. The absolute pressure at a point in a liquid is thus the sum of the atmospheric pressure and the pressure due to the supernatant liquid.

absolute refractive index. *See* REFRACTIVE INDEX.

absolute temperature. Another name for THERMODYNAMIC TEMPERATURE.

absolute zero. The zero point of thermodynamic temperature, i.e. 0 K. *See also* ZERO POINT ENERGY.

absorbed dose. *See* DOSE (def. 1).

absorber. Any body which absorbs radiation.

absorptance. Symbol α. The ratio of the FLUX absorbed by a body to the flux incident on it.

absorption. The conversion of all or part of the energy incident on a substance into some other form of energy within the substance; for example, part of the energy of an incident beam of light may be used in exciting the atoms or molecules of the absorbing substance.

absorption coefficient. Symbol α. A quantity defined by the equation

$$I = I_0 e^{-\alpha x}$$

where I_0 and I are respectively the incident and transmitted radiation intensities and x is the thickness of the matter traversed. The quantity x may be expressed as a length, or a mass per unit area, or in moles per unit area or in atoms per unit area; the corresponding absorption coefficients are respectively described as linear, mass, molar and atomic.

absorption spectrum. A SPECTRUM produced by the absorption of electromagnetic radiation by matter. To obtain an absorption spectrum, incident radiation with a continuous range of wavelengths is used. The beam emerging from the sample is dispersed using a PRISM or DIFFRACTION GRATING. It is found that at some wavelengths radiation has been absorbed; the energy at these wavelengths has been used to raise the energy of atoms or molecules of the absorbing medium to higher levels. Thus when visible radiation is incident, the absorption spectrum consists of dark lines or bands on a bright continuous background. *See also* BOHR THEORY: FRAUNHOFER LINES. *Compare* EMISSION SPECTRUM.

absorptivity. Former name for ABSORPTANCE.

abundance. (1) Symbol C. The ratio of the number of atoms of a given ISOTOPE to the total number of atoms in a mixture of isotopes. Abundance for a naturally occurring isotopic mixture of an element is known as *natural abundance.*
(2) The concentration of a specified substance in the Earth's crust (sometimes called *terrestrial abundance*) or in the universe (sometimes called *cosmic abundance*).

ac Abbrev. for ALTERNATING CURRENT.

acceleration. Rate of change of velocity with time.

acceleration due to gravity. Symbol g. The acceleration of a body falling freely, without air resistance or buoyancy, at a particular point close to the Earth's surface. It is caused by the gravitational attraction of the Earth and varies with latitude because the Earth is not completely spherical and also because of the Earth's rotation. At the poles g is 9.832 metre per second^2 (m s^{-2}); at the equator it is 9.78 m s^{-2}; the Earth's rotation accounts for 0.034 m s^{-2} of the difference. The standard value of g (symbol g_n) is taken as 9.806 65 m s^{-2}

acceleration of free fall. Another name for ACCELERATION DUE TO GRAVITY.

acceleration vector. A line representing acceleration in magnitude and direction.

accelerator. A machine for increasing the kinetic energy of charged ions and ELEMENTARY PARTICLES. Its main use is in fundamental research in nuclear and particle physics and in radiotherapy. *See also* BETATRON; COCKCROFT-WALTON GENERATOR; CYCLOTRON; ELECTROSTATIC GENERATOR; LINEAR ACCELERATOR; SYNCHROCYCLOTRON; SYNCHROTRON; VAN DE GRAAFF GENERATOR.

acceptor. An atom in an extrinsic SEMICONDUCTOR that accepts an electron, thus producing a positive HOLE in the conduction band. *See* BAND THEORY.

access time. The time necessary for information to be supplied from a computer store for processing.

accommodation. *See* EYE.

accumulator. A collection of series-connected secondary cells. The total amount of electricity stored is called the *charge* on an accumulator; it is usually measured in ampere-hour. *See* CELL (def. 1). *See also* LEAD ACID BATTERY; NIFE CELL.

accuracy. The closeness of agreement between an experimentally determined value and the true value. It is affected by operator mistakes, instrument errors and random errors. Care and patience will elimate the first; the second is inherent: the limitations of any instrument should always be considered; the third can be minimized by taking as large a number of measurements as practicable.

achromat. Another name for ACHROMATIC LENS.

achromatic colour. A colour without hue, ie. a grey.

achromatic lens. A compound lens consisting of a biconvex crown glass lens cemented by Canada balsam to a plano-concave flint glass lens so as to form a planoconvex combination. For the combination to image red and blue rays from an axial object point at the same axial image point it is necessary that

$$f_F/f_C = -\omega_F/\omega_C$$

where f and ω refer to FOCAL LENGTH and DISPERSIVE POWER respectively, and suffixes C and F to crown and flint glass respectively. See also ACHROMATIC SEPARATED LENSES.

achromatic prism. A compound prism which eliminates DISPERSION between two colours, usually red and blue. For incident white light dispersion of the other colours in the light remains, but most of the colour effect is eliminated since blue and red are at the two extremities of the visible spectrum. Two small angle prisms, one made of crown and the other of flint glass, are in contact as illustrated in fig. A1. The condition for no red-blue dispersion by the combination (i.e. for blue and red rays to emerge parallel as shown) is

$$(n_{CB} - n_{CR})A_C = (n_{FB} - n_{FR})A_F$$

where A and n refer to prism angle and refractive index respectively, suffixes F and C to flint and crown glass respectively and suffixes B and R to blue and red light respectively.

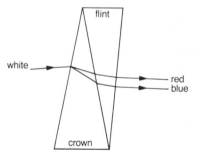

A1 Achromatic prism

achromatic separated lenses. When two convex lenses of the same material and of focal lengths f_1 and f_2 respectively are a distance d apart, the combination will be achromatic provided

$$d = (f_1 + f_2)/2$$

aclinic line. Another name for MAGNETIC EQUATOR.

acoustic impedance. See IMPEDANCE (def. 2).

acoustic measurements. The determination of the variation in the pressure, density, temperature and particle position of a medium transmitting SOUND, the rate of energy transmission, absorption and dissipation in the medium and the SPEED OF SOUND in it.

acoustics. (1) The study of SOUND. *Linear acoustics* is concerned with small-amplitude sounds, i.e. sounds below the pain threshold. *Nonlinear acoustics* is concerned with large-amplitude sounds, for example sounds resulting from explosive and shock waves. *Physiological acoustics* is the analytical assessment of the reception of sound by the ear, and the processing of the resulting signals by the nervous system.
(2) The characteristics of a room or building, determining the fidelity with which music and other sounds can be heard within it.

acoustoelectric effect. The appearance of an electrostatic field when an acoustic wave passes through a medium containing mobile electric charges.

actinic light. Light of a wavelength capable of producing a photographic effect.

actinide element. An element with ATOMIC NUMBER in the range 89 to 103.

actinometer. A RADIOMETER mainly used to measure solar and terrestrial radiation.

actinometry. (1) The measurement of light absorbed by a system during a photochemical change.
(2) The measurement of the energy of solar radiation.

action. The product of momentum and distance.

action potential. An electric impulse passing along a NEURON. After its passage there is a short time interval before another impulse can pass.

action reaction pair. The equal and opposite forces which bodies exert on one another, in accordance with the third of NEWTON'S LAWS OF MOTION.

activate. (1) To supply sufficient energy to an atom or molecule to make it reactive.
(2) To make a material radioactive, for example by neutron bombardment.

activation analysis. A method of chemical analysis depending on the identification and measurement of radioactive isotopes usually formed by irradiating the sample in a nuclear reactor. The technique is very sensitive.

activation energy. The energy required by an atomic or molecular system to enable a particular process to occur, examples are the energy required to raise an electron to the conduction band and the energy needed by a molecule to participate in a chemical reaction.

activity. Symbol A. The number of nuclear disintegrations occurring in a given radioactive substance in unit time.

acute angle. An angle whose value lies between 0° and 90°.

addend. One of a set of numbers to be added.

additive colorimetry. See COLORIMETRY.

additive process. A process in which a coloured light is produced by direct combination of suitable quantities of lights of different colours. See also COLORIMETRY. Compare SUBTRACTIVE PROCESS.

adiabatic approximation. An approximation used when the HAMILTONIAN of a quantum system varies slowly with time.

adiabatic bulk modulus. BULK MODULUS of elasticity when no heat enters or leaves the system, equal to the product of the

RATIO OF SPECIFIC HEATS and the pressure of a gas.

adiabatic curve. A plot of pressure against volume for a given mass of gas undergoing an ADIABATIC PROCESS. The curve for 1 kilogramme of air is illustrated (fig. A2), and for comparison an ISOTHERMAL for 273 K and one for 473 K are also shown. It can be seen that the adiabatic through a point is always steeper than the isothermal.

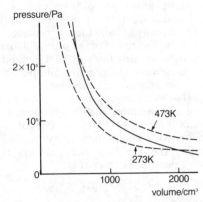

A2 Adiabatic curve

adiabatic demagnetization. A technique for producing temperatures close to absolute zero by
(a) demagnetizing a paramagnetic salt: the magnetized salt is first cooled to liquid-helium temperature and then isolated thermally from its surroundings; on removal of the magnetizing field, the sample becomes demagnetized and cools still further. Temperatures as low as 10^{-3} K have been obtained.
(b) demagnetizing a substance with nuclear magnetic moments: the nuclear spins are magnetically aligned isothermally at a temperature of about 10^{-2} K and then adiabatically demagnetized. Temperatures as low as 10^{-6} k are anticipated.

adiabatic equation. The equation

$$p\,V^{\gamma} = \text{a constant}$$

where p and V are respectively the pressure and volume of a gas undergoing an

ADIABATIC PROCESS and γ is the RATIO OF SPECIFIC HEATS of the gas.

adiabatic process. A process in which no heat enters or leaves the system. Usually the temperature changes. For example, in the adiabatic compression of a gas, work is done on the system and the gas temperature rises. *Compare* ISOTHERMAL PROCESS.

admittance. Symbol Y. The reciprocal of electric IMPEDANCE.

adsorption. The formation of a layer of substance, usually gas, on a solid surface. In contrast to ABSORPTION, there is no penetration of the solid. If the layer is held by covalent bonds, it is one atom or one molecule thick and the process is known as *chemisorption*. If the binding is by van der Waals forces, the layer may have a thickness of several atoms or molecules and the process is known as *physisorption*.

aerial. A conductor or series of conductors, usually comprised of wire. It is sited in an elevated position. The purpose of an aerial is to transmit or receive radio waves: in transmission, electromagnetic waves are produced by accelerating charges in the aerial; in reception, electromagnetic radiation induces small varying currents in the aerial. Many types of aerial exist, suitable for use in particular frequency ranges. The *dipole aerial*, for example, is used at frequencies below about 30 megahertz. It is a straight conductor whose length is usually half a wavelength, but may be a whole wavelength. Connection is made at the centre point.

aerial array. A system of two or more coupled AERIALS, so orientated as to have particular directional radiating or receiving properties.

aerodynamics. The study of gases in motion, expecially the relative motion of solid objects and gases, for example the movement of bodies in air.

aerofoil. A body which when moving through a fluid experiences a lift force much greater than the drag force; the most efficient shape is shown in fig. A3. Due to viscosity there is a velocity difference between the fluid layers reaching the trailing edge B via the upper and lower surface routes; this results in a vortex at the trailing edge accompanied by counter circulation round the aerofoil which produces the lift force. The flight of an aeroplane depends on the use of aerofoils for wing and tail structures.

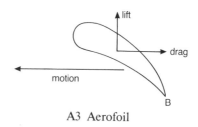

A3 Aerofoil

aerosol. A fine dispersion of liquid or solid particles in a gas; particle radii lie in the range 1 nanometre to 50 micrometre. Examples are hazes, mists, fogs, clouds, smokes, dusts, living bacteria, viruses and moulds. Aerosols feature in atmospheric electricity, cloud formation, precipitation processes, atmospheric chemistry, air pollution, visibility and radiation transfer. Commercial aerosols are used for many purposes and are produced by using a gas, often chlorofluoromethane, under pressure to disperse liquids or solids into the atmosphere. The widespread use of chlorofluoromethane has caused anxiety about its accumulation in the stratosphere; this could produce a reduction of ozone, leading to more solar ultraviolet radiation at the Earth's surface.

aether   Former spelling of ETHER.

afterglow. Another name for PERSISTENCE.

aggregation. A cluster or group of particles held together in a gas or liquid by INTERMOLECULAR FORCE.

agonic line. A line on the Earth's surface joining points of zero magnetic declination, i.e. points at which a compass indicates true north. The two main agonic lines are one passing through America and another following an irregular path through Eastern Europe, Arabia, Asia and Australia.

air. A mixture of gases which, when dry, normally has the following composition by volume:

nitrogen	78.08 %
oxygen	20.94 %
argon	0.9325 %
carbon dioxide	0.03 %
neon	0.0018 %
helium	0.0005 %
krypton	0.0001 %
xenon	0.000 009 %
radon	6×10^{-18} %

Values of other properties of dry air are:
   specific heat capacity
       at constant volume 718 J kg^{-1} K^{-1}
       at constant pressure 1006 J kg^{-1} K^{-1}
   ratio of specific heat capacities 1.403

air breakdown. A phenomenon which usually occurs as a CORONA DISCHARGE at a point of a charged conductor where the electric field strength is highest.

air cell method. A method of measuring the REFRACTIVE INDEX of a liquid relative to air. The cell consists of two thin plane parallel glass plates cemented together so as to contain a film of air of constant thickness. The liquid under investigation is contained in a glass vessel with thin plane parallel sides. Light from an external monochromatic source passes through the container into the liquid, where it strikes the immersed air cell normally. The cell is then rotated about a vertical axis until no light is transmitted by it, i.e. until the light is totally internally reflected. The angle of rotation, θ, is measured on the scale provided; it can be shown that $1/\sin\theta$ is the required refractive index.

air wedge fringes. An example of INTERFERENCE of light waves. A thin wedge of air is illuminated by a monochromatic light source and the light reflected at the upper and lower glass slides, which contain the

wedge, is focused by the eye. The FRINGES are equally spaced and parallel to the wedge edge; if there are N bright bands in a distance l, then the wedge angle is given by

$$N\lambda/(2l)$$

where λ is the wavelength of the light used.

Airy disc. The image of a point source of light that would be produced by an ABERRATION—free optical system; the disc is a DIFFRACTION pattern and decreases in size as the aperture of the optical system increases.

albedo. The ratio of the amount of light scattered from a surface to the amount of incident light.

alcohol thermometer. A type of thermometer similar in construction to the MERCURY IN GLASS THERMOMETER except that alcohol, usually coloured for ease of viewing, replaces the mercury. Since alcohol freezes at $-130°$ C compared to $-39°$ C for mercury's freezing point, the alcohol thermometer is much more useful at low temperatures; alcohol boils at 78° C so that its upper range is limited compared to that of mercury, whose boiling point is 459° C.

Alfvén waves. Magnetohydrodynamic waves propagated through a PLASMA under certain conditions. The waves travel in the direction of the applied magnetic field and the plasma particles oscillate in a direction perpendicular to this.

algebra. A branch of mathematics in which variable quantities and numbers are represented by symbols. Relations between symbols are expressed by equations which are then arranged into convenient forms for solution according to a set of logical rules.

algebraic sum. The result of adding a set of numbers with due regard to sign; thus the algebraic sum of 15 and -10 is 5.

algorithm. A specific sequence of operations which will, in principle, yield the solution to a given problem. There is no limit to the length of the sequence.

alpha decay. A type of radioactive decay in which an unstable nucleus ejects an ALPHA PARTICLE, yielding a nucleus of MASS NUMBER 4 less than the parent's and of ATOMIC NUMBER 2 less than the parent's; for example radium-226, atomic number 88, decays to radon-222, atomic number 86.

alpha iron. A form of iron which is stable below 906° C; it is ferromagnetic up to its CURIE TEMPERATURE of 768° C.

alpha particle. A helium nucleus consisting of two neutrons and two protons; it is frequently emitted during radioactive decay.

α particle. Abbrev. for ALPHA PARTICLE.

alpha rays. A stream of ALPHA PARTICLES.

alpha ray spectrometry. The measurement of the distribution of either the KINETIC ENERGY or MOMENTUM of alpha particles from a specified radioactive nuclide. Usually the nuclide emits a series of single-energy groups of alpha particles, one of which is predominant. The resulting *alpha ray spectrum* therefore consists of 'lines' reflecting this energy distribution.

alpha wave. An outstanding feature obtained in ELECTROENCEPHALOGRAPHY. The wave frequency is in the range 5 hertz to 13 hertz, being lowest when the eyes are closed and highest when they are open; the wave shape is regular sinusoidal. The alpha wave amplitude is greatest near the occipital region which is associated with vision. Focusing the eyes at first suppresses the alpha wave, but later it returns. On sleeping the wave disappears, to be replaced by a wave of frequency about 1 hertz; any stimulus during sleep results in an alpha wave train.

alternating current. An electric current which periodically reverses its direction.

The commonest type has a sinusoidal wave form and is represented by the equation

$$I = I_0 \sin 2\pi f t$$

where I is the current at time t, I_0 the maximum current value and f the frequency of alternation. The alternating current passes through its cycle of values once in every period, over which the average value is zero.

alternating series. An infinite series whose terms are alternately positive and negative.

alternator. A device for producing an alternating current. *See* GENERATOR (def. 1).

altitude. (1) The vertical distance of a point above sea level.

(2) The height of a geometric figure or solid measured perpendicular to its base.

(3) A celestial co-ordinate specifying the angular distance of a celestial body above the horizon. *See also* CELESTIAL SPHERE.

AM. Abbrev. for AMPLITUDE MODULATION.

ammeter. An instrument for measuring electric current. *See* MOVING COIL INSTRUMENT; MOVING IRON INSTRUMENT; HOT WIRE AMMETER.

amount of substance. Symbol n. The product of the AVOGADRO CONSTANT and the number of entities comprising the sample under consideration; for example, the amount of substance of a sample of light is the number of photons present multiplied by the Avogadro constant. The unit is the MOLE.

ampere. Symbol A. The unit of electric current in the SI UNITS system. It is defined as the constant current that when passed through two parallel conductors of infinite length and negligible cross section situated one metre apart in vacuo, produces a force between the conductors of 2×10^{-7} newton per metre of length. This unit replaced the international ampere, which was defined as the current which when passed through silver nitrate solution under specified conditions deposited silver at a rate of 0.001 118 gramme per second; one international ampere equals 0.999 850 ampere.

Ampère balance.   An apparatus used for absolute current measurement. A laboratory form is illustrated (fig. A4). Initially, with the current switched off, the zero screw is adjusted so that plane ABCD is horizontal. The current to be measured is then switched on and kept constant throughout the experiment by slight adjustment (if necessary) of the rheostat for a constant meter reading. The current flows through ABCD and EFGH in series and so FG repels CB. The mass m kilogramme necessary to restore balance is then measured; since the distances of CB and the scale pan from the pivot are equal, the force between the conductors is mg newton, where g is the acceleration due to gravity. The equal lengths (l metre) and separation (r metre) of FG and CB are measured. The current I ampere is then given by

$$[mgr/(2 \times 10^{-7}\, l)]^{1/2}$$

The method is absolute since it is based on the definition of the ampere.

ampere hour.   A unit of quantity of electricity equal to 3600 COULOMB.

Ampère's law.   The formula

$$dB = (\mu_0 I \sin\theta\; dl)/4\pi r^2$$

where dB is the elemental MAGNETIC INDUCTION produced by a current I at a point at a distance r from a conductor element dl. θ is the angle which a line from the point to the conductor makes with the current direction and μ_0 is the MAGNETIC CONSTANT.

Ampère's rule.   The magnetic field due to a current in a wire appears directed clockwise to an observer looking in the direction of the current.

Ampère's theorem.   The formula

$$I = \oint H \cos\theta\; dl$$

where I is the total current enclosed by a loop of length l, and θ is the angle between an element dl of the loop and the magnetic field H due to the current; the symbol $\oint$ indicates that the integral is taken completely round the closed loop. In applying the theorem to a straight wire conductor, the loop, is taken as a circle centred on the wire and plane perpendicular to it; H is

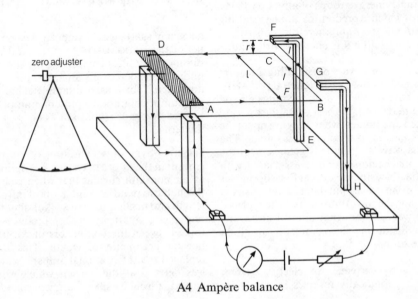

A4 Ampère balance

then constant and lies along the loop and θ is always zero. Hence

$$I = \oint H \, dl = 2\pi rH$$

where r is the loop radius, and so

$$H = I/(2\pi r)$$

ampere-turn. The MAGNETOMOTIVE FORCE produced when a current of one ampere flows through one turn of a coil.

amplifier. A device for increasing the strength of an electric signal by drawing energy from a source independent of the signal. Amplifiers may be classified by the frequency range over which they are designed to work, for example AUDIO FREQUENCY, RADIO FREQUENCY, ULTRA HIGH FREQUENCY. Alternatively they may be classified by the types of TRANSISTOR or VALVE used. *See also* CLASS A, CLASS AB, CLASS B, CLASS C, CLASS D AMPLIFIER.

amplitude. (1) The maximum numerical value of an alternating quantity.
(2) The angular distance of a celestial body, measured on the horizon, from the east or west point.
(3) The quantity $+(a^2 + b^2)^{1/2}$ for the complex number $(a + ib)$.

amplitude modulation. MODULATION in which the carrier amplitude is increased or diminished as the signal amplitude increases or diminishes.

amu. Abbrev. for atomic mass unit. *See* UNIFIED ATOMIC MASS UNIT.

analog computer. *See* COMPUTER.

analyser. A device which transmits only that component of light which is polarized in a particular plane. *See* NICOL PRISM; POLAROID.

analytic geometry. A form of geometry in which lines and curves are represented by equations, and their properties deduced by algebraic reasoning.

anaphoresis. ELECTROPHORESIS towards the anode.

anastigmat. A three- or four-component camera lens designed to exhibit minimum ASTIGMATISM, CHROMATIC ABERRATION, COMA, CURVATURE OF IMAGE and SPHERICAL ABBERATION even for large apertures and wide fields

Anderson bridge. A modification of MAXWELL'S BRIDGE. It is illustrated in fig. A5. Direct current balance is first obtained using a direct voltage supply and a moving coil galvanometer as detector (D). The condition is

$$R_1/R_2 = R_3/R_4$$

The direct supply is then replaced by an alternating one and detector D by for example a microphone; a new balance is obtained by adjusting R only. The condition is

$$L/R_2 = C[R_3(R_4 + R)/R_4 + R]$$

The advantage of this bridge over Maxwell's for obtaining a value for the ratio of a capacitance to an inductance is that the alternating balance in the Anderson bridge does not disturb the direct balance as may happen in Maxwell's bridge; thus the tedious process of trying to find a direct balance which is also an alternating balance is eliminated for the Anderson bridge.

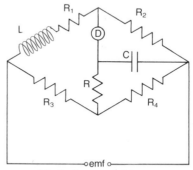

A5 Anderson bridge

Andrews' experiments. Experiments concerned with the behaviour of carbon dioxide when subjected to high pressures at various temperatures. Andrews developed the concepts of CRITICAL PRESSURE and CRITICAL TEMPERATURE.

anechoic chamber. Another name for DEAD ROOM.

anelasticity. A property of a solid such that there is not a unique relationship between STRESS and STRAIN in the preplastic range.

anemometer. A device for measuring the speed of a fluid, especially wind. The momentum of the wind is used to rotate a set of small cups mounted on a spindle; the instrument is calibrated to give the wind speed directly from a dial. *See also* HOT WIRE ANEMOMETER; PITOT TUBE; VENTURI METER.

aneroid. Not containing a liquid

aneroid barometer. *See* BAROMETER.

angle. The figure formed between two intersecting lines; the amount of rotation between the lines is measured in degrees or radians, 180° being equal to π radian: 2π radian is thus a complete turn. The angle between two plane surfaces is the angle between the perpendiculars in each plane to the line of intersection of the planes at a point on it.

angle of contact. The angle, measured in the liquid, between the wall of a vertical capillary tube and the meniscus of a liquid in the tube; if the angle is acute the liquid rises in the capillary tube, whereas if the angle is obtuse the liquid falls. *See* JUVIN'S RULE.

angle of deviation. The angle through which an incident ray is deviated by interaction with different media, for example by reflection of refraction.

angle of dynamic friction. The angle whose tangent is the coefficient of dynamic FRICTION. It is the angle to the horizontal of an inclined plane down which a body will slide without acceleration.

angle of incidence. (1) The angle between a ray falling on a surface and the line perpendicular to the surface (i.e. the normal) at the point of incidence; the point of incidence is the spot where the ray strikes the surface.

(2) The angle between a wavefront and a surface which it strikes.

angle of reflection. (1) The angle between a ray leaving a reflecting surface and the line perpendicular to the surface (i.e. the normal) at the point of leaving; this point coincides with the point of incidence.

(2) The angle between a wavefront and a reflecting surface which it leaves.

angle of refraction. The angle between a refracted ray and the line perpendicular to the refracting surface (i.e. the normal) at the point of entry; this point coincides with the point of incidence.

angstrom. Symbol Å. A unit of length equal to 10^{-10} metre; it was formerly used for wavelength and interatomic distance measurements but has now been replaced by the nanometre, which is equivalent to 10 Å.

angstrom unit. Another name for ANGSTROM.

angular acceleration. The time rate of change of ANGULAR VELOCITY.

angular displacement. The angle through which a point, line or body is rotated about a specified axis in a specified direction.

angular frequency The product of 2π and the frequency in hertz.

angular magnification. Another name for MAGNIFYING POWER.

angular momentum. Symbol L. The product of the ANGULAR VELOCITY of a body and its MOMENT OF INERTIA about the axis of rotation. *See also* CONSERVATION LAW.

angular velocity. Symbol ω. The angle in radians through which a body rotates in 1 second.

anharmonic motion. The motion of a body due to a restoring force which is not directly proportional to the displacement of the body from a fixed point in its line of motion.

anion. A negatively charged ION. Compare CATION.

anisotropic. Not ISOTROPIC.

anisotropy. The variation of physical properties with direction.

annealing. The process of heating a solid and then slowly cooling it in order to remove strain and crystal imperfections.

annihilation. A collision between a particle and its antiparticle resulting in the conversion of the particles into electromagnetic radiation; this radiation is known as *annihilation radiation*. For example an electron and a positron can collide to yield two gamma ray photons. The total energy of the electromagnetic radiation so produced is equal to the sum of the masses and kinetic energies of the particles, in accordance with the CONSERVATION LAW of mass-energy.

annihilation radiation. *See* ANNIHILATION.

annual parallax The maximum angle subtended at a particular star by the Earth's mean (i.e. average of greatest and smallest) orbital radius.

annular eclipse. *See* ECLIPSE.

annulus. (1) A plane figure bounded by two concentric circles.
(2) A thick-walled hollow cylinder.

anode. An electrode in, for example, a cell or valve; the charge on an anode is positive and it is the electrode by which electrons leave a system. *Compare* CATHODE.

anode rays. Positive ions emitted from the anode in a glow discharge.

anode saturation. A phenomenon which occurs when electrons are no longer attracted to an ANODE due to a build-up of electrons around it.

anodizing. The process of forming an oxide layer on aluminium by ELECTROLYSIS; the aluminium oxide layer is porous and can be coloured by certain dyes.

anomalous dispersion. A discontinuity in the regular variation of refractive index with DISPERSION which occurs in the region of an absorption band. The refractive index becomes high on the long wavelength side and low on the short wavelength side of the absorption band.

anomalous expansion of water. The decrease in volume of a given mass of ice on melting and the continued decrease in volume of the resulting water with increase of temperature up to a temperature of 4° C. *See also* HOPE'S EXPERIMENT.

anomalous viscosity. VISCOSITY for which Newton's law of viscosity is not obeyed. Anomalous viscosity is demonstrated by all fluids consisting of two or more phases at the same time, for example every COLLOIDAL SOLUTION.

anomalous Zeeman effect. *See* ZEEMAN EFFECT.

anomaly. An angle defining the position of a planet in its elliptical orbit round the Sun. The angle subtended at the Sun by the line joining the planet and PERIHELION is known as the *true anomaly;* it is shown in fig. A6. The *mean anomaly* is defined as the angle subtended at the Sun by a line joining the perihelion to an imaginary planet moving at constant speed in the real planet's orbit and having the same period. The *eccentric anomaly* is as illustrated. The major axis of the orbit is the diameter of the semicircle

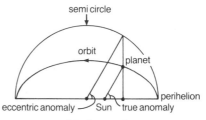

A6 Anomaly

antenna. Another name for AERIAL.

antiderivative. A function whose derivative is a given function. For example x^2 is the antiderivative of $2x$ since the derivative of x^2 is $2x$.

antiferromagnetism. A property of some inorganic compounds, for example manganese oxide. It results from an antiparallel arrangement of adjacent magnetic dipole moments. Antiferromagnetic materials have low positive MAGNETIC SUSCEPTIBILITY which shows a temperature dependence similar to that encountered in FERROMAGNETISM. The susceptibility increases with temperature up to the *Néel temperature* and then decreases with temperature in accordance with the CURIE WEISS LAW, i.e. the substance is paramagnetic above the Néel temperature. *Compare* CURIE TEMPERATURE.

antilog. Short for ANTILOGARITHM.

antilogarithm. The number whose logarithm is a given number; thus if x is the logarithm of y, then y is the antilogarithm of x to the same base.

antimatter. A hypothetical substance whose atoms are composed of ANTIPARTICLES. Atoms of antimatter would therefore have negatively charged nuclei made up of antiprotons and antineutrons; antielectrons, i.e. positrons, would orbit the nucleus. Contact between ordinary matter and antimatter would result in ANNIHILATION. Antimatter has not yet been detected in the universe.

antinode. A position of maximum disturbance in a STANDING WAVE. *Compare* NODE.

antiparallel. Parallel but pointing in opposite directions.

antiparticle. An elementary particle having the same mass as a given elementary particle and having CHARGE, BARYON NUMBER, ISOSPIN quantum number and STRANGENESS of identical magnitude but opposite sign to those of the given particle.

antiprincipal points. Conjugate points of a CENTRED OPTICAL SYSTEM for which the lateral MAGNIFICATION is -1.

aperture. The area of an optical system perpendicular to its axis and available for the passage of light.

aperture synthesis. A method of accurately mapping radio sources in outer space by computer synthesis of the outputs of more than one small adjustable RADIO TELESCOPE for various positions of the telescopes.

aphelion. The point at which a body orbiting the sun is furthest from the sun. *Compare* PERIHELION.

aplanatic. Free from SPHERICAL ABERRATION and COMA.

apochromatic lens. A compound lens corrected for the secondary spectrum of an ACHROMATIC LENS. *See also* CHROMATIC ABERRATION.

apogee. The point at which a body orbiting the Earth is furthest from Earth. *Compare* PERIGEE.

apothecaries' units of mass. *See* Table 6D.

apparent change of weight. (1) The difference in sensation of weight experienced when the reaction between a person and the support is changed. Thus in an orbiting space station the reaction vanishes giving rise to a feeling of weightlessness; in a rocket at liftoff the reaction increases to such an extent that, although lying down, a sensation of greatly increased weight is experienced resulting in difficulty in breathing and in raising the arms or legs.
(2) The upthrust on a body immersed wholly or partially in a fluid. For example a body weighed in air and then in a denser medium such as water appears to suffer a loss of weight. *See also* ARCHIMESDES' PRINCIPLE.

apparent depth. The apparent change in depth of an object when the medium in which it is immersed is changed. Thus the depth of a swimming pool appears to increase when the pool is emptied of water. The phenomenon is due to the REFRACTION of light on passing from one transparent medium to another.

apparent power. *See* POWER FACTOR.

Appleton layer. *See* IONOSPHERE.

apsides. Plural form of APSIS.

apsis. Either of two points at the extremities of the major axis of an elliptical orbit.

aqueous humour. *See* EYE.

arc. (1) A luminous electrical gas discharge having high current density and low potential gradient. The electrodes are heated by the discharge and their evaporation helps maintain it. *See also* CONDUCTION (in gases); GAS DISCHARGE TUBE.
(2) An open segment of a curve.
(3) The inverse of a circular or hyperbolic function; thus the *arc sine*, written arc sin or sin^{-1}, of a number is any angle whose sine is the given number.

Archimedes' principle. When a body is wholly or partially immersed in a fluid at rest it experiences an upthrust equal to the weight of the displaced fluid. *See also* BUOYANCY.

arc lamp. A lamp whose light comes from an electric arc, mostly from the incandescent crater formed at the positive electrode; the light is of high intensity.

arc sine. *See* ARC (def. 3).

are. Symbol a. A metric unit of area equal to 100 square metre.

area. A measure of a two-dimensional surface. The unit is the square of any unit length, for example the square metre.

Argand diagram. A representation of a COMPLEX NUMBER as a point on a plane. Using CARTESIAN CO-ORDINATES the complex number

$$x + iy, \text{ where } i = (-1)^{1/2}$$

is represented by the point (x, y), as illustrated in fig. A7. In POLAR CO-ORDINATES the number, written as

$$r \cos \theta + ir \sin \theta$$

is represented by (r, θ) where

$$r = +(x^2 + y^2)^{1/2} \text{ and } \theta = \tan^{-1}(y/x)$$

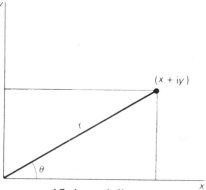

A7 Argand diagram

argon. Symbol A. A gas used in FLUORESCENT LAMP and electric light bulb manufacture. The gas is obtained from the atmosphere as a by-product of the liquefaction of AIR.

argument. (1) A sequence of logical propositions based on a set of premises and leading to a conclusion.
(2) The angle θ for the COMPLEX NUMBER represented by the point (r, θ) in the ARGAND DIAGRAM.

arithmetic mean. The result of dividing the sum of a given set of numbers by the number of numbers in the set.

arithmetic progression. A SEQUENCE in which each term differs from the preceding one by the same number, i.e. it is of the form

$$a, a + d, a + 2d, ..., [a + d(n-1)]$$

where n is the number of terms and d is

known as the *common difference*. The corresponding arithmetic SERIES has the sum

$$[2a + d(n - 1)]n/2$$

Compare GEOMETRIC PROGRESSION.

armature. (1) Any moving part in an electrical machine in which a voltage is induced by a magnetic field. Examples are the rotating coils in an ELECTRIC MOTOR and the ferromagnetic bar attracted by an electromagnet in a relay.
(2) Another name for KEEPER.

Arrhenius' theory. The theory that an electrolyte ionizes as soon as it dissolves with the result that ions are not produced by the passage of current but are present as such in solution before current is passed.

artery. A vessel through which blood passes from the heart to other parts of the body.

artificial pacemaker. A device whereby electric impulses can be applied to a heart to maintain its normal functioning. The earliest pacemakers were implanted in the patient and an operation was required every 18 months to 2 years in order to replace the mercury cells which charged up the capacitor whose discharge stimulated the heart. The next development was the DEMAND PACEMAKER, followed by the EXTERNAL PACEMAKER.

artificial radioactivity. *See* RADIOACTIVITY.

artificial satellite. *See* SATELLITE (def. 1).

asdic. Another name for SONAR.

aspherical surface. A surface which is not part of a sphere; usually an aspherical surface is part of an ellipsoid or of a hyperboloid, or of a paraboloid. Lenses and mirrors with aspherical surfaces show reduced aberrations, especially SPHERICAL ABERRATION, compared to those with spherical surfaces.

associative operation. An operation which is not affected by the way in which the terms operated on are grouped. An example is

$$(x + y) + z = (x + z) + y$$

astatic galvanometer. A sensitive form of moving-magnet galvanometer in which two similar magnets are suspended as illustrated (fig. A8). The net couple due to an external magnetic field is small since the couples on each magnet are nearly equal but opposite; the current through the coil surrounding one of the magnets thus produces a greater rotation for a given current than if only a single magnet were used. In an alternative arrangement, each magnet is surrounded by a coil, the two coils being wound in opposite directions so that even greater sensitivity is achieved.

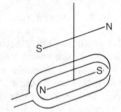

A8 Astatic galvanometer

asteroid. A small planet of diameter generally less than about 1000 kilometre. Most asteroids have near-circular orbits situated between the orbits of MARS and JUPITER. Some small asteroids however have highly elliptic orbits and approach much nearer the Sun than do the majority.

astigmatic lens. A lens used to correct ASTIGMATISM of the eye. It may be plano-cylindrical, sphero-cylindrical or toroidal, the choice depending on the defect.

astigmatism. An aberration of optical systems such that a point off the axis of the system is imaged as two perpendicular lines, the tangential (i.e. horizontal) one being closer to the system than the sagittal (i.e. vertical) one.

astrolabe. An ancient astronomical

instrument used for plotting angular distances on the CELESTIAL SPHERE. It incorporated horizontal and vertical graduated scales.

astrometric binary star. *See* BINARY STAR.

astrometry. The measurement of the positions of the celestial bodies on the CELESTIAL SPHERE.

astronomical telescope. *See* TELESCOPE.

astronomical unit. Symbol au. A unit of length equal to

$$1.495\ 979 \times 10^{11}\ \text{metre}$$

i.e. the mean distance between Sun and Earth. The unit is usually used for distance measurements within the solar system.

astronomy. The study of the universe beyond the Earth's atmosphere. The subject is subdivided into ASTROMETRY, ASTROPHYSICS and CELESTIAL MECHANICS. *See also* GAMMA RAY ASTRONOMY; INFRARED ASTRONOMY; RADIO ASTRONOMY; X RAY ASTRONOMY.

astronaut. A person who voyages in space, beyond the Earth's atmosphere.

astrophysics. The study of the physical processes associated with celestial bodies and with the intervening regions of space. A main concern is stellar energy and the relationship between this energy and evolution.

asymptote. A tangent to a curve at infinity. For example the asymptotes of the rectangular hyperbola $xy = c^2$ are the x and y axes.

athermancy. The property of opaqueness to infrared radiation. *Compare* DIATHERMANCY.

atherosclerotic plaque. An internal lesion in an artery causing it to narrow, so leading to reduced blood pressure in the constricted region. *Compare* VENTURI METER.

atmosphere. (1) The AIR.
(2) A unit of pressure equal to 101 325 pascal. The acutal pressure of the atmosphere fluctuates about this value. *See also* ATMOSPHERIC LAYERS.
(3) Any gaseous medium.

atmospheric circulation. The pattern of wind velocity over the Earth in relation to latitude, longitude and height.

atmospheric convection. Approximately vertical and relatively small-scale movement of the atmosphere.

atmospheric electricity. The general electrical properties of the atmosphere. Mean fine weather values of these properties at about sea level are as follows:

air-earth current
density 2×10^{-14} A m^{-2}
direction of field downwards
potential gradient 130 V m^{-1}
small ion mobility 1.4×10^{-4} m^2 V^{-1} s^{-1}
total conductivity 3×10^{-4} S m^{-1}

Charges arise in the Earth's atmosphere due to ionization by radioactive material and by cosmic rays. During a LIGHTNING flash the electrical properties of the atmosphere are appreciably changed.

atmospheric layers. The gaseous layers into which the Earth's atmosphere is divided according to direction of temperature variation. The vertical extents of the layers vary over the Earth's surface and also show seasonal and diurnal changes at the same place. The approximate altitude limits of each layer and the temperature variation within the layers are illustrated in fig. A9. The lowest temperature occurs at the MESOPAUSE.

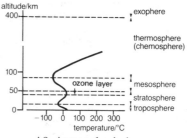

A9 Atmospheric layers

atmospheric optics.   The study of phenomena associated with the scattering, reflection, refraction and diffraction of light by the atmosphere or by ATMOSPHERIC POLLUTION.

atmospheric pollution.   Undesirable substances present in the atmosphere, for example dust and noxious gases. Such pollution may obscure vision, harm plants and animals and erode buildings.

atmospheric pressure.   The pressure at sea level due to the air above the Earth's surface. *See also* ATMOSPHERE (def. 2).

atmospherics.   Short-wave trains of electromagnetic radiation arising from natural electrical atmospheric disturbances. The radiation causes interference with radio reception.

atmospheric windows.   Gaps in atmospheric absorption through which some radiations from space penetrate the Earth's atmosphere. The ranges of wavelengths permitted passage are 300–900 nanometre (the *optical window*), 8 millimetre to 20 metre (the *radio window*) and several narrow *infrared windows* at micrometre wavelengths.

atom.   The smallest part of an element that can take part in a chemical reaction. Originally the atom was thought to be indivisible but later work led to the concept of the atom as a microplanet, with NUCLEUS consisting of a specific number of protons and neutrons at the centre and a number of electrons orbiting the nucleus. The size of an atom, i.e. an atomic diameter, is of the order of 10^{-10} metre; the size of the nucleus is of the order of 10^{-15} metre. *See* BOHR THEORY; CHADWICK'S EXPERIMENT; GEIGER AND MARSDEN'S EXPERIMENT.
 WAVE MECHANICS and QUANTUM MECHANICS introduced further refinements into the conception of the atom so that the pattern of electron energy levels is given by four quantum numbers: the azimuthal quantum number, the magnetic quantum number, the principal quantum number and the spin. These numbers determine

the CONFIGURATION of the atom of a particular element. *See* ATOMIC ORBITAL.

atom bomb.   *See* NUCLEAR WEAPON.

atomic clock.   *See* CLOCKS

atomic energy.   Energy obtained by nuclear FISSION or FUSION.

atomic energy level.   *See* ENERGY LEVEL. *See also* BOHR THEORY.

atomic heat.   The heat capacity of one mole of a substance. *See also* DULONG AND PETIT'S LAW.

atomicity.   The number of atoms in a molecule of a given element. For example the atomicity of sulphur dioxide is 3.

atomic mass unit.   *See* UNIFIED ATOMIC MASS UNIT.

atomic nucleus.   Another name for NUCLEUS.

atomic number.   Symbol Z. The number of protons in the nucleus of a given atom. The atomic number is also the number of electrons in the atom and so determines the chemical properties.

atomic orbital.   An allowed WAVE FUNCTION of an electron in an atom, obtained by a solution of SCHRÖDINGER'S WAVE EQUATION. The square of the modulus of the wave function at a particular point is proportional to the probability of finding an electron at that point. Each allowed wave function has an associated energy value and is characterized by three quantum numbers: the *principal quantum number*, the *azimuthal quantum number* and the *magnetic quantum number*, respectively represented by the letters n, l and m. The principal quantum number, n, can take values 1, 2, 3, etc. in order of increasing energy, respectively corresponding to the ELECTRON SHELLS designated K,L,M, etc. Each shell can contain a maximum of $2n^2$ electrons; thus the K shell is full when it contains 2 electrons, the L shell when it has 8 electrons and the M shell when it has 18 electrons. The shells have subshells characterized by the azimuthal quantum number, l, which gives the orbital angular momentum of the electron and has values

$$0, 1, 2, 3, ..., (n - 1)$$

for each value of n; the electron states corresponding to these values of l are respectively designated

$$\text{s, p, d, f, ..., subshells}$$

For a given value of l the magnetic quantum number, m, can take the values

$$-l, -(l - 1), ..., 0, ..., (l - 1)$$

The K shell thus has only one subshell, i.e. the subshell is the same as the shell, which is an s subshell. The L shell has two subshells: an s subshell and a p subshell. For an s subshell, $m = 0$; for a p subshell, $m = -1$, 0 and 1, i.e. there are three p orbitals within this subshell; similarly there are five d orbitals and seven f orbitals Usually the energies of the orbitals of a particular subshell only differ in the presence of an external magnetic field. *See* LARMOR PRECESSION.

A fourth quantum number M characterizes the direction of electron SPIN. M can have the values $+\frac{1}{2}$ and $-\frac{1}{2}$. Each orbital can contain a maximum of two electrons with opposing spins.

Orbitals are conventionally represented by a surface enclosing a volume in which there is an arbitrarily decided probability (say 95%) of finding the electron. As illustrated in fig. A10, an s orbital is spherical but orbitals with $l > 0$ have angular dependence; p orbitals are also shown.

The four quantum numbers lead to an explanation of the PERIODIC TABLE. *See also* PAULI EXCLUSION PRINCIPLE.

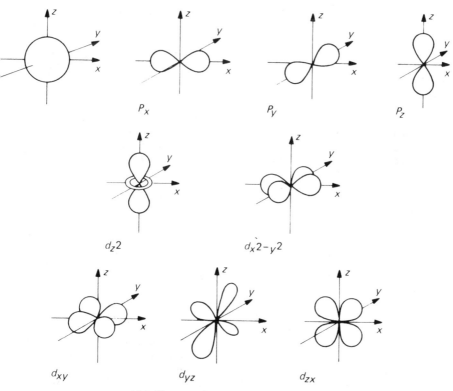

A10 Shapes of atomic orbitals

atomic physics. The study of the physical properties of atoms, regarding the atom as a whole. Atomic physics is thus distinct from NUCLEAR PHYSICS although there is some overlap.

atomic pile. Former name for NUCLEAR REACTOR.

atomic ratio. The ratio of the numbers of different atoms present in a given sample, as opposed to the ratio of masses or volumes of the different constituents.

atomic unit of length. Symbol a_0. The radius of the first orbit of the hydrogen atom according to the BOHR THEORY; it is

$$5.29 \times 10^{-9} \text{ metre}$$

atomic unit of mass. The rest mass of the electron, i.e.

$$(9.1084 \pm 0.003) \times 10^{-31} \text{ kilogramme}$$

Compare UNIFIED ATOMIC MASS UNIT.

atomic volume. The volume in the solid state of one mole of an element.

atomic weight. *See* RELATIVE ATOMIC MASS.

attenuation. (1) A loss of intensity of a signal, electromagnetic wave, sound wave, beam of particles etc. in passing through a medium. The loss may result from ABSORPTION or SCATTERING.
(2) A drop in current or voltage experienced by a signal as it passes through a circuit.

atto−. Symbol a. A prefix meaning 10^{-18}.

au. Symbol for ASTRONOMICAL UNIT.

audibility limits. The frequency limits between which a sound can be heard. They are around 20 hertz, corresponding to a low rumble, and 20 000 hertz, corresponding to a shrill whistle. The upper limit falls appreciably with increasing age. *See also* SENSITIVITY OF THE EAR.

audio frequency. A frequency within the AUDIBILITY LIMITS.

audiometer. An instrument for measuring the acuity of hearing for either speech or tone signals.

auditory heart warning system. An amplified-loudspeaker system whose input is from electrodes attached to a patient's chest in the region of the heart. Pulses of sound are heard as long as the heart is working so that there is noticeable and immediate warning of any trouble.

Auger effect. The spontaneous ejection of an electron by an excited singly charged positive ion to form a doubly charged ion. The excitation of the singly charged ion can arise by absorption in an electron shell of a gamma ray emitted by the nucleus. Alternatively excitation can be induced by an external stimulus, for example electron or photon bombardment.

Auger shower. A shower of elementary particles produced by a primary COSMIC RAY entering the atmosphere. Such a shower may extend over 1000 square metre.

auricle. Either of two chambers in the heart; the left-hand one receives blood from the lungs and the right-hand one receives blood from the general circulation.

aurora. An atmospheric phenomenon consisting of green, red or yellow rays, streamers, luminous arcs etc. It is produced by a stream of charged particles from the Sun entering the upper atmosphere. The phenomenon is most frequently seen in an angular region of about 20° from either pole; when nearest the north pole it is known as *aurora borealis* and when nearest the south pole as *aurora australis*.

autoionization. An effect analogous to the internally produced AUGER EFFECT except that an optical photon rather than a gamma ray photon is involved in the excitation.

autoradiography. A technique for studying the distribution of a radioisotope introduced into a thin specimen. For example to study carbohydrate distribution in leaves, a leaf of a growing plant is kept in a container through which carbon dioxide gas made from radioactive carbon is passed. After a suitable time the leaf is placed between photographic plates, which are affected by beta particles from the radioactive carbon atoms; the regions of greatest concentration of the radioactive carbon in the leaf can then be identified from the developed plates.

autumnal equinox. *See* EQUINOX.

avalanche. A burst of ions, all of which are produced from a single primary ion as in a GEIGER COUNTER.

avalanche breakdown. A breakdown in a semiconductor DIODE due to the cumulative multiplication of free charge carriers in a strong electric field. Some free carriers acquire sufficient energy to liberate new electron-hole pairs by collision.

average. *See* ARITHMETIC MEAN. *See also* GEOMETRIC MEAN.

Avogadro constant. Symbol N_A or L. The number of entities in one MOLE of substance; its value is $6.022\ 52 \times 10^{23}$.

Avogadro's hypothesis. Equal volumes of all gases at the same temperature and pressure contain equal numbers of molecules. In other words, the volume occupied at a given temperature and pressure by a mole of gas is the same for all gases. At standard temperature and pressure this volume is 22.4×10^{-3} cubic metre. *See* IDEAL GAS.

Avogadro's number. Former name for AVOGADRO CONSTANT.

avoirdupois units of mass. A system of mass units based on the POUND. *See* Table 6D.

axial vector. Another name for PSEUDO VECTOR (def. 1).

axiom. A proposition assumed to be true, or considered self-evident, and used as a basis for logical deduction.

axis. One of a set of reference lines relative to which the positions of points are specified. *See also* CARTESIAN CO-ORDINATES; POLAR CO-ORDINATES.

axis of rotation. A line about which rotation occurs.

axis of symmetry. A line about which a figure, curve or body is symmetrical.

axon. A threadlike part of a nerve cell responsible for transmitting an ACTION POTENTIAL.

azimuth. *See* POLAR CO-ORDINATES; CELESTIAL SPHERE.

azimuthal quantum number. *See* ATOMIC ORBITAL.

B

back electromotive force. An electromotive force opposing the flow of current in a circuit.

(1) *of induction*: the back electromotive force which is self-induced when the current in a circuit changes. *See* ELECTROMAGNETIC INDUCTION

(2) *of motor*: the electromotive force opposing motion which is induced in a conductor moving in a magnetic field.

(3) *of polarization*: a reduction of a cell's natural electromotive force which occurs when POLARIZATION produces a back electromotive force. *See also* LENZ'S LAW.

background. In general, unwanted effects in physical measurements above which a phenomenon must show itself in order to be measured. The term is often restricted to naturally occurring effects in distinction to those arising from human activities, for example the effect of COSMIC RAYS in studies of RADIATION.

Bailey's beads. A necklace appearance surrounding the dark body of the Moon just before and after totality during a total solar ECLIPSE. The phenomenon is the result of sunlight shining into valleys in the apparent edge of the Moon.

balance. (1) A condition of partial or complete equilibrium.

(2) An instrument used in weighing. There are various types.

In the *beam balance* a rigid beam is balanced on a FULCRUM at its midpoint and similar pans are suspended from each end of the beam. The object to be weighed is placed in one pan and standard masses are added to the other until the beam is horizontal; the sum of the standard masses then equals the unknown mass. A sensitive beam balance is provided with a rider, i.e. a small mass which can be moved along the top of the balance beam on which is marked a calibrated scale. To minimize friction, the beam of a sensitive balance rests on agate knife edges and, to reduce the effects of air currents and temperature changes, it is enclosed in a glass-fronted case. A vertical pointer attached to the midpoint of the beam enables the position of equality to be detected; an accuracy of 0.001 gramme is obtainable. By marking the beam with a calibrated scale and providing a choice of masses to slide along it, one pan of the beam balance is made redundant, yielding a *single* pan balance. *See also* HYDROSTATIC BALANCE; MICROBALANCE.

In the *spring balance* the mass under investigation is attached to the lower end of a calibrated vertically coiled spring. The weight of the mass is proportional to the extension of the spring, provided the ELASTIC LIMIT is not exceeded. The reading of this type of instrument, in contrast to that of other types of balance, thus depends on the value of the ACCELERATION DUE TO GRAVITY at the place of use.

In the *torsion balance* the force to be measured is applied horizontally to one end of a light horizontal lever; the lever is suspended from a vertical torsion wire (or fibre) and so the wire is twisted. The wire's other end is firmly attached to a knob carrying a pointer which can move over a horizontal circular scale. The amount of rotation of the knob necessary to restore the lever to its original position is a measure of the force applied. Alternatively the angle of a twist may be found by measuring the deflection of a beam of light reflected off a small mirror attached to the wire.

balancing column method. The technique employed using HARE'S APPARATUS.

ballistic galvanometer. An instrument used to measure electric CHARGE. It is essential that the time in which the charge passes through the galvanometer be short compared with the PERIOD of natural OSCILLATION of the galvanometer; the charge then passes before the galvanometer starts to deflect. The initial deflection (throw) of a MOVING COIL INSTRUMENT is then proportional to the charge passing.

ballistic pendulum. A pendulum used to measure the velocity of a projectile striking it. The mass, m, of the freely suspended pendulum is very much greater than the mass, m', of the projectile. If u' is the velocity with which the projectile strikes the pendulum and assuming the projectile embeds in the pendulum, then by the CONSERVATION LAW for momentum

$$m'u' = (m + m')u$$

where u is the velocity with which the pendulum starts to move. By the conservation law for energy

$$0.5(m + m')u^2 = g(m + m')h$$

where h is the observed vertical height of rise of the centre of mass. Hence u' can be found.

ballistics. The study of the motion and propulsion of projectiles. *Internal ballistics* is concerned with the phenomena before launch, *external ballistics* with the phenomena after launch.

ball lightning. *See* LIGHTNING.

balloon sonde. Another name for SOUNDING BALLOON.

Balmer series. A series in the SPECTRUM of the hydrogen atom. It is defined by the equation

$$1/\lambda = R(1/2^2 - 1/n^2)$$

where λ is the wavelength of a line, n is any integer greater than or equal to 3 and R is the RYDBERG CONSTANT. All lines of this series are in the visible part of the spectrum.

band spectrum. A type of SPECTRUM, characteristic of molecules, in which bands occur. The bands are each formed of a number of closely spaced lines. One edge of a band is sharply defined and is known as the *band head*. On one side of the head the intensity is zero but on the other it falls off gradually. If the slow fall-off is on the shorter (longer) wavelength side, the band is said to be degraded towards the violet (red). The band head wavelength is an important clue to the identification of the molecule. Each band corresponds to a transition between a high and low electron ENERGY STATE; different VIBRATIONAL ENERGY LEVELS of the molecule give rise to the lines within a band.

band theory. A theory which results from the application of QUANTUM MECHANICS to the energies of electrons in crystalline solids. In an isolated atom an orbiting electron occupies a discrete ENERGY STATE. In the solid state the atoms are situated very close together; although this has little effect on the tightly bound inner electrons the behaviour of the VALENCE ELECTRONS is completely altered. The orbits of these electrons no longer belong to individual atoms but are shared by the whole lattice and thus help to hold the atoms together in the solid. In the solid state each discrete ENERGY LEVEL is therefore broadened into a band of allowed energy; every band is made up of closely spaced energy levels each of which can accommodate a pair of electrons. The broadening is greatest for the most loosely bound levels. The allowed energy bands are separated by bands of forbidden energy as schematically illustrated in fig. B1. In the resting state of the solid the highest allowed energy band, known as the *conduction band*, is unfilled.

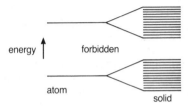

B1 Energy bands

The main success of the band theory is an explanation of the differences in electric conductivity of different solids. Conductivity is due to electron movement in the presence of an electric field; electron movement implies that an electron is gaining energy from the field. By the QUANTUM THEORY this gain can occur in discrete amounts only and equals the difference in energy between two allowed states. In metals and good conductors the forbidden band may not exist, i.e. the allowed bands overlap, and so the energy gain is easily made. In electrical insulators it is postulated that the forbidden band is too wide for the electron to jump. For an INTRINSIC SEMICONDUCTOR it is suggested that the forbidden band is narrower so there is a possibility of an electron crossing it. In an IMPURITY SEMICONDUCTOR the atoms of the impurity provide energy levels in the forbidden zone (band) and these provide stepping stones for the electrons. *See also* n-TYPE SEMICONDUCTOR; p-TYPE SEMICONDUCTOR.

band width. The frequency range over which a specified characteristic of an electrical device falls between specified limits.

banking of track. A means used to increase the safe speed of cornering. When a vehicle moves round a circular path, the necessary force towards the centre of the circle is provided by the friction force at the places of contact with the track. The required force increases with speed and if friction is insufficient the vehicle will move away from the centre, i.e. it will skid. The highest safe speed may be increased by inclining the track to the horizontal (i.e. banking) so that the NORMAL reaction at each place of contact has a COMPONENT towards the centre which assists the friction. This component is proportional to the sine of the angle of banking. It is for this reason that racing-track curves have saucer-shaped tracks; a driver can then move towards a part of the track which is sufficiently steep to prevent side slip while maintaining high speed. A cyclist instinctively further increases stability by leaning towards the centre.

bar. A unit of pressure. *See* Table 6F.

Barkhausen effect. The occurrence of a series of discontinuous steps in the amount of FERROMAGNETISM produced by a continuously changing magnetizing field. The steps are interpreted as the alignment of MAGNETIC DOMAINS; they may be demonstrated by winding two coils on an iron core and steadily increasing the current through one coil. The discontinuities in the INDUCED ELECTROMOTIVE FORCE in the other coil may be displayed on a CATHODE RAY OSCILLOSCOPE.

barn. Symbol b. A unit of area equal to 10^{-28} square metre. In experiments on SCATTERING; atomic nuclei cross sections are usually expressed in barn.

barograph. A recording BAROMETER usually of the aneroid type. Generally the pointer carries a pen which is in contact with a moving chart. A continuous record of the variation of atmospheric pressure with time is therefore obtained.

barometer. An instrument for measuring atmospheric pressure. There are several types.

In the *aneroid barometer*, shown in fig. B2, changes in atmospheric pressure cause movements in the thin corrugated top of an evacuated sealed metal box. These movements are transmitted to a pointer by a train of levers and the pressure is read

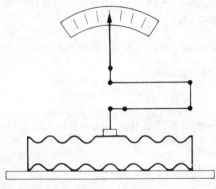

B2 Aneroid barometer

directly on a graduated scale over which the pointer can move. Most domestic barometers are of this type.

The *mercury barometer*, shown in fig. B3, consists of a long glass tube sealed at one end. It is filled with mercury and inverted in an open reservoir of mercury. The atmospheric pressure is often expressed in terms of the vertical height h of mercury that the atmosphere will support. The height is usually around 760 millimetres.

B3 Mercury barometer

barometer conventions. The temperature and gravitational acceleration values at which a mercury BAROMETER is adjusted to read directly in accepted pressure units; the values are 273 kelvin and 9.806 65 metre per square second respectively.

barometric reduction. The correction of barometeric readings to the value they would have at mean sea level. The purpose is to make meaningful comparisons between pressure readings obtained from stations located at different altitudes.

barrel distortion. *See* DISTORTION (def. 2).

barrier potential difference. A potential difference occurring at a p-n JUNCTION between two semiconductors. It opposes the flow across the junction of electrons to the p-type semiconductor and of positive HOLES to the n-type semiconductor. Its maximum value is about 0.1 volt.

baryon. Any ELEMENTARY PARTICLE built from three QUARK particles. Apart from the proton, baryons are unstable; they yield decay products which always include a proton.

baryon number. Symbol B. A quantum number whose value is 1 for every BARYON, –1 for every antibaryon and 0 for all other particles. The total baryon number is conserved in all particle interactions.

base. (1) The region of a TRANSISTOR between the EMITTER and the COLLECTOR.
(2) The bottom face of a geometric solid.
(3) The horizontal straight line at the bottom of a geometric figure.
(4) The number of units grouped together in a positional number system and represented by 1 in the next position. For example the decimal system has a base of ten and the binary system has a base of two.
(5) *See* LOGARITHM.

base units. *See* COHERENT UNITS. *See also* Table 1.

battery. An electric CELL, or two or more such cells connected and used as a single unit. The LEAD ACID BATTERY, extensively used in cars, is a multi-cell example. Single LECLANCHÉ CELL dry batteries are much used in portable radios, torches etc.

beam balance. *See* BALANCE.

beat frequency. The frequency difference between two vibrations which produce BEATS when sounding together.

beats. A phenomenon which occurs when two sound wave trains of the same amplitude but slightly different frequency are superimposed. INTERFERENCE occurs between them giving a resultant vibration whose amplitude alternates between zero and twice the amplitude of either parent wave train. The ear thus registers a succession of maxima; these are the beats. Since the frequency difference between the two parent vibrations is small, it is possible for two ULTRASONIC waves to produce audible beats.

Beckmann thermometer. A mercury thermometer designed for measuring small changes of temperature. As illustrated in fig. B4, the capillary tube has a scale of 5 K and is connected to a reservoir at the top of the thermometer. The scale is very finely graduated. The thermometer is used for measuring differences of temperature (rather than actual temperatures) in the range 0–100° C. To do this, it is necessary to use the reservoir to adjust the amount of mercury in the bottom bulb to a suitable quantity for the temperature level under study.

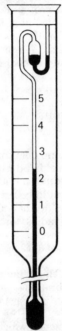

B4 Beckmann thermometer

becquerel. *See* RADIATION UNITS.

bel. *See* DECIBEL.

bending of light rays. A prediction of the theory of RELATIVITY concerning the effect of a gravitational field on electromagnetic radiation.

Bernoulli equation. An equation which applies to a nonviscous fluid in steady flow and states that along a STREAMLINE

$$p + \rho(gh + \tfrac{1}{2} v^2) = \text{constant}$$

p, ρ and v are the fluid pressure, density and velocity respectively; h is the height above a fixed reference line, g is the acceleration due to gravity. The equation is of fundamental importance in all discussions of fluid flow; it only applies to gases if the velocity of the gas is small compared with that of sound.

beta decay. A type of RADIOACTIVE DECAY in which a POSITRON and a NEUTRINO, or an electron and an antineutrino, are ejected from an unstable NUCLEUS. This changes the ATOMIC NUMBER of the nucleus concerned by ±1 but does not change its MASS NUMBER. *See also* RADIOACTIVITY.

beta iron. Iron in a transitional state of magnetism between FERROMAGNETISM and PARAMAGNETISM. *See also* ALPHA IRON.

beta particle. An electron of either positive charge (β^+) or negative charge (β^-) emitted by a NUCLEUS or NEUTRON in the process of BETA DECAY.

beta rays. A stream of BETA PARTICLES.

betatron. A device for accelerating electrons using MAGNETIC INDUCTION. The electron beam is contained in a torus-shaped evacuated chamber which is held between the pole pieces of a ring-shaped magnet. The magnetic field is arranged to increase as the energy of the electrons increases, so that the electrons follow a fixed orbit. The electron momentum is proportional to the magnitude of the magnetic field. The upper limit of energy attainable is about 500 million electrovolts and is set by radiation losses of the electrons. The instrument is used in nuclear physics research, RADIOGRAPHY and RADIOTHERAPY.

betatron synchrotron. Another name for SYNCHROTRON.

beta waves. Waves occurring in ELEC-TROENCEPHOLOGRAPHY. They are usually spindle-shaped and of frequency 14–50 hertz; their amplitude is smaller than that of ALPHA WAVES. During dreaming the beta waves take over completely. The main usefulness of studying the waves is in diagnosis, for example of epilepsy, tumours and accident damage.

bi. A prefix indicating two.

biaxial crystal. A crystal with an OPTICAL AXIS in each of two directions. *See also* DOUBLE REFRACTION.

bichromate cell. *See* DICHROMATIC CELL.

biconcave lens. A LENS with both surfaces concave in shape, i.e. curving inwards.

biconvex lens. A LENS with both surfaces convex in shape, i.e. bulging outwards.

bifocals. Spectacles whose lenses are in two parts. Each upper part is a DIVERGING LENS which corrects MYOPIA when the wearer is looking ahead at distant objects. Each lower part is a CONVERGING LENS which corrects PRESBYOPIA when the wearer is looking down at close work.

bifurcation. A division into two branches. The term is used to describe the behaviour of the equation

$$y_{n+1} = ay_n(1 - y_n)$$

where y_n can take values between 0 and 1 and y_{n+1} is then substituted for y_n to give y_{n+2}, and so on; a is a constant between 0 and 4 for a given set of substitutions. For $a = 2$, the limit of y_n as n tends to infinity is 0.5; for a lying between 2 and 3 the limit changes to a different fixed value. However, for a just greater than 3, alternate values of y_n as n tends to infinity have different limits, i.e. bifurcation has occurred. As a is further increased the two limits become four limits, then eight and so on, leading to a CHAOTIC SYSTEM.

big bang theory. The theory that all matter and radiation in the universe was formed in an explosion of a superdense collection of high-energy matter, which happened at a definite time some 10 to 20 thousand million years ago. The first phase, lasting about 10^{-4} second, saw the creation of ELEMENTARY PARTICLES and their various ANTIPARTICLES. After about 1 second most particles had suffered ANNIHILATION with their antiparticles to produce a PHOTON cloud; the remainder formed the matter destined to be the present material universe. It is considered that DEUTERIUM and HELIUM were synthesized after about 100 second. Down the ages the matter continued to fly apart, slowing down and cooling as it moved outwards and eventually leading to GALAXY and STAR formation. Expansion and consequent cooling of the original photon cloud accounts for the present-day MICROWAVE BACKGROUND radiation of temperature 3 K. The theory explains not only the microwave background but also the cosmic abundance of helium and the EXPANDING UNIVERSE.

It has been suggested that the universe will eventually stop expanding and start to contract until it becomes superdense again, when the whole process may restart. This is the *oscillating universe* model.

bimetallic strip. Two pieces of different metals of unequal COEFFICIENT OF EXPANSION, for example brass and iron, roll welded together. As shown in fig. B5, the strip bends when the temperature changes because of unequal amounts of expansion. Such strips can be used to indicate temperature. They are also used in THERMOSTATS and safety devices to cut off an electric circuit at a preset temperature.

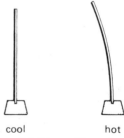

cool   hot

B5 Bimetallic strip

binary. Having two parts or components.

binary arithmetic. A system of representing numbers using BASE two and therefore only the digits 0 and 1. The binary number 1101110 is thus equal to

$$1 \times 2^6 + 1 \times 2^5 + 0 \times 2^4 + 1 \times 2^3 + 1 \times 2^2 + 1 \times 2 + 0$$

which in decimal notation is equal to

$$64 + 32 + 8 + 4 + 2 = 110$$

Similarly the binary fraction 11.001 is equal to

$$1 \times 2 + 1 + 0 \times 2^{-1} + 0 \times 2^{-2} + 1 \times 2^{-3}$$
$$= 2 + 1 + 0.125 = 3.125$$

in decimal notation. The main application of binary arithmetic is in digital COMPUTERS and arises from the fact that the digits 1 and 0 may be represented by the 'on' and 'off' state of an electric circuit.

binary notation. The representation of numbers using only two digits, 0 and 1. *See* BINARY ARITHMETIC.

binary star. A pair of stars each of which rotates about their common CENTRE OF MASS under the influence of their mutual gravitational attraction.

An *astrometric binary star* contains one component which is too faint to be observed directly. Its presence is inferred from perturbations in the motion of the visible component.

A *close binary star* has components which are too close together to be observed separately. Their orbits may deviate from the usual two-body orbits.

An *eclipsing binary star* consists of components each of which passes in front of the other during one orbital period, as observed from Earth. This produces a brightness variation of the system.

A *spectroscopic binary star* is detected by the cyclic variations in the DOPPLER EFFECT caused by the alternate approach and recession of the components.

A *visual binary star* has components which are sufficiently separated to be observable either by the naked eye or by telescope.

binding energy. (1) The net energy required to remove a particle from its parent system, for example an electron from an atom.

(2) The net energy required to decompose a system into its constituent particles, for example a nucleus into NUCLEONS. *See also* MASS DEFECT.

binocular microscope. *See* MICROSCOPE.

binoculars. A pair of terrestrial TELESCOPES for use with both eyes simultaneously, generally with focusing tubes controlled by a common screw adjustment. *Prism binoculars* incorporate a prism system in each telescope as shown in fig. B6, to produce an upright image; in addition the tube length is shortened compared to that of an ordinary telescope of equal power.

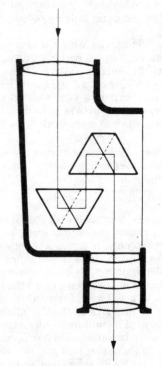

B6 Prism binocular

binocular vision. The simultaneous use of both eyes. Since the eyes are separated, two images from slightly different viewpoints are obtained. The observer's brain synthesizes the two images to give an impression of depth.

binomial theorem. If n is a positive integer,

$$(a + b)^n = \sum_{r=0}^{n} \frac{n!}{r!(n - r)!} a^{n-r}b^r$$

If n is not a positive integer the series

$$\sum_{r=0}^{\infty} \frac{n(n - 1) \dots (n - r + 1)x^r}{r!}$$

converges to the sum $(1 + x)^n$ provided $-1 \langle x \langle 1$. The coefficients of terms in the series are known as binomial coefficients. *See also* FACTORIAL.

bioluminescence. LUMINESCENCE associated with living things, including certain bacteria, fireflies, glow-worms and many deep-sea animals.

biophysics. The application of the methods and principles of physics to the investigation of biological systems. It can be concerned with the physics of the systems themselves, with the biological effects of physical agents and with the use of physical methods in the study of biological problems.

Biot and Savart law. The magnetic field due to a current in a long straight conductor is inversely proportional to the distance of the point of observation from the conductor and directly proportional to the current.

bipolar junction transistor. *See* TRANSISTOR.

biprism. *See* FRESNEL BIPRISM.

birefringence. Another name for DOUBLE REFRACTION.

bisector. A line or plane dividing an angle, line, solid etc. into two equal parts.

bistable circuit. Another name for FLIP-FLOP.

bit. The minimum amount of information necessary to define either of two alternative states. For example, in BINARY ARITHMETIC each of the two digits 0 and 1 is a bit. *See also* BYTE.

black body. A body which absorbs completely any radiation reaching it and reflects none. For equilibrium at any temperature the amount of incident radiation equals the amount of emitted radiation. The spectral distribution of the emitted radiation is described by the PLANCK RADIATION FORMULA. In practice no substance is an ideal black body; the closest practical approximation to black body radiation is the radiation emerging from a small hole or slit in a constant temperature enclosure. *See also* STEFAN'S LAW; WIEN DISPLACEMENT LAW.

black body temperature. The temperature of a body as measured with a radiation PYROMETER. It is generally appreciably less than the true temperature of the body.

black hole. The hypothetical state of a star which has exploded, i.e. become a SUPERNOVA, leaving a core of mass exceeding about three times that of the Sun. The effect of gravitational forces on such a core is to produce high compression of the core, whose gravitational field is then sufficiently large to prevent any matter or radiation escaping from the core. There is some observational evidence for the black hole theory from work on the BINARY STAR Cygnus X1, which is a strong source of X RAYS; this system comprises a SUPERGIANT and a small invisible companion (black hole?) of mass roughly ten times the Sun's. A tentative explanation of the X ray emission is that the presumed black hole is continually attracting matter from the supergiant. In the complex process of entering the black hole, this matter would be compressed and its temperature raised to high values, resulting in the emission of the observed X rays. Behaviour within a black hole is governed by the theory of RELATIVITY. The existence of supermassive black holes, which could be the central powerhouse in quasars etc., has also been postulated.

blanket. *See* NUCLEAR REACTOR TYPES.

blind spot. *See* EYE.

blink comparator. A viewing apparatus by which two photographs, taken at different times, of the same part of the sky can be alternately displayed to the viewer. A change in position of an image appears as an apparent movement, while a pulsation in brightness indicates the presence of a variable star. The apparatus is thus very suitable for detecting position and brightness differences without resort to exact measurements.

Bloch's functions. Solutions of SCHRÖDINGER'S WAVE EQUATION for an electron moving in a potential field varying periodically with distance. The functions are used in the mathematical treatment of the BAND THEORY of solids.

Bloch's wall. A transition layer of finite thickness between adjacent MAGNETIC DOMAINS which are magnetized in different directions. It enables the spin direction to change smoothly rather than abruptly from one orientation to another.

blood flow measurement. Nowadays, usually an electromagnetic measurement in which a magnetic field of square or sawtooth wave form is applied to a bared artery and the potential difference induced across the artery walls by the blood flow is recorded. The induced electromotive force due to the blood flow is of the same form as the magnetic field, whereas the induced electromotive force in the leads is of the form of the rate of change of the magnetic field. Therefore, in contrast to a sinusoidal magnetic field, the flow and lead electromotive forces have different wave forms with a square or sawtooth magnetic field wave form, and so the flow electromotive force can be readily distinguished. In contrast to older methods of blood flow measurement, the electromagnetic method does not require an incision in the artery.

blood pressure. The pressure of the blood on the walls of the arteries. It depends on various physical quantities such as the elasticity of the artery walls, the resistance in the capillaries and the volume and viscosity of the blood. To measure the pressure a SPHYGMOMANOMETER is used.

Board of Trade unit. Symbol BTU. A practical unit of energy equivalent to one kilowatt hour. It is the unit used on some electricity bills.

Bode's law. An empirical relationship between the mean distances of the planets from the Sun. It is of the form

$$x = 0.4 + 0.3 \times 2^n$$

where x is the planet distance in ASTRONOMICAL UNITS and $n = -\infty, 0, 1, 2, ..., 8$. The law yields the results given in the table. It will be seen that the ASTEROID belt is treated as a single planet and that the law fails for Neptune and Pluto.

Bode's Law

planet	n	Bode's law distance an	true mean distance an
Mercury	$-\infty$	0.4	0.39
Venus	0	0.7	0.72
Earth	1	1.0	1.00
Mars	2	1.6	1.52
Asteroids	3	2.8	2.65*
Jupiter	4	5.2	5.20
Saturn	5	10.0	9.54
Uranus	6	19.6	19.20
Neptune	7	38.8	30.10
Pluto	8	77.2	39.50

*Average value

body-centred. Having or involving a crystal structure in which the UNIT CELL has one atom at the centre of the lattice and one at each corner. A unit cell of a body-centred cubic lattice is illustrated in fig. B7.

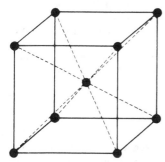

B7 Body-centred cubic unit cell

body force. A force acting throughout a body, for example gravity.

Bohr magneton. The moment of a single electron spin, taken as the unit of atomic magnetic moment and equal to $eh/(4\pi mc)$, where e is the electronic charge, h the PLANCK CONSTANT, m the rest mass of the electron and c the speed of light. The value is

$$9.27 \times 10^{-24} \text{ joule per tesla}$$

Bohr-Sommerfeld theory. *See* BOHR THEORY.

Bohr theory. A combination of QUANTUM THEORY ideas with the concept of an atom as a positively charged nucleus surrounded by electrons moving in circular orbits centred on the nucleus. Bohr postulated that possible electron orbits are restricted to those whose angular orbital momentum mvr equals nh/π, where m is the mass, v the velocity and r the orbital radius of the electron, n is a positive integer and h is the PLANCK CONSTANT. In such orbits the electron does not radiate. It can be shown that

$$r = n^2h^2\varepsilon_0/(Ze^2\pi m)$$

where ε_0 is the ELECTRIC CONSTANT, e the electronic charge and Z the atomic number.

The electron energy, E, is given by

$$E = -Ze^4m/(8n^2h^2\varepsilon_0^2)$$

In the unexcited atom the electrons are in the smallest orbits, i.e. in the lowest energy states. Absorption of a quantum of radiation causes an electron to move to a larger radius orbit. When the electron returns to a smaller radius orbit a quantum of radiation is emitted of energy hv, where v is the frequency of the emitted radiation; hv is the energy difference (i.e. E value difference) between the two orbits. The theory provides an explanation of SPECTRAL SERIES and predicts a value of the RYDBERG CONSTANT for hydrogen in good agreement with experiment.

Later, in conjunction with Sommerfeld, Bohr modified his theory to the *Bohr-Sommerfeld theory* in which the electronic orbits are considered to be ellipses rather than circles. The modification is successful in explaining fine structure in the lines of the hydrogen spectrum. The theories do not explain more complicated atomic phenomena and so have been superseded by WAVE MECHANICS.

boiling. The conversion of liquid to vapour, accompanied by the formation of bubbles in the liquid.

boiling point. The temperature at which the SATURATED VAPOUR PRESSURE of a liquid equals the external pressure. It depends on the intermolecular forces: the stronger these are, the higher the boiling point. The boiling point also rises with increase in external pressure. The value of the boiling point when the external pressure is one standard atmosphere is known as the *normal boiling point*.

boiling water reactor. A NUCLEAR REACTOR in which water is used as both moderator and coolant. *See also* NUCLEAR REACTOR TYPES.

bolide. A very bright and large meteor which sometimes explodes loudly.

bolometer. A highly sensitive instrument for the measurement of small amounts of radiant heat. The radiant heat falls on the blackened side of a thin platinum strip producing a temperature rise in the strip. This temperature rise results in an increase in the strip resistance, which is measured using an electrical bridge.

Boltzmann constant. Symbol k. The UNIVERSAL GAS CONSTANT per atom (or per molecule in the case of a molecular gas). It is thus the universal gas constant per MOLE divided by the AVOGADRO CONSTANT. Its value is

$$1.380\ 54 \times 10^{-23} \text{ joule per kelvin}$$

Boltzmann distribution. The law

$$f = A \exp(-E/kT)$$

where f is the *distribution function*, i.e. the number of particles having positions and speeds within well-defined limits at a given time, in a system consisting of a large number of independent particles in statistical equilibrium; A is a normalization constant, E the particle energy, T the thermodynamic temperature and k the BOLTZMANN CONSTANT.

Boltzmann equation. The equation

$$\partial f/\partial t + V.\ \text{grad}\ f + F.\ \text{grad}_v f = (\partial f/\partial t)_c$$

where f is the distribution function for a system of monatomic particles not in equilibrium; V and F are respectively velocity and acceleration, $\text{grad}_v f$ is the gradient of f in velocity space and $(\partial f/\partial t)_c$ is the rate of change of f due to collisions. *See also* BOLTZMANN DISTRIBUTION.

bomb calorimeter. A calorimeter mainly used to measure HEAT OF COMBUSTION. A specimen of known mass is placed in a small capsule in a steel vessel which can be sealed. Oxygen at approximately 20 atmospheres pressure is then admitted into the sealed vessel which is placed in a lagged vessel containing a known mass of water at a known temperature. The specimen is ignited electrically and the maximum rise in temperature of the water observed. Then using the statement 'heat lost by hot bodies equals heat gained by

cold bodies', the heat of combustion may be calculated.

bond. A union between atoms resulting from their tendency to favour electronic configurations in which at least the outer ELECTRON SHELL is complete. *See also* CO-ORDINATE BOND; COVALENT BOND; ELECTROVALENT BOND; METALLIC BOND.

bond angle. The angle between two BOND LENGTHS in the same molecule.

bond length. The displacement between the nuclei of a pair of atoms between which a BOND exists.

bond strength. The energy required to break a specified BOND in a molecule.

bootstrap theory. A theory of ELEMENTARY PARTICLES in which logical consistency is the ultimate requirement.

Born approximation. A method of calculating cross sections in scattering problems, treating the interactions as perturbations of free particle systems.

Bose-Einstein condensation. The condensation of a vapour, whose molecules obey BOSE-EINSTEIN STATISTICS, to a state in which some of the molecules have a momentum of nearly zero rather than having their momenta spread over a large range of values. The process is thought to be related to the transition between the two forms of liquid HELIUM.

Bose-Einstein statistics. QUANTUM STATISTICS applied to the distribution of similar particles between various allowed energy values when any number of particles can occupy each energy state. At sufficiently high temperature, i.e. when a large number of energy levels are excited, the statistics lead to the classical MAXWELL-BOLTZMANN DISTRIBUTION LAW.

boson. A particle which obeys BOSE-EINSTEIN STATISTICS and does not adhere to the PAULI EXCLUSION PRINCIPLE. Examples include ALPHA PARTICLES,

PHOTONS, PIONS and all nuclei of even MASS NUMBER.

bottom quark. *See* QUARK.

boundary layer. The layer of incompressible fluid of thickness in the range 50 micrometre to 100 micrometre immediately next to a stationary object past which the fluid flows. The behaviour of this layer is important for the study of lift and drag.

Boyle's law. At constant temperature, the pressure of a given mass of gas is inversely proportional to its volume. To some extent the law fails for all real gases but is more closely obeyed the lower the pressure.

Boys' method. A method for the measurement of the RADIUS OF CURVATURE of the surface of a convex LENS. In the diagram (fig. B8), O represents a bright object whose position is such that an image due to reflection at B on the back face of the lens is formed alongside O. Light is therefore reflected normally at B, and the produced path BI of the reflected light is the radius of curvature, *r*, of the back face of the lens. Since I is the virtual IMAGE of O due to refraction by the lens and *u* is measured, the radius of curvature can be calculated from the usual LENS FORMULA, assuming the focal length of the lens is known. The radius of curvature of the other lens face can be measured likewise.

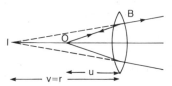

B8 Boys' method for radius of curvature

Brackett series. A series in the SPECTRUM of the hydrogen atom. It is defined by the equation

$$1/\lambda = R(1/4^2 - 1/n^2)$$

where λ is the wavelength of the line, *n* is any integer greater than or equal to 5 and *R* is the RYDBERG CONSTANT. The lines are in the infrared part of the spectrum.

Bragg's law. The condition for maxima in the DIFFRACTION pattern of X RAYS produced by crystal planes is

$$n\lambda = 2d \sin \theta$$

where λ is the X ray wavelength, *d* the distance between parallel crystal planes, θ the angle between the crystal planes and the incident beam and *n* is an integer known as the *order of diffraction*. The equation is based on the assumption that the X rays are reflected from atomic planes and that INTERFERENCE occurs between X rays reflected from different planes. The equation is also valid for ELECTRON DIFFRACTION, NEUTRON DIFFRACTION etc. *See also* X RAY CRYSTALLOGRAPHY; X RAY DIFFRACTION.

breakdown voltage. That voltage at which a given insulator fails to withstand the voltage and starts to conduct. The term is often used in connection with the passage of electric current through a gas.

breaking strain. The strain on a body when BREAKING STRESS is applied.

breaking stress. The tensile STRESS at which a body fractures, or continues to deform with decreasing load.

breeder reactor. A NUCLEAR REACTOR which produces more fuel than it consumes. *See also* NUCLEAR REACTOR TYPES.

bremsstrahlung. The electromagnetic radiation associated with a change in velocity of charged particles. A well known example is the continuous X RAY SPECTRUM emitted by an X RAY TUBE.

Brewster angle. *See* BREWSTER'S LAW.

Brewster's fringes. FRINGES due to INTERFERENCE observed when white light is viewed through two plane parallel plates of glass of nearly equal thickness.

Brewster's law. When light is incident on the reflecting surface of a refracting medium, the reflected light undergoes partial plane POLARIZATION def. (1); the plane-polarized component is largest – almost 100% of the reflected light – for an angle of incidence of $\tan^{-1}n$, where n is the refractive index of the reflecting medium. This angle is known as the *polarizing angle* or *Brewster angle*. Obviously for this angle, the angle between the reflected and refracted rays is 90°. ELECTROMAGNETIC RADIATION in general behaves in a similar way.

bridge.   Any of a variety of electrical four-terminal networks. An e.m.f. is applied to the input terminals and a detector is connected across the output terminals. By adjusting the values of one or more of the network components, the detector reading can be brought to zero. Specific relations then exist between the values of the components so that one component may be measured in terms of the others. The WHEATSTONE BRIDGE is used for measuring resistance as is also the CAREY-FOSTER METHOD; for comparing low-resistance standards the KELVIN DOUBLE BRIDGE, which has eight terminals, is employed. The CAPACITANCE BRIDGE is used to measure capacitance. For comparison of self-inductance with capacitance, the ANDERSON BRIDGE, OWEN'S BRIDGE or MAXWELL's BRIDGE is suitable; for comparison of mutual inductance with capacitance the CAMPBELL BRIDGE may be used.

brillouin zone.   A zone of allowed electron energy levels in a crystal. *See* BAND THEORY.

Brinell number.   *See* BRINELL TEST.

Brinell test.   A test to determine the hardness of a metal by forcing a small hard steel ball into its surface. The hardness is expressed by the ratio of the load on the ball to the area of the depression produced; the ratio is known as the *Brinell number*.

British thermal unit.   Symbol Btu. A unit now taken as having the value

$$1055.055\,853\ \text{joule}$$

It was previously defined as the energy required to raise the temperature of one pound of pure water through 1° F at normal atmospheric pressure.

brittle substance.   A substance which breaks immediately after its ELASTIC LIMIT is reached.

Brownian movement.   Small irregular random movements of small particles suspended in a fluid, arising from bombardment of the particles by the fluid molecules. The theory of Brownian movement is applicable to other problems involving random motion, for example random electron movements leading to NOISE in electrical networks.

brush.   An electrical contact on an ELECTRIC MOTOR or GENERATOR.

brush discharge.   An electric discharge in a gas near high-voltage sharp points. The necessary voltage is less than that required for a SPARK DISCHARGE or an ARC. The discharge appears as short luminous streamers at the tip of the point.

bubble chamber.   An instrument for making visible the tracks of IONIZING RADIATION. It consists of a vessel filled with a liquid (often hydrogen or helium) under pressure at a temperature above the boiling point. SUPERHEATING is achieved by suddenly lowering the pressure on the liquid for a few milliseconds, during which time a photograph is taken. The pressure is reimposed before the liquid boils. Particles passing through the liquid during the superheating cause a track of small bubbles along their path.

It is possible to investigate the DECAY and INTERACTION of particles using a bubble chamber. Large numbers of photographs can be taken from different positions. By applying electric and magnetic fields, the particle tracks can be bent. From a knowledge of the strengths of the electric and magnetic fields and by measuring the amount of curvature, the particles can be identified and their energy evaluated. Since the liquid density is much greater than that of air, the tracks are shorter than in air and so there is a better chance of observing the whole of a reaction than there is when using a CLOUD CHAMBER. Moreover, bubble chamber photographs are clearer than cloud chamber ones.

bulk modulus. The ratio of bulk stress to bulk strain. *See* STRESS; STRAIN.

bulk strain. *See* STRAIN.

bulk stress. *See* STRESS.

bumping. The violent boiling of a liquid which occurs when it is heated to a temperature above its natural boiling point. The bubbles thus have a pressure higher than atmospheric.

Bunsen grease spot photometer. An instrument used to compare the values of LUMINOUS INTENSITY of two light sources. The grease spot is formed on an opaque white screen and is held between the two sources. When juxtaposed mirror images of the two sides of the spot appear identical, the energy received at the screen from each source is the same. Hence by applying the INVERSE SQUARE LAW the source intensities may be compared.

buoyancy. The upward force on a body which is wholly or partially immersed in a fluid. *See also* ARCHIMEDES' PRINCIPLE.

Burgers' vector. The translation vector associated with a particular dislocation in a crystal. Its magnitude indicates the strength of the dislocation.

bwr. Abbrev. for BOILING WATER REACTOR.

byte. A unit of eight BITS which can be stored and handled within a COMPUTER. The capacity of a computer memory is often quoted in bytes.

C

c. Symbol for the SPEED OF LIGHT.

cadmium cell. Another name for WESTON CELL.

cadmium sulphide cell. A layer of cadmium sulphide sandwiched between two electrodes to form a PHOTOCELL. The cell is much more sensitive than a SELENIUM CELL and is used in exposure meters.

caesium clock. *See* CLOCKS.

calcite. A natural transparent form of calcium carbonate which manifests DOUBLE REFRACTION.

calculus. A branch of mathematics concerned with the DIFFERENTIATION and INTEGRATION of functions.

calculus of variations. The study of the maximum and minimum properties of definite integrals. Examples of physical problems for which the method is applicable are the determination of conditions of equilibrium from the LEAST ENERGY PRINCIPLE, the determination of the path of a light ray from FERMAT'S PRINCIPLE OF LEAST TIME and the solution of dynamical problems using the HAMILTON PRINCIPLE.

calendar year. *See* TIME.

calibration. The process in which the readings of an instrument for standard or known quantities are taken, thus enabling arbitrary readings of an instrument to be converted to absolute values.

Callier coefficient. The ratio of the density of a photographic negative as measured with parallel light to that measured with totally diffuse light. The average value is about 1.5.

calomel electrode. A mercury electrode in contact with a solution of potassium chloride saturated with mercurous chloride, i.e. calomel. It is used as a reference electrode.

caloric theory. A theory based on the concept of heat as a weightless fluid. It fails to account for the production of an unlimited amount of heat by friction and was superseded in the mid-19th century when heat was shown to be a form of energy.

calorie. The amount of heat required to raise the temperature of one gramme of water from 14.5° C to 15.5° C at standard pressure. The unit is now obsolete in scientific calculations.

calorific value. The amount of heat liberated by the complete combustion of unit mass of a substance, any steam formed being assumed to condense. The measurement can be made in a BOMB CALORIMETER.

calorimeter. An apparatus for making quantitative thermal measurements. The simplest form, as used for the METHOD OF MIXTURES, consists of a copper can provided with a lid through which pass a thermometer and stirrer; the can is wrapped with insulating material such as cotton wool and placed inside a larger can.

calorimetry. The measurement of quantity of heat, for example SPECIFIC HEAT CAPACITY, CALORIFIC VALUE of a fuel, HEAT OF ATOMIZATION or HEAT OF COMBUSTION.

camera. An apparatus for producing a photographic image. Basically it is a light-tight box with a shuttered convex lens in one side; when the shutter is released the lens focuses a real image on to a photographic film or plate behind the lens. Winding on the film or changing the plate leaves the camera ready for the next exposure.

camera obscura. A box or chamber within which an image of an external scene is formed using a convex lens or a small aperture. Such a device was used in early SUNSPOT studies.

camera tube. *See* TELEVISION.

Campbell bridge. An electrical BRIDGE for measuring capacitance (*C*) in terms of mutual inductance (*M*). The circuit is shown in fig. C1. The detector, D, may be a microphone or oscilloscope. The values of the variable resistors R, R_1, and R_2 are adjusted so that there is zero signal from the detector. Then

$$L/M = (R + R_1)/R$$

and

$$M/C = RR_2$$

where *L* is the self-inductance as indicated.

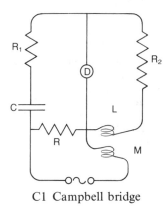

C1 Campbell bridge

canal rays. Streams of positive ions which pass through small holes bored in the cathode of a GAS DISCHARGE TUBE.

candela. Symbol cd. The SI UNIT of LUMINOUS INTENSITY, defined as the luminous intensity in the direction of the normal of a black-body surface, area 1/600 000 square metre, temperature 2040 K

under a pressure of 101 325 pascal. This unit is sometimes called the *new candle.*

candle. An obsolete unit of LUMINOUS INTENSITY. One candle is about 1.02 CANDELA.

candle power. Former name for LUMINOUS INTENSITY.

cannula. A type of VENTURI METER suitable for measuring the flow rate of arterial blood.

canonical. Denoting a standard form of equation, function or rule, especially one of simple form.

canonical equations. Equations representing the motion of a set of particles in classical mechanics. They are

$$dq_n/dt = \partial H/\partial p_n \text{ and } dp_n/dt = -\partial H/\partial q_n$$

where p_n and q_n are respectively the *n*th generalized momentum and position co-ordinate at time *t* and *H* is the HAMILTONIAN.

capacitance. Symbol *C*. The ratio of the electric charge acquired by a body to the resultant change of its potential. The unit is the farad, equal to one coulomb per volt. The capacitance of a given body is constant; for an isolated conducting sphere of radius *a*, the *sphere capacitance* is $4\pi\varepsilon_0 a$, where ε_0 is the absolute PERMITTIVITY for a vacuum.

capacitance bridge. A type of WHEATSTONE BRIDGE arrangement adapted for the measurement of capacitance. The circuit is shown in fig. C2. C_1 is the CAPACITOR of unknown capacitance and the values of capacitance of capacitors C_2, C_3 and C_4 are known, that of C_2 being adjustable. The capacitance of C_2 is varied until the quadrant electrometer, E, shows no deflection when the key, K, is depressed. Then

$$C_1/C_2 = C_3/C_4$$

and so the capacitance of C_1 can be found. (It is usual to use the same symbol for a capacitor as for its capacitance.)

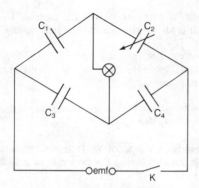

C2 Capacitance bridge

capacitor. A device for storing electric charge; it consists of one or more pairs of insulator-separated conductors. The conductors are called *electrodes* or *plates* and the insulator, which may be solid, liquid or gaseous, is called the DIELECTRIC.

The charge and discharge of a capacitor can be achieved using the circuit shown (fig. C3). Capacitor C is charged when Z is connected to X and discharged when it is connected to Y. At the commencement of charge, electrons flow from the negative terminal of the battery to capacitor plate a and electrons leave capacitor plate b at the same rate; thus a becomes negatively charged and b positively charged with the result that the electron flow is opposed and eventually stopped. During discharge, electrons flow from plate a to plate b until both plates are uncharged. The charge on a capacitor is the charge on its positive plate. The energy stored in a charged capacitor is equal to $CV^2/2$ where C is the CAPACITANCE of the capacitor and V the potential difference between its plates.

The capacitance of a capacitor depends on the absolute PERMITTIVITY of the dielectric and on the shape and size of the conductors.

A *parallel plate capacitor* consists of two parallel conducting plates, each of area A, separated by dielectric of thickness d and absolute permittivity ε. The capacitance is $\varepsilon A/d$. A *variable capacitor* is a capacitor constructed with a set of fixed plates and a set of movable plates so that it is effectively a parallel plate capacitor of variable plate area.

A *coaxial cylindrical capacitor* consists of two hollow coaxial conducting cylinders, radii of cross section a and b respectively and each of length l, separated by a coaxial hollow cylinder of dielectric of absolute permittivity ε. The capacitance is

$$2\pi\varepsilon l/\ln(b/a)$$

A *concentric spherical capacitor* consists of two hollow concentric conducting spheres, radii a and b, separated by a concentric spherical shell of dielectric of absolute permittivity ε. The capacitance is

$$\pi\varepsilon ab/(b-a)$$

An *electrolytic capacitor* is a parallel plate capacitor made by passing a direct current through an electrolyte separating two parallel sheets of aluminium foil. A very thin insulating film of aluminium oxide, which acts as dielectric, then forms on the anode plate. This plate must be connected

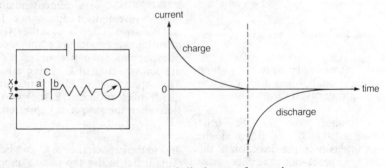

C3 Charge and discharge of capacitor

to the positive side of any circuit in which the capacitor is used; failure so to do results in breakdown of the oxide film. Since the dielectric is thin, the capacitor, although of small volume, can give a large capacitance.

A *mica capacitor* has tin-foil plates and mica dielectric. It is constructed as shown in fig. C4. Two terminals, labelled t, are joined to the plates.

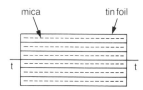

C4 Mica capacitor

A *paper capacitor* has tin-foil plates and paraffin-waxed paper dielectric. It is arranged as shown in fig. C5. A terminal is attached to each piece of foil at opposite ends of the roll.

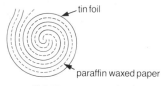

C5 Paper capacitor

Symbols used to represent various types of capacitor – an electrolytic capacitor, a variable capacitor, and one of fixed capacitance – are given in fig. C6.

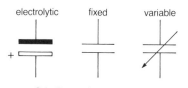

C6 Capacitor symbols

Capacitors *in parallel* are connected together as shown in fig C7 so that the potential difference between the plates of each capacitor is the same; the effective

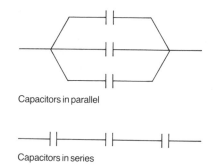

C7 Capacitors in parallel and in series

capacitance is then the sum of the individual capacitances.

Capacitors *in series* are connected together as shown in fig. C7 bottom so that the potential difference between the first and last plate of the capacitor bank is the sum of the potential differences across individual capacitors; the reciprocal of the effective capacitance is then the sum of the reciprocals of the individual capacitances.

capacitor microphone. A MICROPHONE in which a diaphragm forms one plate of a CAPACITOR, the other plate being fixed. Diaphragm movements caused by sound-wave pressure variations result in capacitance variations. The microphone is connected in series with a high resistance across a steady potential difference of about 300 volt; changes in capacitance therefore produce corresponding changes in potential difference across the capacitor. The microphone is free of background noise and has a good frequency response; it suffers however from directional effects and from low output.

capillarity. The study of the influence of SURFACE TENSION on the behaviour of the liquids in vertical capillary tubes, small pores etc. *See also* JUVIN'S RULE.

capillary. Any long hole or pore. For example a capillary tube is a thick-walled glass tube of narrow bore.

capture. Any process whereby an additional particle is acquired by an atom, ion, molecule or atomic nucleus. For example in *radiative capture* a nucleus captures a neutron and then emits a gamma ray; in *K capture* a nucleus acquires a K shell electron and then emits a neutrino, followed perhaps by an electron from an outer shell filling the K shell vacancy, accompanied by emission of an X ray photon.

carat. (1) One of the TROY UNITS OF MASS, equal to 2×10^{-4} kilogramme.

(2) The number of parts by mass of gold in a gold alloy of 24 parts. 24 carat gold is thus pure gold; 20 carat gold has 20/24 of its mass as pure gold and 4/24 as other metals.

carbon cycle. A sequence of six nuclear reactions resulting in the formation of one helium nucleus from four hydrogen nuclei, the carbon effectively acting as a catalyst:

$$^1_1\text{H} + {}^{12}_6\text{C} \rightarrow {}^{13}_7\text{N}$$
$$^{13}_7\text{N} \rightarrow {}^{13}_6\text{C} + {}^0_1\text{e}$$
$$^1_1\text{H} + {}^{13}_6\text{C} \rightarrow {}^{14}_7\text{N}$$
$$^1_1\text{H} + {}^{14}_7\text{N} \rightarrow {}^{15}_8\text{O}$$
$$^{15}_8\text{O} \rightarrow {}^{15}_7\text{N} + {}^0_1\text{e}$$
$$^1_1\text{H} + {}^{15}_7\text{N} \rightarrow {}^{12}_6\text{C} + {}^4_2\text{He}$$

Some stars are believed to derive their energy by this cycle.

carbon dating. A RADIOACTIVE DATING technique for organic remains. In living plants and organisms most of the carbon is ^{12}C but a small amount of the radioactive ^{14}C isotope, of HALF LIFE 5760 years, is also present. In living tissue the ratio of ^{14}C to ^{12}C is constant since ^{14}C is renewed by intake from the environment. In dead tissue no renewal of radioactive carbon occurs and so the isotope ratio and therefore the radioactivity exhibited falls exponentially with time, i.e.

$$\Phi = \Phi_0 e^{-\lambda t}$$

where Φ_0 is the radioactivity of a mass of carbon from a similar but present-day sample and Φ the radioactivity of an equal mass of carbon from the object to be dated; t is the time for which the object has been

dead and λ the decay constant. Hence

$$t = \ln (\Phi_0/\Phi)\lambda = \tau \ln (\Phi_0/\Phi)/\ln 2$$

where τ is the half life of ^{14}C. The values of t so obtained are fairly accurate up to a value of about 4000 year.

carbon microphone. A MICROPHONE which depends for its action on the variation of the electric resistance of carbon with pressure changes in sound waves incident on the microphone.

cardinal points. *See* CENTRED OPTICAL SYSTEM.

Carey-Foster method. A method of finding the difference of resistance between two nearly equal resistors using a METRE BRIDGE. The two resistances form two ratio arms of the bridge and the balance point is found in the usual way; the resistances are then interchanged and the new balance point determined. It can be shown that the distance between the first and second balance points is proportional to the difference between the resistances. The absolute value of the difference can be found by calibrating the bridge wire.

Carnot cycle. A reversible cycle of four operations applied to the gas in a steam engine. The graph of pressure against volume is shown in fig. C8. A — B represents an adiabatic compression, B — C an isothermal expansion at temperature T_1, C — D an adiabatic compression and D — A an isothermal compression at temperature T_2; the pressure, volume and temperature thus return to initial values. The

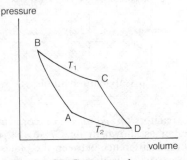

C8 Carnot cycle

Carnot cycle is the cycle of operations occurring in an ideal heat engine.

Carnot's theorem. No engine can be more efficient than a reversible engine working between the same temperatures. It follows that all reversible engines working between the same temperatures are equally efficient, i.e. the efficiency depends only on the temperatures and is independent of the nature of the working substance.

carrier. (1) An electron or HOLE moving through a metal or semiconductor, enabling charge to be transported through the substance and so giving rise to its conductivity.

(2) The normal form of a chemical compound to which the radioactive form is added in order to introduce it into a system for TRACER studies.

carrier wave. A radio wave of constant amplitude and frequency that undergoes MODULATION by the signal in radio transmission.

Cartesian co-ordinates. Co-ordinates used to locate the position of a point with respect to two intersecting lines (axes); each co-ordinate is the distance from one axis measured parallel to the others. In a plane, the x and y axes (shown as rectangular, i.e. meeting at 90°) are as indicated in fig. C9. The co-ordinates of point P are written (a, b); the x co-ordinate, a, is know as the *abscissa* and the y co-ordinate, b, as the *ordinate*. The point where the axes meet is

known as the *origin*, O. In three dimensions, a third axis, the z axis, is introduced; it is perpendicular to the x and y axes at the origin. The co-ordinate of a point P' situated vertically above P outside the plane of the paper are thus (a, b, c) where $c = $ PP'. A positive direction is assigned to each axis, usually as indicated by the arrows in the figure. Distances measured from the origin in the opposite direction are then considered negative.

Cartesian sign convention. *See* OPTICS SIGN CONVENTIONS.

cascade. A process or piece of equipment arranged in a series of consecutive stages so that the output of one stage is the input of the next, as in a cascade of electronic circuits.

cascade liquefaction. A method of liquefaction in which a gas of high CRITICAL TEMPERATURE is liquefied by increase of pressure, and then allowed to evaporate under reduced pressure thus lowering the temperature. This cooling effect is used to reduce the temperature of a second gas below its critical temperature, which is less than that of the first gas. The second gas in turn undergoes the compression-evaporation process thus cooling a third gas, and so on. As illustrated in fig. C10, liquid oxygen may be produced by this process. The process is not suitable for

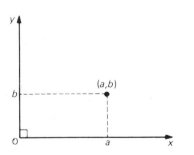

C9 Cartesian co-ordinates

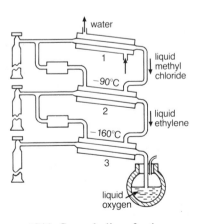

C10 Cascade liquefaction

either hydrogen or helium since their critical temperatures of 33 K and 5.2 K respectively are too low.

cascade shower. The successive production of electrons and high-energy photons by BREMSSTRAHLUNG and PAIR PRODUCTION.

Cassegrainian telescope. An astronomical reflecting telescope of the form shown in fig. C11. The incident light strikes a large concave mirror with a hole at the centre, through which the light is viewed after reflection first at the concave mirror and then at a small convex mirror.

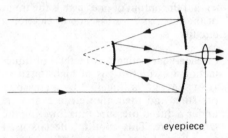

eyepiece

C11 Cassegrainian telescope

catadioptric system. An optical system using mirrors. *Compare* DIOPTRIC SYSTEM.

cataphoresis. ELECTROPHORESIS towards the cathode.

cathetometer. An instrument for measuring vertical heights. It consists of a vertical scale along which a horizontally mounted microscope or telescope can be moved.

cathode. An electrode with a negative electric charge. *Compare* ANODE.

cathode follower. A radio circuit with the load in the cathode circuit; usually the load is in the anode circuit. Although the follower cannot act as a voltage amplifier, it will amplify current over a wide frequency range. The input and output voltages are in phase; the input impedance is high and the output impedance is low.

cathode glow. A small luminous region close to the cathode in a GLOW DISCHARGE.

cathode ray oscilloscope. An instrument whereby electrical signals can be visually displayed on a CATHODE RAY TUBE. The signal under investigation is applied, after amplification, to the vertically deflecting plates of the tube; the amount of deflection of the spot is proportional to the signal strength. The spot is also moved horizontally across the screen by a TIME BASE generator incorporated in the oscilloscope. The resultant trace observed on the screen is therefore a composite of the two voltages and is thus a graph of signal against time. The commonest type of time base is of adjustable frequency and has a SAWTOOTH WAVE FORM so that after traversing the screen at constant speed the spot flies back almost instantaneously to its starting point.

cathode rays. Streams of electrons emitted from the cathode of a GAS DISCHARGE TUBE, CATHODE RAY TUBE etc.

cathode ray tube. A funnel-shaped electronic vacuum tube, shown in fig. C12,

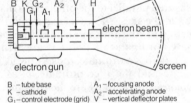

B − tube base	A₁ − focusing anode
K − cathode	A₂ − accelerating anode
G₁ − control electrode (grid)	V − vertical deflector plates
G₂ − accelerating electrode	H − horizontal deflector plates

(1) Electrostatic focusing and deflection

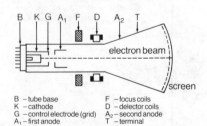

B − tube base	F − focus coils
K − cathode	D − delector coils
G − control electrode (grid)	A₂ − second anode
A₁ − first anode	T − terminal

(2) Electromagnetic focusing and deflection

C12 Cathode ray tube

which enables electrical signals to be visually observed. An electron beam from the cathode eventually strikes a fluorescent screen where it is converted to visible light; the beam intensity and therefore the brightness of the display is controlled by the potential on a grid. The electron beam may be focused and deflected either electrostatically or electromagnetically, or by a combination of both methods. Electrostatic deflection is generally used for the display of high-frequency waves, for example in cathode ray oscilloscopes. Electromagnetic deflection is more suitable for high-velocity electron beams which are required for bright displays as in television and radar receivers.

cation.   An ion with a positive charge and which therefore moves towards the cathode of an electrolytic cell.

Cauchy dispersion formula.   A formula relating the refractive index, n, of a material to the wavelength, λ, of the light passing through it. The formula is

$$n = A + B/\lambda^2 + C/\lambda^4$$

A, B and C are constants. Experimental results for many substances are in agreement with the formula over limited regions of the spectrum.

causality.   The principle that every effect has a definite cause or causes. Since the HEISENBERG UNCERTAINTY PRINCIPLE implies that like individual subatomic particles cannot be identified, the concept of causality is replaced in QUANTUM THEORY by probabilities that specific subatomic particles exist in specific positions and take part in specific events.

caustic.   The curve to which nonaxial rays, after reflection or refraction at a spherical surface, are tangential. The cusp, i.e. the apex of the caustic, is the focal point for paraxial rays.

Cavendish's experiment.   An experiment designed to determine the GRAVITATIONAL CONSTANT, G. A long thin beam, length l, was suspended horizontally by a wire;

each end of the beam carried a small lead sphere, as shown in fig. C13. Near each small sphere was a larger lead sphere located so that the maximum rotation of the rod, due to the gravitational attraction of the spheres, was obtained. The large spheres were then moved so that the maximum deflection in the other direction was obtained, and the average deflection, θ, found. A value of the gravitational constant was obtained using the formula

$$G = cr^2\theta/(mMl)$$

where m and M are respectively the masses of the small and large spheres, r the separation of their centres and c the TORSIONAL CONSTANT. Subsequently the method was refined.

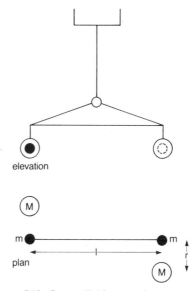

C13 Cavendish's experiment

cavitation.   The formation of small vapour-filled bubbles in the high-velocity regions of a liquid, for example in pumps, close to propellers etc. The bubbles, which may cause surface pitting, collapse when they reach a region of higher pressure.

cavity resonator.   A closed conducting surface which is excited by external means so that an oscillating electromagnetic field is maintained within it. The device has several resonant radio frequencies whose values depend on the cavity dimensions. The main uses are at frequencies greater than 300 megahertz.

celestial equator. The circle of projection on the CELESTIAL SPHERE of the Earth's equator.

celestial mechanics. The branch of ASTRONOMY dealing with the relative motion of celestial objects due to gravitational forces.

celestial meridian. *See* CELESTIAL SPHERE.

celestial poles. The two points in the sky where the Earth's axis would intersect the CELESTIAL SPHERE.

celestial sphere. A sphere concentric with the Earth and of infinite radius; it rotates once in every 24 hours of SIDEREAL TIME and is used for positional astronomy. The main features are shown in fig. C14. EQ is the CELESTIAL EQUATOR and N and S the north and south CELESTIAL POLES; EC is the ECLIPTIC; γ and ≏ are the points of intersection of EQ and EC, i.e. the vernal and autumnal EQUINOXES; HO is the *horizon*, the great circle having an observer's ZENITH, Z, and NADIR, Z_0, as poles; n is the *north point* of the horizon, the point of intersection of ZN extended and HO, and s is the *south point* diametrically opposite; ZNn marks the *celestial meridian* for an observer on Earth.

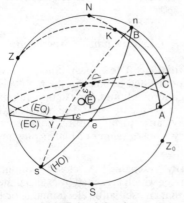

C14 Celestial sphere

K is a celestial object, γA and AK are a pair of co-ordinates giving the position of K. γA being the *right ascension* of K (measured in hours, minutes and seconds anticlockwise from γ) and AK the *declination* of K (measured in degrees above or below EQ); right ascension and declination are widely employed for astronomical observations. nB and BK are another pair of co-ordinates, nB being the *azimuth* of K (measured in degrees east of the north point), BK being the *altitude* of K (measured in degrees above or below HO).

cell.   (1) A device for converting chemical energy directly into electric energy. It consists of two electrodes dipping into an electrolyte; ions are produced or discharged at each electrode so that one electrode becomes positively charged and the other negatively charged. When the electrodes are connected through a circuit, current flows; if this happens as soon as the cell is assembled it is called a *primary cell*. In contrast a *secondary cell* requires charging after its assembly, i.e. a current must be passed through it in the reverse direction to its discharge; the chemical actions in this cell are reversible.

Cells connected so that all their positive terminals are joined, and likewise all their negative ones, are said to be connected *in parallel*. If the cells are identical their combined electromotive force is the electromotive force of any one of them. When the negative terminal of one cell is connected to the positive terminal of the next cell, and so on, the cells are connected *in series* The combined electromotive force is then the sum of the individual electromotive forces. *See also* ACCUMULATOR.

(2) Any of the various devices similar to the above device (def. 1) and used for ELECTROLYSIS, study of electrochemical processes etc.

Celsius scale. The official name for the CENTIGRADE SCALE. The term has been recommended since 1948, but centigrade remains in widespread use.

cent. The interval equal to 1/1200 of the frequency difference of notes which differ by one octave.

centi-. Symbol c. A prefix meaning 10^{-2}.

centigrade scale. A temperature scale in which the melting point of ice is 0° and the boiling point of water 100°. The degree centigrade, symbol ° C, is a hundredth of this temperature interval and is equal in magnitude to the KELVIN.

centimetric waves. Microwaves of wavelength in the range 1–10 centimetre.

centrad. One hundredth of a radian. The unit is used to specify angles of deviation produced by small-angled prisms.

central force. A force which acts on a moving body and is always directed towards a fixed point, or towards a point moving according to known laws; the body is said to be in a *central orbit.*

central processing unit. The part of a digital COMPUTER in which operations are performed on the data.

centred optical system. An optical system composed of a number of spherical reflecting or refracting surfaces having their centres on a common axis. To calculate the position, nature and size of an image formed by rays close to the axis in any such system, the positions of only six special axial points need be known; the points are called *cardinal points.*

Two of these cardinal points are known as *focal points:* the first focal point is the axial point imaged by the system at infinity; the second focal point is the image formed by the system of the axial point at infinity. Another pair of cardinal points, known as *principal points,* are conjugate, i.e. one is the image of the other; they are associated with identical object and image, i.e. unit lateral MAGNIFICATION, and no inversion. The remaining pair of cardinal points are the *nodal points,* which are conjugate points for which the angular magnification, i.e. the MAGNIFYING POWER, is unity. If the medium on the object side of the system is the same as that on the image

side, the nodal points coincide with the principal points.

Extensions of the constructions given in the dictionary entry IMAGE – in detail for an image in a curved mirror and in brief for an image formed by a thin lens –permit the graphical location of the image formed by a centred optical system for different object positions, making use of the cardinal point locations. If object distance is measured from the first principal point and image distance from the second principal point, the elementary MIRROR FORMULA applies; the elementary LENS FORMULA becomes

$$1/f = 1/v - 1/u$$

since it is more convenient to use the Cartesian sign convention rather than the 'real is positive' convention when dealing with complex optical systems (*see* OPTICS SIGN CONVENTIONS). The mirror formula is the same in both systems.

centre of area. Another name for CENTROID.

centre of buoyancy. The CENTRE OF MASS of the fluid which would occupy the space of the immersed part of a body floating in the fluid.

centre of curvature. The centre of a circle whose curvature is in the same direction as the curve at the point under consideration on the curve, and which shares a common tangent with the curve at that point and has a radius equal to the RADIUS OF CURVATURE of the curve at that point.

centre of figure. Another name for CENTROID.

centre of gravity. The point through which acts the resultant of all the forces acting on a body due to its presence in a gravitational field. If the gravitational field is uniform, the gravitational forces acting on each particle of the body are parallel; their resultant therefore passes through a point fixed relative to the body, i.e. through the CENTRE OF MASS, with which the centre of gravity then coincides. If the gravitational field is nonuniform, the resultant forces on a body are in general equivalent to a single force and a couple; the single force does not generally pass through a point fixed relative to the body for all positions of the body and so the centre of gravity does not exist. If, however, the matter in the body is distributed with spherical symmetry, then the couple reduces to zero and the force always passes through the centre of mass, i.e. the body has a centre of gravity. Hence in common usage, centre of mass and centre of gravity are taken to be synonymous, since if a centre of gravity exists it coincides with the centre of mass.

centre of mass. A point such that the sum, for all particles making up a body of the product of particle mass and particle perpendicular distance from any plane through the point is zero. *See also* CENTRE OF GRAVITY.

centrifugal force. *See* CENTRIPETAL FORCE.

centrifugation separation. A technique for separating different phases of the same material, or different materials in the same phase, using a CENTRIFUGE.

centrifuge. A bar or flywheel mounted on a vertical axle about which it can rotate rapidly. Tubes containing the specimens under investigation are attached to the extremities of the wheel or bar in such a way that they are free to pivot outwards at high speed. The forces produced greatly exceed the gravitational force and so the settling of suspensions can be speeded up. The device is also used for the measurement of particle mass and density. *See also* SEDIMENTATION; SEDIMENTATION DENSITY GRADIENT METHOD; SEDIMENTATION EQUILIBRIUM METHOD; ULTRACENTRIFUGE.

centripetal force. A force causing a body, mass m, to move in a circular path of radius r. The force is directed towards the centre of the circle and is of magnitude mv^2/r, where v is the speed of the particle. The body exerts a *centrifugal force*, of equal magnitude but opposite direction to the centripetal force, on whatever provides the centripetal force.

centroid. A point associated with a geometric shape. It coincides with the CENTRE OF MASS of a homogeneous solid body of shape congruent with the geometric one. For example the centroid of a plane figure is the centre of mass of a thin sheet of material of the same shape and dimensions as the figure.

Cerenkov radiation. The bluish light emitted by a beam of high-energy charged particles when travelling through a medium of refractive index n at a speed v which is greater than the speed of light in the medium. The bluish light is emitted in a conical wave front whose axis lies in the beam direction. The cone's apex angle is 2θ where
$$\cos \theta = c/(nv)$$
c is the speed of light in empty space; v may thus be found by measuring θ. The Cerenkov radiation is an electromagnetic shock wave, i.e. it is the optical analogue of the SONIC BOOM.

CGS units. A system of units based on the centimetre as unit of length, the gramme as unit of mass and the second as unit of time. The system was extended to cover thermal measurements by the addition of the inconsistently defined CALORIE. Two methods of further extending the system to embrace electrical measurements were developed; they involved the introduction of *electromagnetic units* and *electrostatic units*. On the electromagnetic system the absolute PERMEABILITY of free space is unity, whereas on the electrostatic system it is the absolute PERMITTIVITY of free space that is unity. All versions of the CGS system have now been superseded by SI UNITS. *See* Table 6.

Chadwick's experiment. An experiment concerned with finding the particle mass of a very penetrating radiation produced when alpha particles were incident on beryllium. Chadwick directed the radiation at various materials. From observation of the energies of the resulting particles he concluded that each particle of penetrating radiation had a mass very close to that of a proton but that it was uncharged. He called the new particle a NEUTRON.

chain reaction. A series of nuclear reactions initiated by a single nuclear FISSION. For example the fission of a uranium-235 nucleus is accompanied by the emission of from one to three neutrons, any of which is capable of causing fission in another ^{235}U nucleus. When each fission causes, on average, one further fission, the reaction is said to be *critical*; below this average the reaction is described as *subcritical* while above it is said to be *supercritical*. *See also* CRITICAL MASS.

change of state. The change of a substance from one state of matter to another, brought about for example by boiling, condensing, freezing, melting or subliming.

channel. (1) A specified band of frequencies, or a particular path, used in the transmission and reception of electric signals.
(2) The region in a FIELD EFFECT TRANSISTOR between source and drain whose conductivity is modulated by the voltage applied to the gate.

chaotic system. A system in which, for any sort of motion, slightly different initial conditions can lead to vastly different outcomes. An example is English weather.

characteristic. *See* LOGARITHM.

characteristic curve. A graph showing the relationship between two variables of importance for a device. One example is the plot of photographic density against logarithm of exposure for specified conditions of processing a particular photographic emulsion; another is the plots of voltage against current for various electrical and electronic devices.

characteristic equation. Another name for EQUATION OF STATE.

characteristic function. One of a set of functions satisfying a particular equation with specified boundary conditions. In WAVE MECHANICS characteristic functions are physically possible solutions of SCHRÖDINGER'S EQUATION for an atomic particle. Characteristic functions also occur frequently in QUANTUM MECHANICS under the name *eigenfunctions*.

characteristics of notes. The distinguishing attributes of musical notes, namely their PITCH, LOUDNESS and QUALITY.

characteristic value. The value of some parameter associated with a CHARACTERISTIC FUNCTION. If there is more than one solution of the differential equation corresponding to a particular characteristic value, the system is said to be *degenerate*. If the characteristic value is associated with an eigenfunction, it is generally known as an *eigenvalue*.

characteristic X radiation. A series of discrete-wavelength X RAYS characteristic of the emitting element and independent of its state of chemical combination.

charge. (1) Symbol Q. The property of certain elementary particles enabling them to attract or repel one another by ELECTROMAGNETIC INTERACTION. Charge is conventionally designated positive or negative such that like charges repel one another and unlike charges attract. The force is thought to result from the exchange of photons between charged particles. The natural unit of negative charge is that possessed by the ELECTRON; the proton carries a positive charge of equal magnitude. The charge on the electron, symbol e, may be determined employing MILLIKAN'S ELECTRONIC CHARGE DETERMINATION method.
The charge of a body or region arises either from an excess or from a deficit of electrons with respect to protons. Quantitatively charge is the integral of electric current with respect to time and is measured in coulomb.
The principle of *conservation of charge* states that charge can neither be created

nor destroyed, but it may be neutralized. Thus when bodies are charged by rubbing for example amber with fur, the amount of negative charge on the amber exactly equals the amount of positive charge on the fur. The principle is an example of CONSERVATION LAW.

(2) The quantity of electricity stored in for example an accumulator.

(3) To store electric charge in a device.

charge conjugation parity. Symbol C. A quantum number associated with elementary particles of zero CHARGE, BARYON NUMBER and STRANGENESS. It is conserved in both STRONG INTERACTION and ELECTROMAGNETIC INTERACTION.

charge coupled device. A type of CHARGE TRANSFER DEVICE.

charge density wave. The ground state of a metal in which the conduction electron charge density is sinusoidally modulated in space.

charged particle. Any particle that carries a positive or negative CHARGE, the electron and proton being examples. When a charged particle moves into a magnetic field whose direction is perpendicular to the direction of motion of the particle, the particle is deflected as indicated in fig. C15,

its path becoming circular while in the field. It will be seen that Fleming's left-hand rule (*see* FLEMING'S RULES) is obeyed, bearing in mind that the conventional current direction is opposite to the direction of motion of negatively charged particles and in the same direction as the motion of positively charged particles. If the particle speed is v, its charge q, and the magnitude of the magnetic induction B, then the magnitude of the force on the particle is Bqv and its direction is perpendicular to the particle path and to the magnetic field. If r is the radius of the resulting circular path, then the CENTRIPETAL FORCE is

$$Bqv = mv^2/r$$

A magnetic field, unlike an electric one, does not change a particle's energy.

charge independence. *See* STRONG INTERACTION.

charge to mass ratio of electron. A quantity first measured by Thomson using the apparatus illustrated in fig. C16. A beam of electrons was subjected to magnetic and electric fields such that they deflected the beam in opposite directions; the strengths of the deflecting fields were adjusted to give zero deflection of the spot where the beam struck the fluorescent screen. It can then be shown that

$$e/m = ED/(B^2Ll)$$

e and m are respectively the charge and mass of the electron; D is the magnitude of the spot deflection produced by either the magnetic field or the electric field acting alone; l is the length of the deflecting plates and L the distance from the centre of the deflecting plates to the screen; E and B are

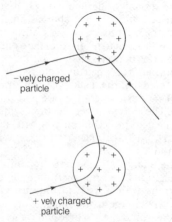

−vely charged particle

+ vely charged particle

+ indicates magnetic field direction into plane of paper

C15 Charged particle in magnetic field

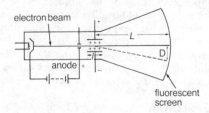

electron beam

L

anode

D

fluorescent screen

C16 Thomson's method for finding e/m

the electric field strength and magnetic induction magnitudes respectively.

charge transfer device. A semiconductor device in which discrete packets of charge are transferred from one location to the next; the device can be used for short-term storage of charge in a particular location.

charging by induction. The process whereby a conductor is charged by a charged body without physical contact between them. The charged body attracts charge of opposite sign to its own on the closest parts of an uncharged conductor to which it is near; charges of the same sign are repelled to the remotest parts of the conductor. Momentary earthing removes the latter charge, and on withdrawing the charged body the remaining charge on the conductor spreads over its surface. The conductor has thus been given a permanent charge of opposite sign to that of the original charged body and without loss of charge of the latter.

Charles' law. The volume of a fixed mass of any gas at constant pressure is directly proportional to the thermodynamic temperature. An alternative statement is that for every change of temperature of 1 K, the volume of a fixed mass of any gas at constant pressure changes by 1/273 of its volume at 273 K. The law is not strictly valid for actual gases, but is the better obeyed the lower the pressure.

charm. A property possessed by any matter which contains a charmed QUARK.

charmed quark. *See* QUARK.

chemical bond. Another term for BOND.

chemical energy. Energy stored in the chemical BONDS of a substance and which is converted to another form when a chemical reaction occurs. For example chemical energy is converted into heat when a substance burns.

chemical hygrometer. A HYGROMETER consisting of drying tubes of known mass through which a known volume of air is drawn; the increase in mass of the tubes gives the amount of water vapour present. The experiment is then repeated for an equal volume of air which has been passed through water before entering the drying tubes. The ratio of the first mass to the second is the relative HUMIDITY of the atmosphere.

chemical shift. A small change in the wavelength of a spectrum peak due to a chemical effect.

chemiluminescence. *See* LUMINESCENCE.

chemisorption. *See* ADSORPTION.

chemosphere. Another name for THERMO-SPHERE.

chief ray. The ray from an object point to the centre of the ENTRANCE PUPIL of an optical system.

chip. A very small piece of SEMICON-DUCTOR containing a component, for example a resistor, transistor etc., or an INTEGRATED CIRCUIT.

chi-squared test. A statistical significance test applied by summing the square of the difference of a result from the expected value, divided by the expected value, for all the results. The sum is denoted by χ^2; the smaller it is, the greater the agreement between observed and expected values.

Chladni's figures. Regular patterns obtained by scattering fine powder on a horizontal vibrating plate; the powder collects along the lines of least vibration.

choke. (1) An inductor which offers impedance to the flow of alternating current in a circuit. It may thus be used to smooth the output of a rectifying circuit.
  (2) A groove cut into the metal surface of a wave guide to a quarter-wavelength depth. It prevents the escape of microwave energy.

chloroplast. A plant cell responsible for PHOTOSYNTHESIS.

48   chord

chord. A straight line joining two points on a curve.

chromatic aberration. An ABERRATION of lenses arising from the dependence of the refractive index of a material on wavelength. The image of a point source of white light therefore has a blue surround at the focus for red light and a red surround at the focus for blue light. At a position between the blue and red foci the image is white circle, known as the *circle of least confusion*.

The distance between the foci of the C (red) and F (blue-green) wavelengths of the hydrogen spectrum, which is chosen as a comparison standard, is known as the *longitudinal chromatic aberration*. The difference between the corresponding powers of the lens is commonly known as the chromatic aberration, and for a thin lens equals ωP where ω is the dispersive power of the glass and P is the power of the lens for the D (yellow) wavelength of the sodium spectrum. The difference in image size for the C and F wavelengths is known as the *lateral chromatic abberation*. When chromatic abberation is corrected for C and F wavelengths (*see* ACHROMATIC LENS) there still remains some residual chromatic aberration for other wavelengths; this is known as the *secondary spectrum*. *See also* APOCHROMATIC LENS.

chromatic colour. A colour other than black, grey or white.

chromaticity. The colour quality of light, definable by two CHROMATICITY CO-ORDINATES.

chromaticity chart. A plane diagram obtained by plotting one of the three CHROMATICITY CO-ORDINATES against either of the other two. A typical $x-y$ plot is shown in fig. C17. The locus of all visible wavelengths is horseshoe shaped. The straight line joining its ends is the *locus of purples*. A point representing any other colour lies within the figure, the point W representing white. The wavelength at which the line from W passing through a point representing any other colour cuts the curve is the

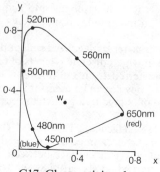

C17 Chromaticity chart

dominant wavelength for that colour. The ratio of the distance of the colour point from W to the distance of the dominant wavelength point from W is known as the *excitation purity* for the colour.

chromaticity co-ordinates. The co-ordinates x, y and z, defined by

$$x = X/(X + Y + Z), y = Y/(X + Y + Z)$$
$$\text{and } z = Z/(X + Y + Z)$$

X, Y and Z are known as *tristimulus values*. They are respectively the amounts of three stimuli, usually red, green and blue lights, required to match exactly the colour under consideration. Since

$$x + y + z = 1$$

any two of the co-ordinates will completely specify a colour. When x, y and z are all equal to 1/3 the colour is white.

chromaticity diagram. Another name for CHROMATICITY CHART.

chromatic resolving power. Another name for RESOLVING POWER of an optical spectrometer.

chromatography. Any technique for separating and analysing mixtures by selective ABSORPTION or ADSORPTION in a flow system.

chromosphere. The inner gaseous atmosphere of the SUN, extending about 15 000 metre beyond the PHOTOSPHERE. *See also* CORONA.

chronon. The time taken by a photon to travel a distance equal to the diameter of the electron; the time is about 10^{-24} second.

chronoscope. An electronic instrument for measuring very short time intervals.

chronotron. A device for electronically measuring the time interval between events.

ciliary body. *See* EYE.

circle. The locus of a point moving so that its distance from another point remains constant. If r is the radius of the circle then its circumference is $2\pi r$ and its area πr^2. An ellipse of zero eccentricity is a circle.

circuit. An arrangement of electrical conductors, for example resistors or inductors, connected together to form a conducting path. When current passes a *closed circuit* is formed; in the absence of current an *open circuit* exists.

circuit-breaker. A device for making and breaking an electric circuit. *See also* CONTACTOR; SWITCH.

circular measure. The measurement of angles in RADIAN.

circular mil. An obsolescent unit of area equal to the area of a circle of diameter 0.001 inch.

circular motion. The motion of a body round a circle. If a body moves with constant speed v round a circle of radius r, it has an acceleration of magnitude v^2/r directed towards the centre of the circle. Since $v = r\omega$ where ω is the body's angular speed, the acceleration magnitude may also be written as $r\omega^2$.

circular polarization. *See* POLARIZATION (electromagnetic).

circumcentre. The centre of the circle passing through all three vertices of a triangle. It is the point of intersection of the perpendicular bisectors of the sides of the triangle.

circumference. The line enclosing any simple closed curve, for example a circle.

circumpolar. Denoting stars or constellations which are always above the horizon when observed from a given geographical latitude.

cladding. (1) The process of bonding one metal to another to prevent corrosion.
(2) The material used to enclose the fuel in a NUCLEAR REACTOR. It is designed to prevent the escape of fission products and to support the fuel.

Clark cell. A voltaic cell whose positive electrode is mercury coated with mercury sulphate paste; the negative electrode is zinc and the electrolyte is a saturated zinc sulphate solution. The electromotive force is 1.4345 volt at 15° C and was used as a standard until the Clark cell was superseded by the WESTON CELL for this purpose.

class A amplifier. An amplifier operating so that output current flows during the whole of the input cycle. It has low distortion and low efficiency.

class AB amplifier. An amplifier operating so that output current flows for between one half and the whole of the input cycle. It tends to operate as CLASS A AMPLIFIER for low input signal levels and as a CLASS B AMPLIFIER for high ones.

class B amplifier. An amplifier operating so that output current flows for only half the input cycle. It is efficient but produces some distortion.

class C amplifier. An amplifier operating so that output current flows for less than half the input cycle. It is efficient but produces appreciable distortion.

class D amplifier. An amplifier operating by means of PULSE MODULATION. The input signal modulates the MARK SPACE

RATIO of a square wave causing it to operate a switching system. The resulting output current is proportional to the mark space ratio and therefore to the input signal. Such amplifiers are very efficient but switches operating sufficiently fast to avoid distortion are generally impracticable.

classical physics. That part of physics excluding relativity, nuclear physics and quantum theory.

Clausius-Clapeyron equation. The equation

$$dp/dT = L/[T(V_2 - V_1)]$$

where L is the specific latent heat for the change of a substance from one phase to another at thermodynamic temperature T; V_1 and V_2 are the specific volumes of the substance in the two different phases and p is the vapour pressure.

Clausius' equation. The equation

$$c_2 - c_1 = Td(L/T)/dt$$

where c_2 and c_1 are the specific heat capacities of a substance for the liquid and vapour phases respectively; L is the specific latent heat of vaporization at thermodynamic temperature T.

Clausius' theorem. For a system undergoing any reversible cycle of changes in which it returns to its initial state,

$$\int \Delta Q/T = 0$$

where ΔQ is the infinitesimal quantity of heat absorbed by the system at thermodynamic temperature T.

cleavage. A splitting of a crystal to yield two smooth surfaces. It occurs along planes of atoms in the crystal.

Clebsch-Gordon coefficient. Another name for VECTOR COUPLING COEFFICIENT.

clinical thermometer. A mercury in glass thermometer used to measure body temperature. Unlike other types of mercury in glass thermometer the clinical ther-

mometer has a constriction in the capillary just above the bulb. This leads to ease of reading, the thread of mercury remaining in position after reaching its maximum value. The mercury may be returned to the bulb by shaking the thermometer. The calibration range is usually 35°–43° C, i.e. 95°–110° F; normal human body temperature is 36.9° C, i.e. 98.4° F.

clocks. Devices for measuring time, generally using a periodic process of constant frequency. The more important types are as follows.

The *pendulum clock* makes use of the fact that the period of a pendulum is a function of its length only; such a clock is accurate to about 0.01 second per day.

The *quartz clock* has a regulating device in the form of an oscillator controlled by a crystal of quartz, which is kept in resonant vibration by means of the PIEZOELECTRIC EFFECT; the clock is accurate to about 0.001 second per day.

The quartz clock is an example of a *crystal clock*.

The *atomic clock* is a device in which the periodic process is a molecular or atomic event associated with a particular spectral line; for example the nitrogen atom in an ammonia molecule can be stimulated by a quartz oscillator to vibrate at a frequency of 23 870 hertz between equivalent positions on either side of the plane of hydrogen atoms. Should the oscillator vary from the required frequency, its energy is no longer absorbed by the ammonia and is used in a feedback circuit to correct the oscillator.

The *caesium clock* is a device in which the frequency is defined by the energy change induced in the caesium-133 nucleus by a magnetic field (*see* NUCLEAR MAGNETIC RESONANCE); the associated radiation frequency is

9 192 631 770 hertz

and is accurate to better than one part in 10^{13}. This clock is used in the definition of the second in SI UNITS.

close binary star. *See* BINARY STAR.

closed circuit. *See* CIRCUIT.

closed circuit television. A television system in which camera, control system and receiver are cable linked so that there are no aerials or open circuits.

closed pipes. *See* VIBRATIONS IN PIPES.

close packed structure. A structure obtained when atoms, assumed to be spheres of equal radius, are packed together so as to occupy a minimum volume. The structure is either face centred cubic or close packed hexagonal or combinations of these (*see* CRYSTAL SYSTEM). Each atom is symmetrically surrounded by 12 others.

cloud. A mass of minute water drops or ice crystals suspended in the atmosphere and appearing as an opaque drifting body. Minute atmospheric particles such as smoke encourage the formation of drops. Clouds are usually classified by their appearance and height.

cloud chamber. An instrument for making the tracks of ionizing particles visible as rows of little droplets, formed by condensation from a supersaturated vapour. The supersaturation may be produced for a short time by rapid expansion of the vapour causing adiabatic cooling. Continuous supersaturation is obtained in the DIFFUSION CLOUD CHAMBER.

Clusius column. An apparatus for separating gaseous isotopes. It consists of a vertical column many metres high with a heated wire running down its axis. A radial temperature gradient builds up between the wire and the tube wall. As a result of THERMAL DIFFUSION heavier molecules collect around the wire and lighter ones diffuse towards the wall; the lighter isotope collects at the top of the tube.

cluster. A collection of stars moving in the same direction. There are two main types. An *open cluster* comprises relatively small numbers of widely spaced young stars and is found in the spiral arms of the Galaxy. A *globular cluster* consists of enormous numbers of closely packed older stars and is found in the spherical halo surrounding the central nucleus of the Galaxy. Neighbouring open clusters are much closer to each other than are neighbouring globular ones.

coax. Short for COAXIAL CABLE.

coaxial cable. A cable consisting of a wire enclosed in an insulating coating several millimetre thick, which in turn is surrounded by a coaxial conducting cylinder which is often earthed. Coaxial cables do not produce external fields and are not affected by them and so are used for the transmission of high-frequency signals.

cochlea. A snail-shaped cavity within the ear. It is where the main process of hearing occurs.

Cockcroft-Walton accelerator. *See* COCK-CROFT-WALTON GENERATOR.

Cockcroft-Walton generator. A high-voltage direct-current machine especially used for proton acceleration.

Coddington lens. A powerful magnifying glass comprising a complete sphere with a central stop.

coefficient. (1) A constant multiplier of a variable in an algebraic expression; for example in $5y^3 + 2y$, 5 and 2 are coefficients.
(2) A constant measuring some physical property.

coefficient of contraction. The ratio of the area of the VENA CONTRACTA of a fluid jet to the area of the orifice through which it is discharging; values lie between 0.5 and 1.

coefficient of coupling. The ratio of the actual mutual inductance between two coils to the maximum possible value.

coefficient of cubic expansion. *See* CO-EFFICIENT OF EXPANSION.

coefficient of dynamic friction. *See* FRICTION (def. 1).

coefficient of expansion. The fractional increase in size of a substance per unit temperature rise; the unit is thus K^{-1} or $°C^{-1}$.

A solid has three coefficients of expansion: $α$, which refers to length and is known as the *coefficient of linear expansion*; $β$, which refers to area and is known as the *coefficient of superficial expansion*; $γ$, which refers to volume and is known as the *coefficient of cubic expansion* or the *volume expansivity*. For linear expansion,

$$l_t = l(1 + αt),$$

where l is the original length and l_t the length after an increase of t in temperature. Analagous expressions apply to superficial and cubic expansion. Since the coefficients are small it can be shown that $β = 2α$ and $γ = 3α$. Coefficients of expansion of solids are measured by a COMPARATOR method or by INTERFERENCE techniques.

Liquids have two coefficients of cubic expansion: a *coefficient of apparent expansion*, for which no correction is made for the expansion of the container, and a *coefficient of absolute expansion*, based on the real expansion of the liquid. The absolute coefficient equals the sum of the cubic coefficient for the container material and the apparent coefficient for the liquid. Apparent coefficients can be measured by DILATOMETER.

In contrast to solids and liquids, the coefficient of cubic expansion of a gas at constant pressure is defined with reference to the gas volume at $0°$ C, i.e.

$$V_t = V_0(1 + γ)$$

where V_t and V_0 are the volume of the gas at $t°$ C and $0°$ C respectively, the pressure remaining constant. Likewise the coefficient of increase of pressure of a gas at constant volume is the ratio of the change in pressure per $°$ C change in temperature to the pressure at $0°$ C, the volume remaining constant. For an ideal gas, both the cubic and pressure coefficients equal 0.003 660 8 per $°$ C.

coefficient of extinction. Another name for LINEAR ATTENUATION COEFFICIENT.

coefficient of linear expansion. *See* CO-EFFICIENT OF EXPANSION.

coefficient of mutual inductance. *See* ELECTROMAGNETIC INDUCTION.

coefficient of restitution. Symbol e. The ratio of the component of the relative velocity before impact, along the line of centres of two spheres, to the component of the relative velocity along the same line, in the opposite direction, after impact. The ratio is constant and depends only on the material of the spheres; for perfectly elastic material the ratio is 1; for absolutely inelastic material it is 0. For other material it lies between 0 and 1; for example for ivory it is very nearly 1 and for putty it is almost 0. The value of e varies slightly at high relative velocity.

coefficient of self-inductance. *See* ELECTROMAGNETIC INDUCTION.

coefficient of static friction. *See* FRICTION (def. 2).

coefficient of superficial expansion. *See* COEFFICIENT OF EXPANSION.

coefficient of viscosity. *See* VISCOSITY.

coercive force. The magnetic field strength required to remove the residual magnetism of a ferromagnetic material. If the substance is initially saturated the coercive force is known as the *coercivity*.

coercivity. *See* COERCIVE FORCE. *See also* HYSTERESIS LOOP.

coherent radiation. Electromagnetic radiation of single PHASE. Most practical radiation sources are not coherent over a significant period of time since each emits large numbers of waves all out of phase with each other. In contrast LASER radiation is coherent. Two sources of electromagnetic radiation are said to be coherent if they produce in-phase wave forms.

coherent scattering. *See* SCATTERING.

coherent units. A system of units in which a *derived unit* is produced by multiplying or dividing *base units* without introduction of a numerical factor; the base units are an arbitrarily defined set of physical quantities. SI UNITS are coherent.

cohesion. A type of force between molecules. For example when a liquid is split, the greater the force of cohesion between the molecules the greater the chance of drop formation (rather than spreading over the receiving surface).

coincidence circuit. A circuit with two or more input terminals, all of which must receive a pulse within a specified time interval if an output pulse is to occur.

colatitude. *See* POLAR CO-ORDINATES.

cold cathode. An electronic-tube cathode which emits electrons by FIELD EMISSION rather than by THERMIONIC EMISSION.

cold trap. A tube cooled with liquid air or frozen carbon dioxide in acetone. The trap is used to condense vapour passed into it.

collapsar. Another name for BLACK HOLE.

collector. The electrode in a TRANSISTOR through which a CARRIER leaves the inter-electrode region.

collector ring. Another name for SLIP RING.

colligative property. A property depending on the number of particles present in a substance rather than on the nature of the particles. Examples are ELEVATION OF BOILING POINT and DEPRESSION OF FREEZING POINT.

collimator. (1) A system for producing a beam of parallel light or other radiation, for example in a spectrometer.
  (2) A small fixed telescope attached to a larger one in order to set the line of sight of the large instrument.

collision. An interaction between free particles (including protons, atoms and nuclei), aggregates of particles and rigid bodies in which they approach sufficiently close to exert a mutual influence, with or without actual contact, involving exchange of energy, momentum or charge.
  An *inelastic collision* is a collision in which kinetic energy is not conserved, but may be changed to heat, excitation or ionization energy.
  An *elastic collision* is either a collision in which kinetic energy is conserved, or is a collision in which a bombarded nucleus scatters a bombarding particle without becoming excited or experiencing fission.

collison density. The total number of collisions of a specified type occurring in unit volume of material in unit time.

colloidal solution. A system consisting of particles of a diameter range 10^{-9} metre to 10^{-5} metre dispersed in a medium. Such particles are detectable by ULTRAMICRO-SCOPE. Solutions whose components can be separated and then regenerated by mixing are known as *reversible* colloidal solutions; others are *irreversible* colloidal solutions.

colloidal state. A half-way stage between a true SOLUTION and a true SUSPENSION. *See* COLLOIDAL SOLUTION.

cologarithm. The logarithm of the inverse of a number.

colorimeter. An instrument used in COLORIMETRY.

colorimetry. The determination of colour specification.
  Additive colorimetry is concerned with the provision of a colour specification in terms of any three suitable radiations, usually red, green and blue monochromatic radiations, which when mixed together and presented to a standard observer match the colour under test. CHROMATICITY CO-ORDINATES are generally used to express the specification.
  Subtractive colorimetry is concerned with

the provision of a colour specification in terms of the amount of light absorbed, i.e. subtracted from, an initially white beam. Generally yellow (blue-absorbing), magenta (green-absorbing) and cyan (red–absorbing) filters of appropriate thickness are used to give a match for the standard observer to the colour under test. The readings may be converted to chromaticity co-ordinates. Mixing paints is also a subtractive colour-mixing process, the resulting colour depending on that of the light not absorbed by the pigment.

colour. (1) A sensation which is normally produced in the eye by light but can also result from applying pressure or electric or magnetic fields to the eye. It may also arise from drug taking and pathological conditions.
(2) A GLUON and QUARK property, similar to electric charge but occurring in three varieties. It is believed to be the source of the STRONG INTERACTION between quarks and permits quark arrangements which appear to violate the PAULI EXCLUSION PRINCIPLE.

colour analyser. A device for providing a colour specification related to a particular process involving colour change. The results are generally not expressed in CHROMATICITY CO-ORDINATES.

colour atlas. A collection of charts each consisting of colour samples systematically arranged to cover as much COLOUR SPACE as practicable.

colour blindness. See COLOUR VISION.

colour centre. A defect introduced into a crystal and resulting in new absorption bands in the light transmitted by the crystal. Colour centres therefore alter the colour of a crystal. Examples of methods of defect introduction are treatment with ionizing radiation and heating in alkali metal or halogen vapours. Many types of centre have been documented.

colour comparator. An instrument to facilitate the comparison of coloured samples.

colour deficiency. See COLOUR VISION.

colour emissivity. The ratio of the energy radiated at a particular wavelength by a body to that which would be radiated by a black body at the same wavelength and temperature.

colour filter. A layer, film or plate which changes the spectral distribution of light passing through it. It operates either by simple absorption or by interference.

colour hue. The attribute of radiation whereby it is identified as resembling the appearance of a particular radiation represented by a point on the CHROMATICITY CHART boundary. At its most sensitive, i.e. in the middle part of the VISIBLE SPECTRUM, the eye can detect hue differences corresponding to 1 nanometre to 3 nanometre wavelength difference; there are about 130 steps of just detectable hue difference across the visible spectrum and an average of about 20 tints along any line drawn from the white point to the boundary.

colour mixture. A process used in COLORIMETRY.

colour photography. See PHOTOGRAPHY.

colour picture tube. A type of cathode ray tube suitable for use in COLOUR TELEVISION receivers. The colour mixing is on the additive principle. Three electron beams, one each for the red, green and blue components of the radiation from an object, are directed towards a screen covered with a regular pattern of phosphor dots; one third of the dots luminesce red, one third green and one third blue. Dots of each kind are evenly distributed over the screen and so arranged that the red-controlled electron beam scans the red luminescing dots, the green-controlled beam the green luminescing dots and the blue-controlled beam the blue luminescing dots, so that a complete coloured image is produced. Tubes differ in their arrangement of beams and phosphor dots. For example the *Colourtron* has a triangular arrangement for both electron guns and dots, whereas the more recent *Trinitron* has a single electron gun with three cathodes horizontally aligned and like dots in vertical lines, a system resulting in a smaller and lighter tube with sharper focusing.

colour pyrometer. *See* PYROMETER.

colour rendering. The appearance of a surface when illuminated by lights of different spectral composition. It is to be noted that lights can appear to be of the same colour yet differ appreciably in their colour rendering.

colour saturation. The difference in appearance of a colour from white or grey.

colour space. A three-dimensional conceptual space described by three colour attributes, for example brightness, hue and saturation.

colour system. Data for defining the colour matching characteristics of a standard observer. The C.I.E. (i.e. Commission Internationale de l'Eclairage) system is the best known.

colour television. A system resulting in the display of coloured images on the screen of a COLOUR PICTURE TUBE. The signal arriving at this tube is made up of *luminance* and *chrominance signals*; the former gives rise to a monochrome image thus allowing compatability with black and white receivers (*see* TELEVISION). The chrominance signal is obtained by the use of a separate camera tube for each primary colour: one camera tube has a phosphor activated by red light, another has a phosphor sensitive to green light and the third has a phosphor responding to blue light.

colour temperature. The temperature of a black body with the same relative spectral distribution as that of the radiation under investigation.

Colourtron. *See* COLOUR PICTURE TUBE.

colour vision. A complex subject for which many theories have been pro-pounded concerning the function of the retinal receptors in the EYE; none is completely satisfactory. At sufficiently low light intensities there is no colour perception. At light intensities sufficiently high to permit colour perception, it only occurs for images formed in a small area of the retina surrounding the centre of the field of view; since cells known as *cones* predominate in this area, it is concluded that they are responsible for colour discrimination.

The effect of any colour can be matched by a mixture of red, green and blue lights in suitable proportions. This fact lead to the suggestion that there are three pigments, responsible for colour vision, present in the eye, and that *colour blindness*, i.e. *colour deficiency*, results from the absence of one or more of these pigments. However only two pigments have been isolated and their absorption spectra do not agree with the theoretical response curves deduced from DICHROMAT studies. In an attempt to bolster the *three-colour theory*, i.e. *trichromatic theory*, it has been suggested that the two pigments discovered are merely concerned with the gross detection of light and so are present in relatively large quantity, while amounts of other pigments, sufficiently small to escape detection, are responsible for the sensation of colour. Another suggestion is that three different types of cone exist, but there is no histological evidence for this. *See also* PHOTOPIC VISION; IODOPSIN; RHODOPSIN.

coma. (1) An ABERRATION of optical systems in which a point object off the axis of the system gives rise to a comet-shaped image.
(2) The diffuse luminous region surrounding the head of a COMET.

combination. A selection of a number, r, of entities from a set of n entities, regardless of the order of selection. It is written ^nC_r and equals

$$2n!/(r!(n-r)!)$$

See also FACTORIAL; PERMUTATION.

combination of thin lenses. If the lenses are in contact, then the power of the combination is equal to the sum of the separate powers of the components. It is possible to produce an ACHROMATIC LENS by suitable choice of the components, one of which must be diverging and the other converging. *See also* ACHROMATIC SEPARATED LENSES.

comet. A celestial object which moves in a path round the Sun and sometimes has a bright head and tail in the Sun's vicinity. The COMA is formed when the comet approaches the Sun and is due to vaporization of the ice which is a major constituent of comets. Part of the coma material emerges in the direction remote from the Sun and forms the comet's tail, which may be several million kilometre long.

common base connection. A mode of transistor operation in which the BASE, which is usually earthed, is common to the input and output circuits. The EMITTER is the input terminal and the COLLECTOR the output terminal.

common collector connection. A mode of transistor operation in which the COLLECTOR, which is usually earthed, is common to the input and output circuits. The BASE is the input terminal and the EMITTER the output terminal.

common denominator. A number which is divisible by the denominator of each of two or more FRACTIONS.

common difference. *See* ARITHMETIC PROGRESSION.

common emitter connection. A mode of transistor operation in which the EMITTER, which is usually earthed, is common to both input and output circuits. The BASE is the input terminal and the COLLECTOR the output terminal.

common fraction. *See* FRACTION.

common impedance coupling. The coupling of two circuits by connecting them across the same reactive component.

common logarithms. *See* LOGARITHM.

common ratio. *See* GEOMETRIC PROGRESSION.

commutative operation. An operation for which the order of the terms does not affect the result, for example addition and multiplication.

commutator. (1) A device for reversing the direction of an electric current.
(2) A device for connecting one section of an armature winding after another to an external electric circuit.
(3) The expression AB–BA where A and B are OPERATORS which are not commutative. *See* COMMUTATIVE OPERATION.

comparator. (1) An instrument for the accurate comparison of lengths. It may work electrically, mechanically, optically or pneumatically.
(2) A circuit, for example a differential amplifier, which produces an output related to the difference between two signals.

compass. A device consisting of a magnet pivoted at its centre so that it can move freely in a horizontal plane over a circular scale which is marked with the cardinal points and divided into degrees. The magnet sets itself along the horizontal component of the Earth's magnetic field and thus points in the direction of magnetic north. *Compare* GYROCOMPASS.

compensated pendulum. A pendulum for which the distance of its centre of mass from the support does not vary with temperature and so neither does its period of oscillation. Such a pendulum may be produced by using a hollow bob containing a suitable volume of mercury.

compensating eyepiece. An eyepiece designed to correct for the lateral CHROMATIC ABERRATION of the objective with which it is designed to be used.

compiler. A computer program for converting information written in a programming language into machine code.

complementarity. The principle that certain entities, for example electrons and photons, have both a particle and a wave nature.

Pairs of complementary colours

wavelength	colour		wavelength	complementary colour
563 nm	yellow-green		433 nm	violet
567 nm	yellow		464 nm	indigo-blue
585 nm	yellow-gold		485 nm	blue
607 nm	orange		489 nm	blue
656 nm	red		492 nm	blue-green

complementary colours. The colours of two spectral radiations which when mixed together yield white light; pairs of complementary colours are shown in the table.. The two points of intersection of a line through the white point with the horseshoe boundary of the CHROMATICITY CHART represent complementary colours. Green does not have a complementary colour but will yield white when mixed with a suitable quantity of purple.

complementary transistors. Two transistors of opposite type, for example an n-p-n and p-n-p bipolar junction TRANSISTOR.

complex fraction. *See* FRACTION.

complex number. A number expressed in the form $a + ib$ where a and b are real numbers and i is $(-1)^{1/2}$. *See also* ARGAND DIAGRAM.

component. Any of two or more VECTORS which when added produce a given vector.

composite number. An integer that is not prime.

compound microscope. *See* MICROSCOPE.

compound nucleus. A highly excited short-lifetime nucleus formed immediately after a nuclear collision.

compound pendulum. *See* PENDULUM.

compressibility. The reciprocal of the BULK MODULUS of a substance. It is very small for solids and for liquids but appreciable for gases.

compression. A phenomenon occurring as sound waves traverse an elastic medium. Since the particles of medium vibrate along the direction of wave travel, local density changes occur due to the positions of the particles. As illustrated in fig. C18, at

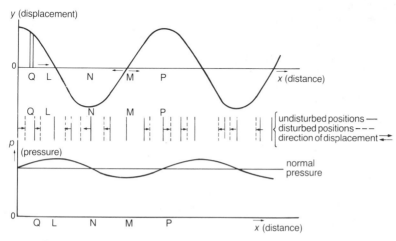

C18 Displacement and pressure variation due to sound wave

points of maximum density such as L there is compression and at points of minimum density such as M there is rarefaction. There is thus a continuous pressure variation in the medium as the wave passes.

compressor.   *See* VOLUME COMPRESSOR.

Compton effect.   The scattering of radiation by free electrons resulting in an increase in the wavelength and a decrease in the energy of the radiation and an increase in electron velocity. The wavelength increase equals

$$(2h/mc) \sin^2 (\phi/2)$$

where h is the Planck constant, m the electron mass, c the speed of light in vacuo and ϕ the angle through which the radiation is scattered. The effect is best observed by passing X or gamma radiation through elements of low atomic number.

Compton electron.   An electron taking part in the COMPTON EFFECT.

Compton recoil.   The recoil of a COMPTON ELECTRON.

Compton scattering.   Another name for COMPTON EFFECT.

computer.   Any of various devices for processing data according to predetermined instructions. The most widely used form is the electronic *digital computer* for which information is stored and processed in digital form. The *analog computer*, which accepts a continuous input, is the other main type. The *hybrid computer* has a continuous input which is converted into digital form for processing and thus has some of the features of both the other two types.

concave.   Having one or two surfaces with a shape identical to part of the interior surface of a sphere, paraboloid etc.

concave lens.   *See* LENS.

concave mirror.   *See* CURVED MIRROR.

concavo-convex lens.   *See* LENS.

concentration cell.   A cell in which two identical metal electrodes are immersed in solutions of different concentration of a salt of the same metal. The metal dissolves in the weaker and is deposited from the stronger solution. The electromotive force is generally a few hundredths of a volt.

concentric.   Having a common centre. For example a number of spheres may be concentric.

condensation.   (1) The change of a gas or vapour into a liquid accompanied by an emission of heat.
(2) The ratio of the instantaneous excess of density above normal value to the normal value at a point in a medium carrying a sound wave.

condensation pump.   Another name for DIFFUSION PUMP.

condenser.   (1) An optical system whereby light from a source is focused on an object, thereby increasing its illumination. It is used for example in microscopes and projectors.
(2) A device for the continuous removal of heat, for example the stream of cold water which is used to remove the latent heat of vaporization when a vapour liquefies, as in distillation.
(3) Obsolete term for CAPACITOR.

conductance.   (1) Symbol G. The reciprocal of the resistance in a direct current circuit. The unit is the siemens.
(2) The real part of the admittance in an alternating current circuit, i.e.

$$R/(R^2 + X^2)$$

where R is the resistance and X the reactance.

conductiometric.   Denoting an experimental technique which involves electrical conductivity measurement.

conduction.   The transmission of electrical, thermal or acoustic energy via a medium without any transfer of mass.
The *conduction of electricity* results from

the action of an applied electric field on charge carriers in a medium. In an ELECTROLYTE charge is carried on positive and negative ions. In solids the carriers are electrons and/or positive holes (*see* BAND THEORY).

Conduction in gases depends on the strength of the field applied between electrodes in the gas, on the electrode separation and on the gas pressure. At small field strength the current passing is due to the presence of natural ions, i.e. ions produced by ultraviolet light, cosmic rays etc., and is therefore small. As the field strength is increased so the current increases up to a constant value; this value is maintained with increasing field strength until a rapid rise commences, and the rise continues with ever-increasing field strength until breakdown potential is reached and an *electric discharge* commences. At atmospheric pressure the discharge is a SPARK DISCHARGE or, if the electrodes are very close, an ARC; at low pressure GLOW DISCHARGE occurs. Immediately after the start of the discharge the current becomes steady. *See also* GAS DISCHARGE TUBE.

conduction band. *See* BAND THEORY.

conduction current. A CURRENT flowing in a conductor.

conduction electron. An electron whose energy lies in the conduction band. *See* BAND THEORY.

cone. (1) A surface obtained by drawing straight lines from a fixed point to every point of a closed curve. If the closed curve is a circle, radius r, and the fixed point lies on the perpendicular to the plane of the circle at its centre, the cone is known as a *right circular cone*: its lateral surface area is $\pi r l$, where l is the length of any of the straight lines; its volume is $\pi r^2 h/3$, where h is the vertical height.
(2) *See* COLOUR VISION.

cone of friction. The locus of the resultant force between two horizontal flat surfaces for different directions of initial relative motion of the surfaces. The axis of the cone

is perpendicular to the surfaces and its semi-apical angle is $\tan^{-1}\mu_d$, where μ_d is the coefficient of dynamic friction. *See* FRICTION.

configuration. The arrangement of electrons orbiting the nucleus in a given atom. For example the configuration of oxygen is

$$1s^2\ 2s^2\ 2p^4$$

where 1s, 2s and 2p denote the SUBSHELL and the superscript gives the number of electrons present in that subshell. The first two shells of oxygen thus contain 2 and 6 electrons. *See also* ATOMIC ORBITAL.

confocal. Denoting conic sections with the same foci.

congruent. Denoting geometric figures identical in all respects.

conic. The locus of a point which moves so that its distance from a fixed point, known as the *focus*, is a constant fraction of its perpendicular distance from a fixed line, known as the *directrix*; the constant fraction is known as the *eccentricity*, symbol e. For a circle $e = 0$, for an ellipse $e < 1$, for a parabola $e = 1$ and for a hyperbola $e > 1$. Any conic may be obtained by taking the appropriate section of a suitable cone.

conical pendulum. *See* PENDULUM.

conjugate. (1) A complex number for which the imaginary part is the negative of the imaginary part of a given complex number; thus

$$a + ib \text{ and } a - ib$$

are conjugate to each other.
(2) One of a pair of things, each of which is interchangeable with respect to the properties of the other. For example if I is the image of O, then I and O are conjugate; if I were made the object, O would be its image.

conjuction. *See* OPPOSITION.

consequent poles. Magnetic poles existing in a body in addition to the two usually found.

conservative law. A law which states that a particular property of a system remains unaltered, although exchange of the property between components of the system may occur. The law applies to such properties as electric CHARGE, ANGULAR MOMENTUM, LINEAR MOMENTUM and mass-energy (*see* MASS-ENERGY RELATION). At speeds small compared to that of light, mass and energy are individually conserved.

conservative field. A field of force in which the work done in moving a particle from one point to another is independent of the path taken between the points.

conservative system. A system in which only a conservative field of force is operative.

consonance. A combination of two or more musical notes which sounds agreeable to most people; the opposite is *dissonance.*

constantan. An alloy of approximately equal masses of nickel and copper. It has high RESISTIVITY and a low temperature coefficient of resistance. It is extensively used in electrical resistance windings and as one member of a THERMOCOUPLE.

constant pressure gas thermometer. An instrument in which the volume changes of a fixed mass of gas at constant pressure are used to measure the temperature of the bath in which the glass bulb containing the gas is immersed. The gas volumes, V_0, V_t and V_{100} at $0°$ C, at the unknown temperature $t°$ C and at $100°$ C respectively are measured. Then

$$t = 100(V_t - V_0)/(V_{100} - V_0) \; ° \text{C}$$

This value requires correction to the scale for an IDEAL GAS since no real gas is ideal.

constants (physical). *See* Table 5.

constant volume gas thermometer. An instrument similar to the CONSTANT PRESSURE GAS THERMOMETER (and described by replacing the word volume(s) by the word pressure(s) and the symbol V by the pressure symbol P in that entry).

constrain. To limit to a predetermined position or path.

constellation. Any of the 88 areas into which the whole of the CELESTIAL SPHERE is now divided.

contactor. A type of switch, designed for frequent use, for making and breaking a circuit.

contact potential. A potential difference of a few tenths of a volt that occurs when two conductors of different materials are placed in contact.

containment. (1) The process of preventing contact between a PLASMA and the walls of the reaction vessel enclosing it.
(2) The prevention of the escape of unacceptable quantities of radioactive material.
(3) The containment system of a NUCLEAR REACTOR.

continuous flow calorimeter. A calorimeter in which heat is supplied at a constant rate to fluid flowing at a constant rate. Eventually a steady state is reached when input and output temperatures may be accurately read. If the rate of heat supply is then adjusted so that the same steady-state temperatures are reached for a different flow rate, it is possible to correct completely for any heat losses to the surroundings and so, for example, obtain an accurate value of the SPECIFIC HEAT CAPACITY of the flowing fluid.

continuous spectrum. An unbroken sequence of wavelengths over a relatively wide range. Such spectra are given by incandescent solids, liquids and compressed gases.

continuum. (1) *See* FOUR-DIMENSIONAL CONTINUUM.
(2) A CONTINUOUS SPECTRUM.

contrast. The light-intensity difference between the image of an object and the image of its surroundings.

contravariant tensor. *See* COVARIANT TENSOR.

control electrode. The valve electrode to which the input signal is applied in order to change the current in one or more of the other electrodes.

control group. A group of people employed in drug testing who receive placebos, i.e. sugar pills, believing them to be drugs, whereas another group actually receives the drug under test. From the responses the significance of the treatment can be statistically evaluated.

control rod. One of several rods which can be moved along its axis in or out of the core of a nuclear reactor in order to control the CHAIN REACTION rate. Each rod usually contains a neutron absorber, for example cadmium or boron.

convection. The process in which heat is transfered through a fluid by motion of the fluid.
Free convection occurs solely as a result of density gradients due to temperature gradients resulting from the presence of a hot body in contact with the fluid. The fluid thus moves under the influence of gravity.
Forced convection occurs when relative motion between the fluid and a hot body in contact with it is maintained by an external agency, the gravity contribution being negligible.

convection current. (1) The flow of fluid due to convection.
(2) The electric current due to the movement of an electrified body. It produces no heat and flows without potential difference or energy change, but can produce a magnetic effect.

convectron. An instrument giving electrical indication of deviation from the vertical. It uses the fact that convection cooling of a heated straight fine wire is much greater for the wire horizontal than for the wire vertical.

conventional current flow. *See* CURRENT.

convergent. Denoting an infinite SEQUENCE whose terms tend to a limiting value, or an infinite SERIES which has a finite sum.

converging lens. A LENS which causes a beam of light to converge.

converging mirror. A CURVED MIRROR which causes a beam of light to converge.

conversion electron. An electron ejected as the result of INTERNAL CONVERSION.

conversion factor. The ratio of the number of fissile atoms produced from the fertile material in a CONVERTER REACTOR to the number of fissile atoms of fuel destroyed in the process.

converter reactor. A nuclear reactor in which FERTILE material is transformed into FISSILE material: *conversion* is said to occur.

convex. Having one or two surfaces with a shape identical to part of the exterior surface of a sphere, paraboloid etc.

convex lens. *See* LENS.

convex mirror. *See* CURVED MIRROR.

convexo-concave lens. *See* LENS.

coolant. A fluid used to transfer heat. For example in a nuclear reactor the coolant transfers heat from the core either to the steam-raising plant or to an intermediate heat exchanger.

cooling curve. The graph of temperature against time for a substance. If such a plot includes the change of a substance from the liquid to the solid state, the temperature for the straight line portion of the graph is the melting point of the substance.

co-ordinate bond. A type of COVALENT BOND in which one of the atoms supplies both the electrons.

co-ordinate geometry. Another name for ANALYTIC GEOMETRY.

co-ordinates. Numbers representing the position of a point with respect to reference lines or points. Two co-ordinates are necessary to specify the position of a point in a plane; in three-dimensional space three co-ordinates are required. The commonest systems are CARTESIAN CO-ORDINATES, POLAR CO-ORDINATES and CYLINDRICAL CO-ORDINATES.

Copernican system. A system of celestial mechanics which laid the foundations of modern astronomy.

copper loss. The power loss in watt due to the electric current flowing in the windings of an electrical machine or transformer.

copper oxide rectifier. A device illustrated in fig. C19. It conducts well when the lead (Pb) is made positive with respect to the copper (Cu), but badly when the lead is negative with respect to the copper. Such rectifiers are frequently used in small battery chargers.

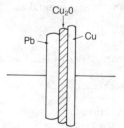

C19 Copper oxide rectifier

core. (1) The part of a nuclear REACTOR in which the reaction occurs.
(2) The magnetic circuit of a transformer, electric motor or similar device. The core is made of ferromagnetic material and is generally laminated to reduce EDDY CURRENT losses.
(3) A small ferrite ring used in a type of computer MEMORY to store one BIT of information.

core loss. Power loss due to EDDY CURRENTS and magnetic HYSTERESIS in a magnetic circuit during cyclic changes of magnetization.

core store. *See* MEMORY.

core-type transformer. A transformer in which most of the core is enclosed by the windings.

Coriolis force. A concept introduced in order to simplify calculations on the motion of bodies observed from a rotating frame of reference, for example the motion of an artificial satellite as observed from Earth.

corkscrew rule. If a corkscrew is advanced in the direction of an electric current, the direction of rotation of the corkscrew is the direction of the lines of magnetic field.

cornea. *See* EYE.

corollary. A result following without further detailed proof from a result which has already been proved.

corona. The outermost region of the sun's atmosphere beyond the CHROMOSPHERE. Its temperature is about 10^6 K.

corona discharge. A luminous electric discharge appearing in the air round a conductor when the potential gradient at its surface rises above a critical value. Such discharges give rise to power loss in transmission lines.

corposant. A CORONA DISCHARGE occurring around the point of a conductor such as a LIGHTING CONDUCTOR.

corpuscular theory. The theory that a luminous body emits small elastic particles which travel in straight lines in an isotropic medium, are repelled on reflection and suffer change of direction on refraction. The theory was discounted for a time since, contrary to the experimental evidence, it required light to increase its speed on entering an optically denser medium. In contrast the WAVE theory correctly predicts the velocity decrease and moreover gives a readier explanation of interference, diffraction and polarization. However, the interactions of light with matter give support to a quasicorpuscular theory involving radiation quanta, i.e. photons. Aspects of both theories are thus now accepted.

correcting plate. A lens system used to correct SPHERICAL ABERRATION in spherical mirrors or to correct COMA in parabolic mirrors.

correlation coefficient. Symbol r. A statistical measure of the degree of relationship between two variables. The value lies between 0 and 1, $r = 0$ indicating no relationship.

correspondence principle. The principle that, for large systems, classical and quantum physics lead to identical conclusions.

cos. Abbrev. for cosine. *See* TRIGONOMETRIC FUNCTIONS.

cosec. Abbrev. for cosecant. *See* TRIGONOMETRIC FUNCTIONS.

cosecant. *See* TRIGONOMETRIC FUNCTIONS.

cosine. *See* TRIGONOMETRIC FUNCTIONS.

cosine law. *See* LAMBERT'S LAW.

cosmic abundance. *See* ABUNDANCE.

cosmic background radiation. Electromagnetic radiation that permeates all regions or large regions of space. It has been found in radio, microwave, infrared, X ray and gamma ray regions of the spectrum.

cosmic rays. Energetic particles from space which bombard the Earth's atmosphere from all directions. They are mainly nuclei of the most abundant elements, i.e. atomic masses up to 56, but all known nuclei are represented; protons are the most abundant. Small numbers of electrons, positrons, neutrinos and gamma ray photons are also present. On entering the Earth's atmosphere most of the so-called *primary cosmic rays* collide with atomic nuclei producing *secondary cosmic rays*, mainly consisting of elementary particles. The intensity of cosmic rays at sea level is about one particle per square centimetre per minute; the intensity varies with latitude because of the influence of the Earth's magnetic field. The origin of cosmic rays remains obscure; it is thought that the higher-energy rays originate outside the solar system and could result from supernova explosions. Large increases in cosmic-ray intensity have been observed to coincide with the appearance of strong solar flares.

cosmogony. The study of the origin and evolutionary development of the universe.

cosmology. The study of the universe in its entirety, including its origin, evolution and present state.

cosmotron. A PROTON SYNCHROTRON at Brookhaven, USA capable of accelerating protons to an energy of 3 giga-electron-volts.

cot. Abrev. for cotangent. *See* TRIGONOMETRIC FUNCTIONS.

cotangent. *See* TRIGONOMETRIC FUNCTIONS.

coulomb. Symbol C. The SI UNIT of electric charge, equal to the charge transported by a current of one ampere in one second.

Coulomb field. The electric field around a point charge.

Coulomb force. The force existing between charged particles. Its magnitude is inversely proportional to the square of the distance between the particles. It may be either attractive or repulsive. *See* COULOMB'S LAW.

coulombmeter. Another name for VOLT-AMETER.

Coulomb scattering. The scattering of a charged particle by the electrostatic field of a nucleus.

Coulomb's law. The magnitude of the force exerted by point charge Q_1 on point charge Q_2 is

$$Q_1 Q_2/(4\pi\varepsilon d^2)$$

where d is the distance apart of the charges and ε is the absolute PERMITTIVITY of the medium in which they are situated. The direction of the force lies along the line joining the particles.

Coulomb's theorem. The magnitude of the electric field strength near a surface whose SURFACE DENSITY OF CHARGE is σ is equal to σ/ε where ε is the absolute PERMITTIVITY of the medium surrounding the surface.

counter. (1) Any device for counting individual particles and photons. Most counters work by multiplication of the number of ions or electrons formed by a single particle or photon; each particle thus gives rise to a pulse of current or voltage. The pulses are electronically counted. *See* CRYSTAL COUNTER; GEIGER COUNTER; IONIZATION CHAMBER; PROPORTIONAL COUNTER; SCINTILLATION COUNTER.
(2) Any electronic circuit that records and counts pulses of current or voltage.

counterglow. Another name for GEGENSCHEIN.

couple. A system of forces comprising, or equivalent to, two forces of equal magnitude acting in opposite directions along parallel lines. The *moment of the couple* is the product of the magnitude of either force and the perpendicular distance between the forces; the moment has the same value

about any axis perpendicular to the plane of the forces.

coupled systems. Two or more mechanical vibrating systems connected so that they react on each other, energy being transferred from one system to the other.

coupled circuits. Circuits so related that alternating-current effects are transferred but steady-state direct-current effects are not. Examples are linking by mutual inductance and linking through a common capacitor.

coupling. (1) *jj coupling*. The interaction of the total angular momentum, i.e. orbital plus spin, of an individual particle in an atom or nucleus with that of other particles.
(2) *LS coupling*. The interaction of the resultant, *L*, of the orbital angular momenta of all the particles in an atom or nucleus with the resultant, *S*, of the spins of all the particles.

covalent bond. A type of bond in which atoms are held together by shared pairs of electrons.

covariant equation. An equation which retains its form under a transformation, such as a LORENTZ TRANSFORMATION, to quantities measured by another observer.

covariant tensor. A TENSOR which is differentiated from a *contravariant tensor* by its transformation behaviour under change of co-ordinates.

CPU. Abbrev. for CENTRAL PROCESSING UNIT.

Crab nebula. A turbulent expanding mass of gas and dust in the constellation Taurus. It is a SUPERNOVA remnant emitting SYNCHROTRON RADIATION.

craters. Circular rocky formations observed on the Moon and on several other bodies in the solar system. They are believed to have originated mainly by meteorite impact, some being later modified by erosion or volcanic processes.

creep. Slow permanent deformation of a metal caused by continuous stress.

critical angle. Symbol c. The smallest angle of incidence at which electromagnetic radiation suffers TOTAL INTERNAL REFLECTION. It is given by

$$\sin c = n'/n$$

where n is the refractive index of the medium in which the radiation is incident and n' (which is less than n) is the refractive index of the medium on the other side of the interface at which total internal reflection occurs.

critical damping. *See* DAMPING.

critical exponent. The power α in the equation

$$Y \simeq (T - T_c)^\alpha$$

for small values of $T - T_c$ where T is the thermodynamic temperature, T_c the CRITICAL TEMPERATURE and Y a thermodynamic variable which most clearly characterizes a phase transition. Thus for the gas-liquid phase change Y would represent the density.

critical isotherm. The isotherm for a gas at its CRITICAL TEMPERATURE.

critical mass. The minimum mass of a fissile material that will sustain a CHAIN REACTION.

critical point. *See* CRITICAL STATE.

critical potential. Another name for excitation energy. *See* EXCITATION (def. 1).

critical pressure. The saturated vapour pressure of a liquid at its CRITICAL TEMPERATURE.

critical reaction. *See* CHAIN REACTION.

critical state. The state of a substance at critical temperature, pressure and volume, i.e. at the *critical point* on its isotherm. The liquid and vapour densities of the substance are then equal.

critical temperature. The temperature above which a gas cannot be liquefied by increase of pressure.

critical velocity. The velocity at which the motion of a flowing fluid changes from laminar to turbulent.

critical volume. The volume occupied by one mole of substance at CRITICAL TEMPERATURE and CRITICAL PRESSURE.

CRO. Abbrev. for CATHODE RAY OSCILLOSCOPE.

Crookes' dark space. *See* GAS DISCHARGE TUBE.

Crooke's radiometer. An instrument for detecting infrared radiation. It consists of an evacuated bulb in which is mounted cross wire holding four vertical thin metal vanes each blackened on one side and polished on the other; the blackened side of one is opposite the polished side of the next. The cross wire can rotate freely about a vertical axis through its centre. The black sides absorb more radiation than the polished sides and so residual gas molecules rebounding from the black sides acquire more momentum than those rebounding from the polished sides. The vane structure therefore rotates under the influence of radiation. The amount of rotation, and hence of radiation, may be measured by observing the deflection of a beam of light by a mirror attached to the cross wire.

crossed Polaroids. Two sheets of POLAROID arranged so that their vibration planes are perpendicular. Any incident light is completely absorbed by the combination since the Polaroid on which any ordinary light is incident renders it plane-polarized; this plane-polarized light is then completely absorbed by the other Polaroid since its vibration plane is perpendicular to that of the light.

crossover network. A filter circuit which passes signals of frequencies above a certain value, the *crossover frequency*, via one path and frequencies below that value by another path. The network is often used in high-fidelity systems.

cross product.   Another name for VECTOR PRODUCT.

cross section.   (1) A plane surface formed by cutting a solid, or the area of this surface.

(2) Symbol σ. The effective area presented by a particular particle to a beam of radiation. It represents the probability that a collision will occur between the beam and the particle.

CRT.   Abbrev. for CATHODE RAY TUBE.

cryogenics.   The study of the production and effects of very low temperatures.

cryometer.   A thermometer suitable for use at low temperatures.

cryoscope.   Any instrument or apparatus for finding the value of a freezing point.

cryotron.   A superconducting switching element which operates in liquid helium.

crystal.   That form of a substance characterized by an orderly arrangement of atoms or positive and negative ions or molecules, repeated more or less perfectly throughout a region large compared with atomic dimensions.

An *ionic crystal* is illustrated in fig. C20; it is the sodium chloride (NaCl) crystal.

crystal analysis.   *See* X RAY ANALYSIS.

crystal base.   The entire contents of the UNIT CELL of a crystal.

crystal clock.   *See* CLOCKS.

crystal controlled oscillator.   An oscillator with very high frequency stability controlled by a quartz crystal. The quartz is set in mechanical vibration using the PIEZOELECTRIC EFFECT.

crystal counter.   A type of COUNTER in which a high potential difference is applied to a crystal of suitable material. A particle of ionizing radiation renders the crystal conducting on striking it and thus produces a pulse of current in it.

crystal detector.   A signal DETECTOR which relies on the rectifying properties of the junction between two crystals of suitable materials, or of the junction beween a metal and a crystal, for example steel and carborundum. It was used in early radio receivers and is now employed for detecting and mixing centimetric waves.

crystal filter.   A FILTER containing one or more PIEZOELECTRIC CRYSTALS each of which forms a resonant or antiresonant circuit in the filter. *See also* RESONANCE (def. 2).

crystal grating.   A crystal acting as a three-dimensional DIFFRACTION GRATING by virtue of the symmetrical arrangement of the crystal atoms in a series of parallel planes. It is suitable for X RAY DIFFRACTION and ELECTRON DIFFRACTION.

crystal habit.   The external appearance of a crystal, governed by the regular internal arrangement of its constituent atoms, ions or molecules and by the way in which the crystal has grown.

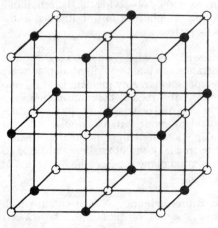

C20 Ionic crystal, sodium chloride

crystal momentum. The product of the Planck constant and a wave vector in a crystal.

crystalline lens. *See* EYE.

crystallography. The study of the form, internal structure, properties and symmetry of crystals. *See also* X RAY CRYSTAL-LOGRAPHY.

crystal microphone. A type of microphone in which the diaphragm vibrations induced by sound waves are used to vibrate a PIEZOELECTRIC CRYSTAL, thus producing an alternating potential difference.

crystal oscillator. An oscillator circuit in which a fixed frequency, constant to 1 part in 2×10^7, is produced by the vibrations of a quartz crystal. Such an oscillator is used in the quartz clock (*see* CLOCKS) and as a frequency standard.

crystal pick-up. A pick-up used on some record players. Vibrations produced by the record groove are converted to a varying potential difference by a PIEZOELECTRIC CRYSTAL.

crystal plane. A plane of atoms, ions or molecules in the lattice of a crystal.

crystal spectrometer. A type of spectrometer for use with X rays, which are dispersed by diffraction from the face of a single crystal. *See also* X RAY DIFFRACTION.

crystal structure. The geometric framework of a crystal and the arrangement, relative to that framework, of the atoms, molecules or ions.

crystal system. Any of the seven groups into which all crystals are classified according to the angles between the edges of the UNIT CELL and the lengths of the edges, as indicated in fig. C21. The seven groups are known as the *cubic, tetragonal, orthorhombic, hexagonal, trigonal* (also called *rhombohedral*), *monoclinic* and *triclinic systems*. Sometimes the trigonal system is included under the hexagonal system.

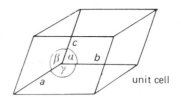

cubic	$a=b=c$	$\alpha=\beta=\gamma=90°$
tetragonal	$a=b\neq c$	$\alpha=\beta=\gamma=90°$
orthorhombic	$a\neq b\neq c$	$\alpha=\beta=\gamma=90°$
hexagonal	$a=b\neq c$	$\alpha=\beta=90°; \gamma=120°$
trigonal	$a=b\neq c$	$\alpha=\beta=\gamma\neq 90°$
monoclinic	$a\neq b\neq c$	$\alpha=\gamma=90°\neq\beta$
triclinic	$a\neq b\neq c$	$\alpha\neq\beta\neq\gamma$

C21 Crystal systems

cube. (1) A regular solid with six congruent square faces.
(2) The third power of a number.

cube root. *See* ROOT.

cubic expansivity. Another name for coefficient of cubic expansion. *See* COEFFICIENT OF EXPANSION.

cubic system. *See* CRYSTAL SYSTEM.

curie. Symbol Ci. A unit of ACTIVITY of a radioactive substance, corresponding to 3.7×10^{10} disintegrations per second; this is approximately the activity of one gramme of radium. *See also* RADIATION UNITS.

Curie constant. *See* CURIE'S LAW.

Curie point. Another name for CURIE TEMPERATURE.

Curie's law. The MAGNETIC SUSCEPTIBILITY of a paramagnetic substance is inversely proportional to its thermodynamic temperature. The constant of proportionality, which is characteristic of the material, is known as the CURIE CONSTANT. *See also* PARAMAGNETISM.

Curie temperature. The temperature above which ferromagnetic material becomes paramagnetic.

Curie-Weiss law. The MAGNETIC SUSCEPTIBILITY of a paramagnetic substance is inversely proportional to the difference of its themodynamic temperature and its CURIE TEMPERATURE. This modification of CURIE'S LAW improves agreement with experiment. *See also* PARAMAGNETISM.

curl. The vector product of the differential vector operator DEL, ∇, and a vector. Thus for vector F,

$$\nabla \times F = \operatorname{curl} F$$

$$= i(\partial F_z/\partial y - \partial F_y/\partial_z)$$

$$+ j(\partial F_x/\partial v - \partial F_z/\partial x)$$

$$+ k(\partial F_y/\partial x - \partial F_x/\partial y)$$

where i, j and k are unit vectors along the x, y and z axes respectively and F_x, F_y and F_z are the components of F along the same axes.

current. Symbol I. The flow of electric charge. The current magnitude is the amount of charge passing in unit time and is measured in AMPERE, one ampere corresponding to a flow of about 10^{18} electrons per second. The direction of conventional current flow is taken to be from a point of higher potential to one of lower potential, i.e. opposite to that of electron flow. Other charge carriers are ions and positive holes.

current balance. Another name for AMPÈRE BALANCE.

current density. Symbol j. The current flowing in a medium per unit of cross section, the cross section being perpendicular to the direction of flow. The medium may be a conductor or a beam of charged particles.

current transformer. An instrument transformer whose primary winding is in series with the main circuit and whose secondary winding is in series with an instrument. such as an ammeter. Uses of such a transformer include extension of the range of an alternating-current instrument, isolation of an instrument from a high-voltage circuit and operation of a safety device in an alternating-current power installation.

cursor. A transparent slider for a slide rule. A hair line on the slider parallel to the scale divisions facilitates reading of the scales.

curvature. The rate of change of direction of a line on a surface. In two dimensions it is given in POLAR CO-ORDINATES by $d\theta/dr$.

curvature of field. An ABERRATION of optical systems in which the images of off-axis points lie on a curved surface, known as the *Petzval surface*, rather than on a plane. Often the effects of ASTIGMATISM are used to offset the curvature of the Petzval surface. In practice this may involve using more than one lens, adjusting the lens spacing and adding suitably positioned stops.

curved mirror. A nonplane mirror. The type most commonly used is the SPHERICAL MIRROR, which may be either CONCAVE or CONVEX. A *concave mirror* and a *parabolic mirror*, viewed on the reflecting side, curve inwards. A *convex mirror* similarly viewed, curves outwards. On account of their wide angle of view, convex mirrors are used as driving mirrors. In search lights and head lamps the reflector is parabolic since it gives rise to a parallel beam of light from a source placed at its focus (*see* PARABOLOID OF REVOLUTION). Shaving mirrors are concave since they give a magnified upright image of an object located inside the focal length. *See also* IMAGE (in a spherical mirror).

cut-off bias. The grid bias voltage on a valve that just reduces the anode voltage to zero, or the bias voltage that just reduces the electron beam current of a cathode ray tube to zero. In both cases the cut-off bias depends on the voltages applied to the other electrodes.

cut-off frequency. The point at which a transmission system passes from a low-attenuation frequency range to one of high attenuation.

cuvette. A parallel-sided glass or quartz container for solutions used in a SPECTROPHOTOMER. Light strikes the parallel sides normally.

cycle. A series of events in which the initial and end conditions are the same. Usually a cycle of events is recurrent.

cycle per second. Former name for HERTZ.

cycloid. The locus of a point on the circumference of a circle as the circle rolls along a straight line. It is a special case of a TROCHOID.

cyclotron. A type of particle accelerator, illustrated in fig. C22. A uniform vertical magnetic field is produced by a powerful electromagnet. An electric field of constant frequency is applied to two hollow D-shaped horizontal metal electrodes, known as the *dees*, which are separated by a small gap. Under the influence of the magnetic field, a charged particle entering the dees describes a semicircular orbit having a radius proportional to the particle velocity. The time taken to complete such a semicircular orbit is equal to

$$\pi m/(Be)$$

where m and e are particle mass and charge

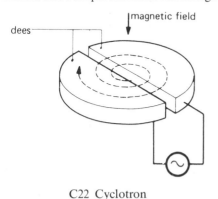

C22 Cyclotron

respectively and B is the magnitude of the magnetic flux density. This time is thus independent of velocity. The orbiting particle keeps in step with the alternating electric field and is accelerated by the electric field, always in the same sense, every time it crosses a dee gap. It therefore describes a spiral as shown. As the particle approaches the dee edge, an auxiliary electric field deflects it from its circular path and it emerges through a thin window.

The maximum particle energy attainable is limited by relativistic effects. The relativistic increase of mass at high particle velocity significantly increases the orbiting time and so the particle passes a gap out of step with the electric field; it therefore experiences less acceleration. The energy limit is around 25 mega-electronvolt. Higher energies are given by the SYNCHRO-CYCLOTRON and the SYNCHROTRON. *See also* RELATIVITY.

cylindrical co-ordinates. Co-ordinates used to locate the position of a point in space by the POLAR CO-ORDINATES in a reference plane of the foot of a perpendicular from the point to that plane and the length of the perpendicular.

cylindrical lens. A lens with one face a portion of the curved surface of a cylinder. A thin cylindrical lens is used for correcting ASTIGMATISM of the eye.

cylindrical winding. A type of winding sometimes used in a transformer. The coil, either single layer or multi-layer, is helically wound. Usually the axial length is several times the diameter.

cystoscope. An instrument for direct viewing of the inside of the bladder. Modern versions use a FIBRE OPTICS system.

cytoplasm. A conducting fluid contained inside an AXON.

D

dalton. *Another name for* UNIFIED ATOMIC MASS UNIT.

Dalton's law of partial pressures. At constant temperature, the total pressure exerted by an ideal gas mixture in a container is the sum of the pressures exerted by each individual gas if it alone occupied the container.

dam. A structure designed to control the flow of water. It can have several uses: raising a water level to assist navigation, diverting water, generating hydroelectric power, storing water, land irrigation, flood control, providing recreation. Concrete dams are usually built with a convex face in contact with the water, thus making use of the arch principle of strengthening. Gravity dams are usually massive structures of earth and rock, presenting a flat face to the water and having a greater thickness at the bottom than at the top.

damping. A decrease in the amplitude of oscillations with time. It is produced by forces resisting the motion, for example friction and electromagnetic forces. If after disturbance there is a large number of oscillations before the undisturbed state is regained, *underdamping* is said to occur. In the case of *critical damping*, the time taken to regain the undisturbed state is a minimum since oscillation just fails to occur. For *overdamping* there is likewise no oscillation, but a longer time is required to regain the undisturbed state than for critical damping. These types of damping are illustrated in fig. D1. In many measuring instruments, critical damping is often intentionally introduced in order to facilitate the observation of readings.

Daniell cell. A primary CELL in which the positive pole is a copper rod which is

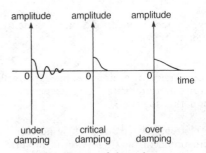

D1 Types of damping

immersed in copper sulphate solution and the negative pole is an amalgamated zinc rod which is immersed in dilute sulphuric acid; the two solutions are separted by a porous partition. The cell reaction is

$$Zn + Cu^{2+} \longrightarrow Zn^{2+} + Cu$$

This yields a fairly constant electromotive force of about 1.08 volt; the internal resistance is a few ohm.

daraf. *See* ELASTANCE.

dark field illumination. A type of illumination sometimes used on a MICROSCOPE specimen to enhance visibility. The specimen is illuminated by light from the side rather than from underneath so that only light diffracted by the object reaches the eye. The technique is only applicable to the examination of small particles or fine lines, which appear as bright images against a dark background.

dark space. A comparatively non-luminous portion of an electrical discharge through a gas. *See also* GAS DISCHARGE TUBE.

dasymeter. An instrument for measuring gas density by finding the upthrust on a tube of known volume.

dating. Abbrev. for RADIOACTIVE DATING.

daughter product. A NUCLIDE resulting from the decay of another nuclide, often called the *parent nuclide*.

Davisson-Germer experiment. An experiment in which an electron beam was directed on to the face of a single nickel crystal in high vacuum. The reflected electrons were found to be diffracted, showing a maximum intensity at one particular angle, and thereby demonstrating the wave nature of the electron.

Davy lamp. A lamp whose flame burns inside a wire mesh cage which cools the hot gases from the flame by conduction, thus preventing ignition by the flame of flammable gases outside the lamp. It was the first lamp which was safe for use in mines. It has now been superseded.

day. The time taken for the Earth to rotate once about its axis. It is measured in various ways. *See* SIDEREAL TIME; SOLAR TIME; TIME.

dc. Abbrev. for DIRECT CURRENT.

dead room. A highly sound-absorbent room. It is made by covering all surfaces with sound-proofing material, structured so as to minimize sound reflection, and by placing absorbent screens at suitable locations so as to reduce the probability of setting up standing waves.

dead time. The post-stimulation time during which an electrical device is insensitive to another stimulus.

de Broglie wave. A wave of wavelength $h/(mv)$, associated with a particle of mass m moving with speed v; h is the PLANCK CONSTANT. Since the square of the amplitude of the wave at a given point represents the probability of finding the particle at that point, the wave is sometimes regarded as a wave of probability. *See also* DAVISSON-GERMER EXPERIMENT.

debye. Symbol D. A unit of electric dipole moment equal to

$$3.335\ 64 \times 10^{-30} \text{ coulomb metre}$$

Debye theory of specific heats. A quantum treatment in which a solid is regarded as a continuous elastic body of atomic structure such that its vibration frequencies stop abruptly at a maximum value. The theory correctly predicts that at low temperature the specific heat at constant volume is proportional to the cube of the temperature.

deca−. Symbol da. A prefix meaning 10.

decay. (1) The spontaneous disintegration of a radioactive nuclide into a daughter product, which may or may not be radioactive. *See also* ALPHA DECAY; BETA DECAY; RADIOACTIVITY.
(2) The decline in brightness of an excited PHOSPHOR after removal of the exciting radiation.

decay constant. Symbol λ. The probability per unit time of the radioactive decay of an unstable nucleus, equal to

$$-N^{-1}\ dN/dt$$

where N is the number of undecayed nuclei present at time t. Integration gives

$$N = N_0\ e^{-\lambda t}$$

where N_0 is the number of undecayed nuclei at $t = 0$.

deci−. Symbol d. A prefix meaning 1/10.

decibel. Symbol dB. A logarithmic unit used to compare two power levels, usually of sound or electricity. Power levels P and P_0 are said to differ by n decibel where

$$n = 10 \log(P/P_0)$$

For sound, P_0 is usually chosen to be at the threshold of AUDIBILITY. One decibel is roughly the smallest detectable change, which explains why the decibel is preferred to the *bel*, which is ten times as large.

decimal system. The common system of notation, using BASE 10 and digits 0–9.

declination. (1) The angle between the horizontal component of the Earth's magnetic field at a point and the direction of true north at that point. *See also* TERRESTRIAL MAGNETISM.
   (2) *See* CELESTIAL SPHERE.

defect. A discontinuity in the regular structure of a crystal. If only one single atom or molecule is involved, the defect is known as a *point defect*; three possibilities are illustrated in fig. D2. A *Frenkel defect* consists of a vacancy interstitial pair, formed when an atom moves to an *interstitial position*, leaving behind a vacancy. A *Schottky defect* occurs either when an atom moves to the surface leaving a vacancy, or when an atom moves from the surface to an interstitial position; surface includes grain boundaries or dislocations. Point-defect formation is encouraged by heat treatment, the number of defects rising exponentially with temperature. Strain and irradiation with ionizing radiation also result in point-defect production.

Frenkel defect   Schottky defects

D2 Frenkel and Schottky defects

Extended departure from structure regularity is known as a *line defect* or a *dislocation*. One example, *edge dislocation*, is illustrated in fig. D3. In this defect one plane of atoms is missing from the lattice.

Edge dislocation

D3 Edge dislocation

defect of eye. *See* COLOUR DEFICIENCY; ASTIGMATISM; MYOPIA; HYPERMETROPIA; PRESBYOPIA.

definition. The sharpness of the image formed by an optical system.

deflector coils. *See* CATHODE RAY TUBE, diagram 2.

deflector plates. *See* CATHODE RAY TUBE, diagram 1.

deformation potential. The electric potential resulting from deformation of the crystal lattice of either a conductor or a semiconductor. *See also* PIEZOELECTRIC EFFECT.

degaussing. The process of neutralizing the magnetic field of an object, for example by applying an equal and opposite field produced by coils of wire carrying a current. Important applications are to ships in order to prevent them from detonating magnetic mines, and to television receivers in order to neutralize the Earth's magnetic field and so improve picture quality.

degeneracy. (1) The existence of an atom, electron or nucleus in more than one possible state coresponding to a given energy level. Degeneracy can be removed by applying a perturbation which results in a separation of the levels for the various states.
   (2) The existence in classical dynamics of two or more independent modes of vibration having the same frequency.
   (3) The state of a gas at temperatures below 20 K, for which the molecular heat is less than $3R/2$ where R is the UNIVERSAL GAS CONSTANT.
   (4) The existence of atoms stripped of all their electrons. This occurs at very high densities, for example in neutron stars.

degenerate semiconductor. A semiconductor so heavily doped as to resemble a metal.

degree. (1) A unit of angular measure.
(2) A unit of temperature difference.
(3) The power to which a variable is raised. If several variables are multiplied together, the overall degree is the sum of the powers; if the variables are added, the degree is that of the term of highest degree. Thus y^4 is of the fourth degree while y^4z^3 is of the seventh degree, as is $y^4 + y^4z^3$.

degrees of freedom. (1) The minimum number of co-ordinates required to specify the motion of each particle in a dynamical system. Any particle has three *translational* degrees of freedom, corresponding to motion in three directions. Molecules have, in addition, *rotational* and *vibrational* degrees of freedom. A linear molecule has two rotational degrees of freedom, corresponding to rotation about two axes perpendicular to its axis; other molecules have three rotational degrees of freedom. The number of vibrational degrees of freedom of a molecule depends on the number of its atoms. By the principle of EQUIPARTITION OF ENERGY, the energy for each degree of freedom is $kT/2$ where T is the absolute temperature and k the Boltzmann constant.
(2) The number of independent variables required to define a system having a given number of phases and components. *See also* PHASE RULE.

de Haas-van Alphen effect. The periodic variation of magnetic susceptibility with magnetic field strength observed in a number of metals, usually at temperatures below 20 K.

dekatron. A type of low-pressure neon ELECTRON TUBE containing ten small cathodes arranged in a circle around a central anode. The circuit arrangement is such that an incoming pulse causes a GLOW DISCHARGE to be transferred sequentially from one cathode to the next, i.e. only one cathode functions at a time. The device can thus be used as a visual counting tube in the decimal system; it can also be used for switching.

del. Symbol ∇. The differential vector operator

$$i\, \partial/\partial x + j\, \partial/\partial y + k\, \partial/\partial z$$

where i, j and k are UNIT VECTORS along the x, y and z axes respectively. *See also* PARTIAL DERIVATIVE.

delocalization. The existence in some molecules of electrons which are not confined to the region of one particular bond or atom but can move over the whole molecule. The molecule is then more stable than it would be if the electrons were localized.

demagnetization. The removal of magnetism. It is best achieved by taking the magnetized material through a series of diminishing HYSTERESIS LOOPS to zero. To do this, the material is placed in a coil of many turns which is connected to an alternating-current supply; the material is then slowly withdrawn.

demand pacemaker. A pacemaker which only stimulates a heart which fails to beat within a fixed period. The danger of ventricular fibrillation (inherent in ordinary pacemakers) should the heart return to normal rhythm is thus avoided. *See also* ARTIFICIAL PACEMAKER.

demodulation. The process of separating the signal from a modulated carrier wave.

denominator. *See* FRACTION.

densitometer. An instrument for measuring the REFLECTION DENSITY or the TRANSMISSION DENSITY of a material. The most common form of the apparatus has a standard light source for illuminating the material and a PHOTOCELL, connected to a microammeter, for measuring the radiation. Readings are taken with and without the specimen in place and the density calculated. An important use is the measurement of density of photographs of spectral lines, X ray diffraction patterns etc, and so deducing the composition of the radiation giving rise to the phenomena.

density. (1) The mass of unit volume of a substance under standard conditions. The SI UNIT is the kilogramme per cubic metre. *See also* RELATIVE DENSITY; VAPOUR DENSITY.

(2) The closeness of any linear, superficial or space distribution, for example CURRENT DENSITY or ELECTRON DENSITY.

(3) *See* REFLECTION DENSITY; TRANSMISSION DENSITY.

dependent variable. A VARIABLE whose value can be calculated when that of another variable, the *independent variable*, is known. Thus in

$$y = 7x^6 + 2x$$

x is the independent and y the dependent variable.

depletion layer. The narrow region separating the n- and p-type components of a SEMICONDUCTOR RECTIFIER. It is denuded of charge carriers and a potential difference acts across it.

depolarization. The reduction of POLARIZATION effects in electrochemical cells. It may be achieved in a variety of ways: increasing the operating temperature, stirring the electrolyte, or chemically as in the LECLANCHÉ CELL.

depression of freezing point. The lowering of the freezing point of a solvent which occurs when a substance is dissolved in it. Relative molecular masses can be determined from the freezing point depressions produced.

depth of field. The range of axial distance on either side of an object whose image is in focus in an optical instrument, within which other objects also appear in acceptable focus.

depth of focus. The conjugate axial distance to the DEPTH OF FIELD, i.e. the range of axial distance on either side of an image over which the image remains in acceptable focus.

derivative. *See* DIFFERENTIATION. *See also* PARTIAL DERIVATIVE.

derived unit. *See* COHERENT UNITS. *See also* Table 2.

desaturated colour. A colour whose chromaticity co-ordinates lie within the boundaries of the CHROMATICITY CHART; the closer the co-ordinates are to those of the white point, the more desaturated the colour.

detached retina. A condition due to disease, injury or degenerative changes in the EYE. If the retina is not promptly reattached to the choroid, blindness will result in the affected eye since alterations in the nutrition and metabolism of the retina when detached result in rapid and irreversible degeneration. A one millisecond flash of light from a LASER focused on the retina will efficiently weld the retina back to the choroid; this procedure can be performed without anaesthetic and without mechanical fixing of the eye and so is suitable for use in the outpatients' department.

determinant. A mathematical expression consisting of a square array of numbers or variables and used in the solution of simultaneous equations.

detector. (1) The circuit used to separate a signal from its carrier wave.

(2) Any device used to detect or measure radiation energy, particle energy etc.

detonation. An explosion, usually small, which initiates a second larger explosion.

deuterium. The hydrogen isotope ^2_1H of MASS NUMBER 2, sometimes represented by the symbol D.

deuterium oxide. *See* HEAVY WATER.

deuteron. The DEUTERIUM nucleus. It has a spin of 1 and a positive magnetic moment.

deviation. (1) The difference between a particular measurement of a quantity and the mean value of all the measurements of the quantity.

The *mean deviation* is the average of the

magnitudes of all the deviations in a set of measurements.

The *standard deviation* is the square root of the average of the squares of all the deviations in a set of measurements.

(2) *See* MINIMUM DEVIATION; ANGLE OF DEVIATION.

Dewar flask. A vessel for maintaining its contents at a temperature different from that of its surroundings. Conductive and convective heat transfer with the surroundings is minimized by using a double-walled vessel and evacuating the space between the walls. To reduce heat transfer by radiation the side of the walls in contact with the vacuum are silvered. To prevent evaporation the vessel is stoppered.

dew point. The temperature below which moisture from the atmosphere condenses on a surface in contact with it, i.e. it is the temperature at which the air is saturated and depends on the humidity.

dextrorotatory. Denoting a substance which rotates the plane of polarization of polarized light in a clockwise direction, as viewed by an observer looking in the direction opposite to that of incidence of the light. *See* POLARIZATION (electromagnetic).

dextrorotory. Another name for DEXTRO-ROTATORY.

diagnostic physics. The study and use of physical instruments and techniques to assist in medical diagnosis. *See* ELEC-TROCARDIOGRAPHY; ELECTROENCEPHAL-OGRAPHY; ELECTROMYOGRAPHY; FLUORES-ENCE ASSAY; MONITORING PILLS; RADIOCARDIOGRAPHY; SPECTROSCOPY; THERMOGRAPHY; ULTRASOUND SCANNING; X RAY DIAGNOSIS.

diamagnetism. The property exhibited by some substances, when subjected to a magnetic field, of becoming magnetized in such a direction as to oppose the field, i.e. to display negative magnetic susceptibility. Diamagnetism is independent of temperature. The phenomenon has been explained in terms of changes in the orbits of electrons. All substances exhibit diamag-

netism, but since it is a weak effect it is frequently masked by stronger PARA-MAGNETISM and FERROMAGNETISM.

diaphragm. An aperture, sometimes variable, whose centre lies on the axis of an optical system and whose plane is perpendicular to this axis. The purpose of the diaphragm is to control the amount of light passing through the optical system.

diaphragm gauge. *See* MICROMANOMETER.

diathermancy. The property of transmitting infrared radiation. *Compare* ATHER-MANCY.

diatomic. Containing two atoms in the molecule. Hydrogen for example is diatomic.

dichroism. (1) The property of exhibiting two colours, usually one colour by transmitted light and the other by reflected light.

(2) The property of some materials which exhibit DOUBLE REFRACTION whereby the ordinary and extraordinary rays are absorbed to different extents.

dichromat. A colour- deficient person all of whose sensations of colour can be stimulated by a mixture, in suitable proportions, of only two PRIMARY COLOURS. There are three types of dichromat: one whose members require red and green primary lights, another whose members require red and blue primaries and the third whose members require green and blue primary lights; the last type is least common. *See also* COLOUR VISION.

dichromate cell. A primary CELL for which poles of carbon and amalgamated zinc are immersed in a solution of potassium dichromate in dilute sulphuric acid. The electromotive force is 2.03 volt.

dielectric. An insulator with electric conductivity less than 10^{-6} siemens.

dielectric constant. *See* PERMITTIVITY (relative).

dielectric heating.   The heating effect produced by rapid alternations of electrostatic charges applied across a dielectric, for example by means of a radio-frequency electric field. Practical applications include cooking and the heat treatment of plastics.

dielectric polarization.   The stress set up in a dielectric due to the presence of an electric field. The field displaces electrons in the atoms of the dielectric with respect to their nuclei, so turning the atoms into electric dipoles and producing the stress. The polarization equals

$$D - \varepsilon_0 E$$

where E is the applied field strength (a vector), D the ELECTRIC DISPLACEMENT (also a vector) and ε_0 the ELECTRIC CONSTANT.

dielectric strength.   The maximum potential gradient that can be applied to a material without causing electrical breakdown; it is usually measured in volt per millimetre.

differential calculus.   The branch of calculus concerned with DIFFERENTIATION.

differential equation.   An equation containing derivatives. The *order* of the equation is the order of the highest derivative, and the *degree* of the equation is the highest power present of the highest derivative. Thus

$$(d^2y/dx^2)^3 + dy/dx + x = 1$$

is a differential equation of the second order and the third degree.

differential steam calorimeter.   Another name for JOLY'S STEAM CALORIMETER.

differentiation.   The process of finding the *derivative* of a function. For the function

$$y = f(x)$$

the derivative is defined as

$$\lim_{\delta x \to 0} [f(x + \delta x) - f(x)]/\delta x$$

and is written as dy/dx. For $y = x^n$ it can be shown that

$$dy/dx = nx^{n-1}$$

where n has any rational value. The value of dy/dx at any point on a curve is the gradient of the curve at that point.

diffraction.   The spreading or bending of waves as they pass through an aperture or round the edge of a barrier. Subsequently the diffracted waves interfere with each other, giving regions of reinforcement and of cancellation. *See also* FRESNEL DIFFRACTION; FRAUNHOFER DIFFRACTION; ELECTRON DIFFRACTION; X RAY DIFFRACTION.

diffraction grating.   A series of parallel lines of separation comparable to the wavelength of light, ruled on a glass or reflecting metal surface; alternatively a plastic replica of such a set of rulings may be used. The grating, whether operating by transmitted or reflected monochromatic light, gives a series of maxima and minima at different angles of scattering. For a plane grating,

$$n\lambda = d(\sin i + \sin \theta)$$

where d is the distance between corresponding points on adjacent lines, i is the angle of incidence and θ the angle which the reflected or transmitted light makes with the normal, and λ is the wavelength of the light; n is known as the *order* and has the values

$$0, \pm 1, \pm 2, \pm 3, \dots$$

Thus for a mixture of wavelengths, several spectra are obtained corresponding to the different orders. For $n = 0$ all wavelengths are superimposed, i.e. no spectrum is formed.

Diffraction gratings are sometimes used in spectrometers in preference to prisms. For ultraviolet work concave reflecting gratings are required since they enable the absorption problems to be overcome. *See also* ULTRAVIOLET SPECTRUM.

diffuse reflection.   The type of reflection occurring at an uneven surface. An incident beam is scattered in all directions by the surface. An example of an almost perfect diffuser is a thick layer of snow

diffusion. (1) The migration of atoms or molecules due to their random thermal motions. The process is very slow in solids but rapid in gases.

(2) The passage of nuclear particles through matter when scattering is much more likely than capture. An example is neutron diffusion in a nuclear reactor.

diffusion cloud chamber. A type of CLOUD CHAMBER where diffusion of a vapour from a hot to a cold surface through an inert gas results in supersaturation. Since the vapour supply is continually replenished by diffusion, the chamber can be almost continuously sensitive to ion tracks. Another advantage is the absence of moving parts.

diffusion pump. A type of vacuum pump in whch gas is removed from a system by a stream of vapour, either oil or mercury. Gas molecules diffuse into the vapour stream, which issues from a nozzle, and are carried away. The vapour is recondensed in a backing pump, for example a ROTARY OIL PUMP, which is always required with a diffusion pump. Pressures as low as 10^{-7} pascal can be reached.

digit. Any of the symbols used in a notation for the integers. The number of different digits used in such a system is the BASE of the system; thus the binary system has two digits and the decimal system has ten.

digital computer. *See* COMPUTER.

digital display. A method of displaying measured values of some quantity, for example voltage or time, in digital form rather than by the reading of a pointer on a scale. *See also* LIGHT EMITTING DIODE; LIQUID CRYSTAL DISPLAY.

digital recording. A method of recording or transmitting sound in which the pressure in a sound wave is sampled about 30 000 times a second and the successive values represented by numbers which are recorded or transmitted. The numbers are then restored to analogue form in the receiver or player. The method gives high fidelity since no distortion or interference accompanies the process.

dilatancy. The phenomenon whereby the viscosity of a fluid increases with its speed. The effect is shown by some suspensions and pastes but the opposite effect, i.e. THIXOTROPY, is more common.

dilatometer. An apparatus for measuring volume changes. In a common form a narrow capillary tube is attached to a glass bulb. Any expansion of the liquid in the bulb produces a marked and therefore easily measurable rise in the liquid level in the capillary tube.

dimensional analysis. A method of checking an equation by analysing the DIMENSIONS in which it is expressed since, if the dimensions on the two sides are not the same, the equation is wrong. If the dimensions are the same, the equation is not necessarily correct but the error is most likely to be arithmetical. The method is also useful for establishing the form of an empirical relationship.

dimensions. The set of powers of basic independent physical quantities in terms of which other physical quantities are defined. For example the dimensions of force in terms of the basic quantities length, L, mass, M, and time, T, are

$$M\ L\ T^{-2}$$

diode. Any electronic device with only two electrodes. The main use of diodes is as rectifiers.

The *semiconductor diode*, symbol $\rightarrow\!\vdash$, consists of a single P-N JUNCTION, the n component usually being earthed as indicated (fig. D4). Current I flows when the potential V is positive; when V is negative the current is very small until breakdown is reached. The main use of the diode is as a SEMICONDUCTOR RECTIFIER.

The *thermionic diode* is a THERMIONIC VALVE which passes current when the anode is at a positive voltage with respect to the cathode, but passes no current when

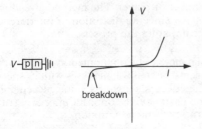

D4 Semiconductor diode

the anode is negative with respect to the cathode.

dioptre. A unit used to express the power of a spectacle lens and equal to the power of a lens of focal length one metre. The power in dioptre is thus the reciprocal of the focal length in metre. The power is taken to be positive for a converging lens and negative for a diverging one.

dioptric system. An optical system using lenses. *Compare* CATADIOPTRIC SYSTEM.

dip. The angle with the horizontal made by the Earth's magnetic field at a particular place. It varies from 0° at the magnetic equator to 90° at the magnetic poles.

dip circle. An instrument for measuring the angle of DIP at a particular place. Basically it consists of a thin magnet suspended so as to be able to rotate freely in the magnetic meridian about a horizontal axis through its centre of mass. The inclination of the magnet to the horizontal is read off on a vertical scale.

dipole. Two electric charges of equal magnitude and opposite sign separated by a very short distance. The product of the magnitude of either of the charges and their separation is the *dipole moment*, symbol *p*. Some molecules behave as dipoles, and by measuring their moments information about the configuration of the molecules may be deduced. *See also* MAGNETIC DIPOLE.

dipole aerial. *See* AERIAL.

dipole moment. *See* DIPOLE.

Dirac constant. Symbol $\hbar$ (called h bar or crossed *h*). The PLANCK CONSTANT divided by 2π.

Dirac equation. A quantum mechanical equation introduced to describe electron behaviour and now thought to be relevant to all elementary particles with spin of ½. The equation embodies the concept of negative energy states from which follows the possibility of POSITRON existence.

direct access device. *See* STORAGE DEVICE.

direct current. An electric current which is unidirectional, continuous and fairly steady. *Compare* ALTERNATING CURRENT.

directrix. *See* CONIC.

direct vision spectroscope. A device for producing DISPERSION with zero ANGLE OF DEVIATION for the mean wavelength of the visible spectrum. It consists of several crown and flint glass prisms cemented together, as illustrated for one pair (fig. D5). The prisms are mounted in a straight tube and viewed through an eyepiece. For the mean (yellow) ray, the angle of deviation in one direction produced by a crown glass prism is balanced by the angle of deviation in the other direction due to a flint glass prism. The instrument is used for the quick examination of the spectrum from a source.

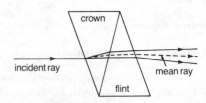

D5 Direct vision spectroscope component

discharge. (1) The removal of electric charge, for example from a capacitor.

(2) The conduction of electricity through a gas or other insulating material, usually accompanied by the emission of light. *See* GAS DISCHARGE TUBE.

(3) The conversion to electrical energy of chemical energy stored in a cell.

disintegration. Any process in which a nucleus breaks into two or more fragments, either spontaneously or as a result of a collision with a high-energy particle or radiation.

disintegration constant. Another name for DECAY CONSTANT.

disc. A direct-access STORAGE DEVICE used in computers. It consists of a flat circular plate coated with a layer of magnetic iron oxide. As the disc spins, data can be stored on a series of concentric tracks by means of a head which moves radially over the disc; the same head can also be used to retrieve the data.

disk. *See* DISC.

dislocation. *See* DEFECT.

disorder. *See* ORDER-DISORDER TRANS-FORMATION.

dispersion. The splitting up by REFRAC-TION or DIFFRACTION of a beam of electromagnetic radiation of mixed wavelengths into its consituent wavelengths. Deviation is measured by the rate of change of the angle of deviation, θ, with wavelength, λ, at a particular wavelength, i.e. by $d\theta/d\lambda$. *See also* ANOMALOUS DIS-PERSION.

dispersion force. A very weak force existing between molecules having no permanent dipole. It basically arises from the perturbation of the electronic orbits of neighbouring molecules. *See also* VAN DER WAALS FORCES.

dispersive power. Symbol ω. A property of a refracting medium over a particular wavelength range. It is defined by the equation

$$\omega = (n_2 - n_1)/(n - 1)$$

where n_1 and n_2 are the refractive indices of the medium at the two extreme wavelengths and n is the average refractive index over the range.

displacement. (1) The vector representing the change in position of a particular particle of a system from its initial position, due to the action of a force.
(2) The quantity of fluid displaced by a body wholly or partially submerged in the fluid.

displacement vector. *See* DISPLACEMENT (def. 1).

disruptive strength. Another name for DIELECTRIC STRENGTH.

dissociation. The breakdown of molecules into smaller molecules, radicals, ions or atoms, often reversibly as when acids are ionized in aqueous solution.

dissonance. *See* CONSONANCE.

distance-time graph. A plot of distance travelled by a body against time. Its slope at any time gives the speed of the body at that time.

distortion. (1) The deviation of the output of a system from what is required. For example an electrical or sound transmission system may introduce unrequired changes in waveform of the input voltage, current or sound. The changes may involve the introduction of features not present in the original, or the suppression or modification of features that were originally present.
(2) *optical distortion.* The lack of geometric similarity between object and image. For example magnification may vary with object size, resulting in either *barrel distortion* or *pincushion distortion* as illustrated in fig. D6. For barrel distortion, magnification decreases with object size; magnification increases with object size for pincushion distortion. Optical distortion may be reduced by the use of suitable stops.

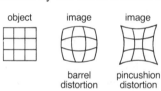

D6 Optical distortion types

div. Abbrev. for DIVERGENCE.

divergence. The SCALAR PRODUCT $\nabla.F$, written div F, where ∇ is the operator DEL and F is a vector. It equals

$$\partial F_x/\partial x + \partial F_y/\partial y + \partial F_z/\partial z$$

where F_x, F_y and F_z are the components of F along the x, y and z axes respectively.

divergent. A description of a sequence or series that is not CONVERGENT.

diverging lens. A LENS which causes a beam of light to diverge.

diverging mirror. A CURVED MIRROR which causes a beam of light to diverge.

division of amplitude. A method of producing optical interference in which the amplitude of a wave is divided into two parts which are later recombined. The fringes formed in a wedge-shaped air film are produced in this way.

division of wave front. A method of producing optical interference in which a wave front gives rise to secondary wave fronts which are later recombined. YOUNG'S FRINGES are formed in this way.

divisor. A quantity by which another quantity is divided; thus for a/b, b is the divisor.

D layer. See IONOSPHERE.

D lines of sodium. Two yellow lines very close together in the emission spectrum of sodium; their wavelengths are 589.6 nanometre for the D_1 line and 589 nanometre for the D_2 line. Since the lines are bright and easily reproduced they are used as reference lines in SPECTROSCOPY.

domain theory. See FERROMAGNETISM.

donor. An atom which donates one or more electrons in an extrinsic SEMICONDUCTOR, thus increasing the number of electrons in the conduction band. See BAND THEORY.

doping. The addition of impurities to a SEMICONDUCTOR in order to achieve a desired n-conductivity or p-conductivity. Doping can be effected by diffusion or b ion implantation.

doping compensation. The addition of a particular type of impurity to a SEMICON DUCTOR in order to modify the effect of an impurity already present.

doping level. The amount of material added to a SEMICONDUCTOR in DOPING. The smaller the amount the higher th resistivity of the semiconductor.

Doppler effect. The change in apparen frequency of a wave source due to relativ motion of source and observer. Various ci cumstances may occur; examples are a follows.
(a) Source moving with speed u_s towards stationary observer. If v is the wave spee and f waves are emitted per second by th source, then f waves occupy a distanc $v - u_s$; the wavelength is therefore

$$(v - u_s)/f$$

and the apparent frequency is

$$vf/(v - u_s)$$

which is greater than f.
(b) Source moving away from a stationar observer with speed u_s; by analogou reasoning to that used in (a), the apparen frequency is

$$vf/(v + u_s)$$

which is less than f.
(c) Observer moving with speed u_o toward a stationary source; the apparent fre quency is

$$f(v + u_o)/v$$

which is greater than f.
(d) Observer moving with speed u_o awa from a stationary source; the apparent fre quency is

$$f(v - u_o)/v$$

which is less than f.
(e) Source and observer both moving; i

moving in the same direction, the apparent frequency is

$$f(v - u_o)/(v - u_s)$$

f moving in the opposite direction the apparent frequency is

$$f(v + u_o)/(v - u_s)$$

Examples of the Doppler effect are the drop in pitch of a whistling locomotive as it passes an observer, and the RED SHIFT of light from receding stars.

Doppler radar. See RADAR.

dose. (1) *absorbed dose.* The energy of ionizing radiation imparted to unit mass of matter at a particular location in it. The unit is the RAD (*See also* RADIATION UNITS).

The *integral absorbed dose* is the integral of the absorbed dose throughout the mass of irradiated matter in a volume of interest.

The *maximum permissible dose* is the upper limit of absorbed dose to be received in a given time, as laid down by the International Commission on Radiological Protection

The *median lethal dose* is the absorbed dose of ionizing radiation that will kill, in a prescribed time, half of a large population of the species of organism under investigation.

(2) *exposure dose.* The energy of ionising radiation to which the body under study is exposed. It is measured in RÖNTGEN. (*See also* RADIATION UNITS).

dosemeter. Any device used to measure absorbed dose or exposure dose of ionizing radiation (*See* DOSE). Photographic films and ionization chambers are some of the devices used.

dosimetry. The measurement of absorbed dose or exposure dose of ionizing radiation (*See* DOSE). In chemical dosimetry, the change in optical density produced by irradiation of a suitable solution is measured; in lithium fluoride dosimetry, the THERMOLUMINESCENCE resulting from irradiation of lithium fluoride is measured

with a PHOTOMULTIPLIER. The apparatus most commonly used for dosimetry is the DOSEMETER.

dot product. Another name for SCALAR PRODUCT.

double refraction. The property possessed by certain substances of forming two refracted rays from a single incident ray. One ray, known as the *ordinary ray*, obeys the normal laws of refraction; the other ray, known as the *extraordinary ray*, has a refractive index (i.e. speed) which varies with its direction in the crystal. Both rays are plane polarized in directions at right angles to each other (*See* POLARIZATION, electromagnetic). The wave fronts of the ordinary ray are spherical whereas those of the extraordinary ray are ellipsoidal. Crystals which show double refraction have either one or two directions along which light is not doubly refracted; each of these directions is known as an *optical axis.* The difference of the maximum and minimum value of the refractive indices is taken as the measure of the double refraction.

doublet. (1) A lens composed of two components, especially an ACHROMAT.
(2) Two close spectral lines resuting from transitions between a single electron energy level and two close energy levels.

down quark. See QUARK.

drain. The electrode in a FIELD EFFECT TRANSISTOR through which carriers leave the inter-electrode region.

D region. Another name for D layer. *See* IONOSPHERE.

drift tube. See. LINEAR ACCELERATOR.

drift velocity. The velocity of charge carriers moving under the influence of an electric field in a particular medium.

driving mirror. A CURVED MIRROR, convex in shape, hence the images formed are always erect and the field of view large.

drum. A direct-access storage device used in a COMPUTER. It consists of a rapidly rotating cylinder coated with a layer of magnetic iron oxide. Fixed electromagnetic read-write heads are arranged around the drum, one for each of the tracks carried on the drum's surface.

dry cell. A type of primary CELL. Most are based on the LECLANCHÉ CELL in which the electrolyte is a paste rather than a liquid; spilling is therefore prevented. A zinc container, the negative electrode, is lined with an ammonium chloride plaster of Paris paste. At the centre is a carbon rod, the positive electrode, surrounded by a mixture of ammonium chloride, powdered carbon, zinc sulphate and manganese dioxide, bound together in a thick paste by glycerine. The electromotive force is about 1.5 volt.

ductility. The property, exhibited by some substances such as copper, of readily being drawn out into a wire without cracking or breaking.

Dulong and Petit's law. The MOLAR HEAT CAPACITY of all solid elements equals approximately $3R$ where R is the UNIVERSAL GAS CONSTANT. The law is valid for elements of simple crystal structure and at normal temperature. At lower temperatures the molar heat capacity is less than $3R$ and is proportional to the cube of the absolute temperature, in accordance with EINSTEIN'S THEORY OF SPECIFIC HEATS.

dust core. A magnetic core made of pulverized magnetic particles held together by a binder. Since there is little EDDY CURRENT loss in such a core, it is suitable for use with high frequencies.

dust tube. *See* KUNDT'S TUBE.

dwarf star. Another name for main sequence star. *See* HERTZSPRUNG-RUSSELL DIAGRAM. *See also* WHITE DWARF.

dynamics. A branch of mechanics concerned with the motion of bodies and the forces producing the motion.

dynamic viscosity. *See* VISCOSITY.

dynamo. Another name for GENERATOR (def. 1).

dynamometer. (1) Another name for TORQUE METER.
(2) Short for ELECTRODYNAMOMETER.

dynamo rule. *See* FLEMING'S RULES.

dynatron. A thermionic valve operated so that the anode current increases as the anode voltage decreases. Its main use is as a generator of oscillations.

dyne. Symbol dyn. The unit of force in CGS UNITS, equal to 10^{-5} newton.

dynode. An electrode giving SECONDARY EMISSION in an electron tube such as a PHOTOMULTIPLIER.

E

e. The irrational number 2.718 28..., defined as the limit of $(1 + 1/n)^n$ as n tends to infinity.

e. The symbol for the charge on an electron, equal to

$$-1.6021 \times 10^{-19} \text{ coulomb}$$

ear. The sense organ responsible for the maintenance of balance and, except for fish, the detection of sound. The ear is divided into outer, inner and middle sections.

earth. An electrical connection between a piece of equipment and the surface of the Earth. Such a connection is made either for safety reasons, so that a device will not become live if a fault develops, or (and) to give an arbitrary zero of electric potential.

Earth. The third nearest planet to the Sun, situated a mean distance of 149 598 000 kilometre from it and moving round it in approximately one YEAR. Earth's mass, diameter, relative density and period of revolution on its axis are about 6×10^{24} kilogram, 12 600 kilometre, 5.6 and 24 hour respectively.

earthshine. The Sun's light reflected by the Earth's surface. It is analagous to moonlight.

ebullition. Another name for BOILING.

eccentric anomaly. *See* ANOMALY.

eccentricity. *See* CONIC.

ECG. Abbrev. for ELECTROCARDIO-GRAPHY.

echelon grating. A type of DIFFRACTION GRATING made by stacking between 20 and 40 accurately parallel glass plates, of equal thickness to within a fraction of a wavelength, in optical contact to form a series of equal steps of about 1 millimetre width. The device may be used as either a reflection or transmission grating and will give high RESOLVING POWER of between 10^5 and 10^6. The device is mainly used to study the fine structure of spectral lines.

echo. (1) A sound repeated as a result of the reflection of sound waves from a surface. It occurs for intervals between the sound and its reflection greater than 0.1 second. A high-pitched sound usually gives a better echo than one of low frequency.
(2) The reflected signal in RADAR.

echo location. A technique such as RADAR or SONAR in which an object is located by its echo.

echo sounding. A technique for measuring the depth of water by transmitting a pulse of sound into it and timing the arrival of the reflected pulse. *See also* SONAR.

eclipse. The temporary blocking of the light from the Sun or Moon, known respectively as a *solar eclipse* and a *lunar eclipse*, as viewed from Earth. An eclipse occurs when Sun, Moon and Earth lie in or nearly in a straight line, as illustrated for a solar eclipse (fig. E1). For case A, observers directly in the Moon's UMBRA see a total eclipse those in the PENUMBRA see only a partial one. If the Earth and Moon are in positions shown in B, the eclipse as viewed from c is said to be *annular*.

eclipsing binary star. *See* BINARY STAR

ecliptic. The circle formed by the intersection of the plane of the Earth's orbit and

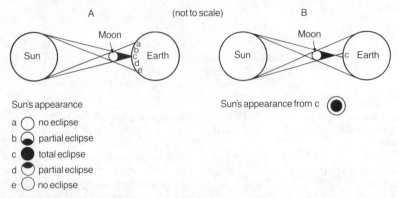

E1 Solar eclipse

the CELESTIAL SPHERE. It marks the apparent annual path of the Sun around the Earth. It is inclined at an angle of 23.5° to the CELESTIAL EQUATOR

eddy current. The current induced in a stationary conductor situated in a changing magnetic field, or in a moving conductor in a fixed field. Such currents occur in the cores of alternating current machinery, such as transformers, producing a loss of useful energy. It is usual to reduce these losses by LAMINATION of the core or by using a FERRITE core. Eddy currents produced in a MOVING COIL INSTRUMENT are often desirable because of the resulting DAMPING, known as *electromagnetic damping.*

edge dislocation. *See* DEFECT.

Edison accumulator. A storage battery with steel grid plates, the positive plate being filled with a metallic nickel/nickel hydrate mixture and the negative plate with iron oxide paste. The electrolyte is potassium hydroxide solution. The accumulator gives about 1.4 volt per cell. This is less than that given by the LEAD ACID BATTERY, which however has the disadvantage of a larger mass.

EELS. Abbrev. for ELECTRON ENERGY LOSS SPECTROSCOPY.

effective resistance. The resistance of an electric circuit element when used with alternating current. It may differ from the direct current value since it is influenced by EDDY CURRENT effects and the SKIN EFFECT.

effective value. *See* ROOT MEAN SQUARE.

effective weight. The weight of an object in vacuuo less the upthrust on it when immersed in a fluid. A floating body thus has zero effective weight.

efficiency. (1) *See* MACHINE.
(2) The ratio of the external work performed by a heat engine to the heat taken in, both quantities being measured in the same units. For a reversible heat engine the efficiency is given by

$$(T_1 - T_2)/T_1$$

where T_1 and T_2 are respectively the temperatures at which heat is taken in and given out. For real engines the efficiency is always less than this value. *See* CARNOT'S THEOREM.

effort. *See* MACHINE.

effusion. The leakage of gas through a fine orifice due to thermal motion of the gas molecules. When the mean free path of the molecules is small compared with the dimensions of the orifice, i.e. at ordinary pressures, GRAHAM'S LAW applies. When the mean free path is greater than the dimensions of the orifice, i.e. at low pressures, the effusion rate is given by

$$s(kT/2\pi m)^{\frac{1}{2}}$$

where s is the orifice area, T the thermodynamic temperature, m the molecular weight and k the BOLTZMANN CONSTANT. Partial separation of gases of different molecular weights can be achieved by effusion.

EHF. Abbrev. for EXTREMELY HIGH FREQUENCY.

Ehrenfest's theorem. The motion of a quantum mechanical WAVE PACKET will be identical to that of the classical particle it represents, provided any potentials acting on it are sensibly constant over the dimensions of the packet.

eightfold way. A method of grouping the ELEMENTARY PARTICLES which reveals regularities in their properties. It is the forerunner of QUARK models.

eigenfunction. *See*   CHARACTERISTIC FUNCTION.

eigenvalue. *See* CHARACTERISTIC VALUE.

einstein. A unit of energy of electromagnetic radiation of a particular frequency, equal to the energy of one mole of photons of that frequency. Its value is $N_A h\nu$ where N_A is the AVOGADRO CONSTANT, h the PLANCK CONSTANT and ν the frequency.

Einstein shift. A slight RED SHIFT in the lines of the spectrum of a star's radiation due to the star's gravitational field.

Einstein's law. The law $E = mc^2$ where E is the intrinsic energy of mass m and c is the speed of light in a vacuum. *See also* RELATIVITY.

Einstein's photoelectric equation. *See* PHOTOELECTRIC EFFECT. *See also* PHOTO-IONIZATION.

Einstein's probability coefficients. Three coefficients respectively representing the probability of spontaneous electron transitions between higher and lower energy levels in atoms or molecules, the probability of induced transitions in either direction, and the probability of the acquisition of energy by absorption. Einstein introduced the coefficients in his derivation of the PLANCK RADIATION FORMULA.

Einstein's theory of specific heats. A theory based on the assumption that a solid is an assembly of bound atoms or molecules each of which vibrates independently as a three-dimensional harmonic oscillator, all the atoms having the same vibration frequency, ν. Applying quantum theory then gives the specific heat at constant volume as

$$3R y^2\, e^y (e^y - 1)^{-2}$$

where R is the universal gas constant and $y = h\nu/kT$ where h is the PLANCK CONSTANT, k the BOLTZMANN CONSTANT and T the thermodynamic temperature. The theory has now been superseded by the DEBYE THEORY OF SPECIFIC HEATS.

Einthoven galvanometer. An instrument consisting of a single conducting filament stretched midway between the poles of a powerful electromagnet. The deflection of the filament, which occurs when a current is passed through it, is measured by means of a high-power microscope let through one of the pole pieces. The galvanometer will detect currents as low as 10^{-11} ampere.

elastance. The reciprocal of CAPACITANCE. The unit, farad^{-1}, is sometimes called a *daraf*

elastic collision. *See* COLLISION.

elasticity. The property which enables a body, deformed by an applied load, to recover its original configuration when the load is removed.

elastic limit. The stress value above which a body no longer deforms elastically, i.e. the stress value at which plastic deformation starts. *See* HOOKE'S LAW; YIELD POINT.

elastic modulus. The ratio of stress to the resulting strain before the ELASTIC LIMIT is reached. There are different moduli corresponding to the various types of strain: examples are BULK MODULUS, RIGIDITY MODULUS and YOUNG'S MODULUS. For substances not obeying HOOKE'S LAW, for example cast metal, concrete, marble and wood, the moduli are defined at a particular stress value as the ratio of a small increase in that stress to the resulting strain increase.

elastic scattering. *See* SCATTERING.

elastomer. A substance, usually synthetic, which has elastic properties similar to those of rubber.

elastoresistance. The change in electric resistance of a material under the action of stress within the elastic limit.

E layer. *See* IONOSPHERE.

electret. A dielectric substance which retains an electric moment after removal of the applied electric field. It is thus the electrical analogue of a permanent magnet.

electrical calorimetry. A method of CALORIMETRY in which the rise of temperature of a known mass of substance, produced by supplying a measured amount of electric energy, is measured. The method is used for measuring SPECIFIC HEAT CAPACITY and SPECIFIC LATENT HEAT. *See also* CONTINUOUS FLOW CALORIMETER.

electrical image. An imaginary charge or dipole introduced in order to facilitate the calculation of potential distribution.

electric arc. *See* ARC (def. 1).

electric charge. *See* CHARGE.

electric conductivity. Symbol K. The reciprocal of the RESISTIVITY of a material.

electric constant. Symbol ε_0 The absolute PERMITTIVITY of free space. It has the value

8.854 18 farad per metre

electric current. *See* CURRENT.

electric current density. *See* CURRENT DENSITY.

electric dipole. *See* DIPOLE.

electric discharge. *See* CONDUCTION (in gases).

electric displacement. Symbol D. A vector whose direction at a given point is that of the electric field at that point and whose magnitude is the produce of the magnitude of the ELECTRIC FIELD STRENGTH and the absolute PERMITTIVITY.

electric double layer. *See* HELMHOLTZ ELECTRIC DOUBLE LAYER.

electric energy. Energy possessed by a charged body due to its position in an ELECTRIC FIELD. It equals QV where Q is the body's charge and V the potential at the position of the body.

electric field. The space surrounding an electric CHARGE and throughout which the charge will exert a detectable force on another electric charge.

electric field of charged conductor. Inside the conductor the field is zero. For a charged spherical conductor the field outside the conductor is the same as if the whole charge of the sphere was concentrated at its centre.

electric field strength. The force experienced by a stationary unit positive charge placed in an electric field at a point where its strength is to be determined.

electric flux. The product of the area under consideration and the component, normal to this area, of the average ELECTRIC FLUX DENSITY over it. It is therefore the surface integral of the ELECTRIC DIS-PLACEMENT normal to the surface.

electric flux density. Another name for ELECTRIC DISPLACEMENT.

electric hysteresis. *See* HYSTERESIS.

electric impedance. *See* IMPEDANCE (def. 1).

electric induction. Another name for ELECTRIC DISPLACEMENT.

electric intensity. Former name for ELECTRIC FIELD STRENGTH.

electricity. The branch of physics concerned with the study of phenomena caused by electric charges. It is known as *current electricity* when concerned with moving charges and as *static electricity* when concerned with stationary charges.

electric motor. A quiet clean device for converting electrical energy to mechanical energy (efficiency 75–95%), using electromagnetic induction. An electric motor may be considered as a GENERATOR which is supplied with current through its coil. Several types of design are available.

electric oscillation. *See* OSCILLATION.

electric polarization. Another name for DIELECTRIC POLARIZATION.

electric potential. Another name for electrostatic potential. *See* POTENTIAL.

electric susceptibility. The ratio of the electric dipole moment per unit volume to the electric field strength.

electrocardiography. The recording of potentials which have spread to the body surface from the heart. Since the potentials are small they are amplified prior to recording. To minimize the presence of nerve and muscle action potentials on the recording, it is made with the patient lying down and as relaxed as possible. The various direct-current potentials, which are of the order of 1 millivolt, are elec-tronically eliminated from the recording. Studies of recordings enable heart desease to be diagnosed early and relatively simply.

electrochemical equivalent. The mass of a given substance which is liberated or deposited from a solution containing it at an electrode by the passage of 1 coulomb of electricity. *See* FARADAY'S LAWS OF ELECTROLYSIS.

electrochemistry. The study of ELECTROLYSIS, electrolytic cells and the general behaviour of ions in solutions.

electrode. A conductor by means of which current is passed into or removed from an electrical system. The positive electrode is the ANODE and the negative one is the CATHODE.

electrodeposition. The coating of one metal on the surface of another metal: the metals are made the electrodes in an electrolytic cell which contains a solution of a salt of the metal to be deposited. The thickness of the deposited layer is governed by the quantity of electricity passed.

electrode potential. The POTENTIAL DIFFERENCE between an electrode and the electrolyte with which it is in contact.

electrodynamics. The study of the mechanical forces existing between neighbouring current-carrying conductors.

electrodynamometer. An electric measuring instrument using the force between two or more current-carrying coils. With appropriate connections it may be used as a voltmeter, ammeter or wattmeter for either alternating or direct current.

electroencephalography. The study of electric potentials on the surface of the head. The potentials are of the order of 100 microvolt. Electrodes are hooked to the scalp at up to 24 locations and the pattern of potential difference between any pair of electrodes can be studied. Prior to recording, background noise and other electrical disturbances in the body are electronically filtered out of the signals; the signals are then amplified and recorded. The main components of the recording are the ALPHA WAVE and the BETA WAVE. The principal use of the recordings is in diagnosis and less frequently in treatment. Epilepsy produces typical patterns; tumours near the brain surface yield a characteristic trace from electrodes sited close to the tumour; brain damage due to accidents can often be assessed from the traces. The alternative to this method of diagnosis is often a difficult and dangerous brain operation.

electrokinetic potential. Another name for ZETA POTENTIAL.

electroluminescence. FLUORESCENCE induced in a substance by bombarding it with electrons.

electrolysis. The passage of an electric current through a liquid containing ions. The current flows in both directions and so differs fundamentally from the electron flow in a metal conductor. An important aspect of the subject is chemical change occurring at the electrodes. See FARADAY'S LAWS OF ELECTROLYSIS.

electrolyte. A solution of a substance, or the substance itself in a fused state, which conducts electricity by the movement of positive and negative ions.

electrolytic capacitor. See CAPACITOR.

electrolytic conductivity. The ratio of the CURRENT DENSITY in an electrolyte to the electric field strength.

electrolytic gas. The gas mixture produced by the ELECTROLYSIS of water, i.e. an explosive mixture of hydrogen and oxygen in the proportions 2:1 by volume.

electrolytic separation. The separation of isotopes by making use of the different rates at which they are released in ELECTROLYSIS. For example, in the electrolysis of water less deuterium than hydrogen pro

rata is formed at the cathode since deuterium is heavier than hydrogen; in time therefore the water becomes enriched with HEAVY WATER.

electromagnet. A coil of insulated wire wound round a soft ferromagnetic core: when current passes through the coil the core is strongly magnetized but it loses its magnetism when the current is switched off. Electromagnets are used for example in electric bells, switches, solenoids and lifting cranes for some metals.

electromagnetic damping. See EDDY CURRENT.

electromagnetic deflection. A method of deflecting an electron beam using an ELECTROMAGNET. The most common application is in the CATHODE RAY TUBE of a television or radar receiver.

electromagnetic field. A field representing the interacton of electric and magnetic forces. For free space, MAXWELL'S EQUATIONS express the relationships between the variables; the POYNTING VECTOR gives the direction and magnitude of the energy flow. According to electromagnetic field theory, an electric field is set up by stationary electric charges and a magnetic field by moving electric charges. The nature of the observed field will therefore depend on the observer's relative motion, so that a field appearing magnetic to one observer may appear electric to another and vice versa.

electromagnetic focusing. The use of a magnetic field to produce convergence of an electron beam. The field is generally obtained by passing direct current through a coil which is usually short and coaxial with the beam. See also CATHODE RAY TUBE.

electromagnetic induction. The production of an electromotive force in a conductor by changing the magnetic flux linked with it. The flux can be varied either by varying the magnetic field or by altering the configuration of a conductor relative to a constant magnetic field: the first method is used to produce the current in the secondary coil of a TRANSFORMER; the second method is employed in producing current in a GENERATOR.

According to the *Faraday-Neumann law*, the magnitude of the induced electromotive force is proportional to the rate of cutting of the magnetic flux. According to *Lenz's law*, the force is in such a direction as to oppose the change. The combined statement of the two laws is

$$V = -L \, dI/dt$$

where V is the induced electromotive force, dI/dt is the rate of change of current I with time t, and L is a constant known as the *coefficient of self-inductance*. The phenomenon is known as *self-inductance*.

When the current in a circuit changes, its associated changing magnetic field will induce an electromotive force in a neighbouring circuit. The phenomenon is known as *mutual inductance.* The induced electromotive force is given by

$$-M \, dI/dt$$

where M is the *coefficient of mutual inductance* and dI/dt measures the rate of change of original current.

The stored energy due to induction is half the product of the appropriate inductance coefficient and the square of the current.

electromagnetic interaction. A type of interaction between ELEMENTARY PARTICLES arising from any electric and magnetic fields associated with the particles. Electromagnetic interaction has about 1/200th of the strength of STRONG INTERACTION. In an electromagnetic interaction there is conservation of ANGULAR MOMENTUM, BARYON NUMBER, CHARGE, CHARGE CONJUGATION PARITY, ISOSPIN, PARITY and STRANGENESS.

electromagnetic levitation. *See* LEVITATION.

electromagnetic moment. *See* MAGNETIC DIPOLE MOMENT.

electromagnetic polarization. *See* POLARIZATION (def. 1).

electromagnetic pump. A simple pump without moving parts and suitable for use with conducting liquids. A strong magnetic field is applied across a diameter of a pipe holding the liquid, through which a current is passed. The liquid therefore experiences an induced force acting along the axis of the pipe and is thus propelled along the pipe. An important application of the pump is the removal of the liquid-sodium coolant used in some NUCLEAR REACTOR TYPES.

electromagnetic radiation. Waves of energy caused by the acceleration of charged particles. The waves involve transversely oscillating electric and magnetic fields at right angles to each other. The waves travel through a vacuum at the speed of light and are described by MAXWELL'S EQUATIONS. An alternative and complementary model treats electromagnetic radiation as PHOTON streams.

electromagnetic spectrum. The distribution of ELECTROMAGNETIC RADIATION throughout the frequency range, as shown in the table overleaf.

electromagnetic units. *See* CGS UNITS.

electrometallurgy. The extraction, refining or shaping of metals by ELECTROLYSIS.

electrometer. An instrument for detecting or measuring potential difference. Originally electrostatic instruments based on the electroscope, such as the electrostatic voltmeter and quadrant electrometer, were used. Apart from special applications these have now been superseded by electronic devices such as an amplifier with very high input impedance drawing negligible current and employing FIELD EFFECT TRANSISTORS. Electrometers can be used to measure currents as low as 10^{-9} ampere by passing the current through a known high resistance and measuring the resulting voltage drop with the electrometer.

electromotive force. Symbol E. The energy supplied by a current source in driving unit

Electromagnetic spectrum

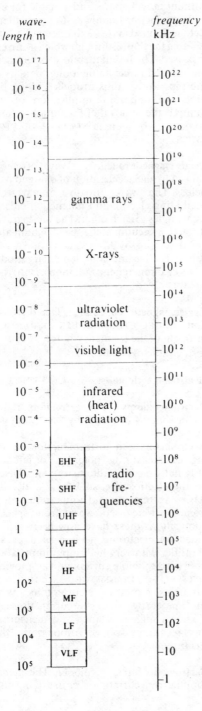

wave-length m		frequency kHz
10^{-17}		
		10^{22}
10^{-16}		
		10^{21}
10^{-15}		
		10^{20}
10^{-14}		
		10^{19}
10^{-13}		
		10^{18}
10^{-12}	gamma rays	
		10^{17}
10^{-11}		
		10^{16}
10^{-10}	X-rays	
		10^{15}
10^{-9}		
		10^{14}
10^{-8}	ultraviolet radiation	
		10^{13}
10^{-7}		
	visible light	10^{12}
10^{-6}		
		10^{11}
10^{-5}	infrared (heat) radiation	10^{10}
10^{-4}		
		10^{9}
10^{-3}		
	EHF	10^{8}
10^{-2}	radio fre-quencies	
	SHF	10^{7}
10^{-1}		
	UHF	10^{6}
1		
	VHF	10^{5}
10		
	HF	10^{4}
10^{2}		
	MF	10^{3}
10^{3}		
	LF	10^{2}
10^{4}		
	VLF	10
10^{5}		
		1

CHARGE around an electric circuit in the direction of the electromotive force. If the current source has internal resistance, the potential difference across it when it is supplying current is less than its electromotive force, which equals the potential difference across it on open circuit.

electromotive series. A series of the metals arranged according to their readiness to form positive ions. The more electropositive an element, the greater its ability to replace less electropositive elements from their salts.

electromyography. The process of registering the electrical activity of muscle. Needle electrodes are inserted into the muscle under investigation and their output is either displayed on a CATHODE RAY TUBE or fed to a loudspeaker. The technique is considered to be an efficient means of diagnosing disease of the lower motor neurons, myoneural junctions and skeletal muscle fibres.

electron. An elementary particle of negative charge, the magnitude of which is

$1.602\ 192 \times 10^{-19}$ coulomb

It has a mass equal to

$9.109\ 56 \times 10^{-31}$ kilogramme

It has a SPIN of ½ and is classified as a LEPTON. Electrons are constituents of all atoms and, when free, are primarily responsible for electric conduction in most substances. Electrons are produced in beta decay, in electric discharges and in thermionic and photoelectric emission. The ANTIPARTICLE of the electron is the POSITRON.

electron affinity. Symbol A. The energy released when an electron is attached to an atom or molecule. It is frequently positive, i.e. the negative ion is often more stable than the neutral species.

electron beam technology. The use of electron beams in a variety of applications. These include the refining, casting and welding of refractory metals, for example tungsten; hole drilling and milling of both metals and ceramics; treatment of deep-seated malignant tumours.

electron bombardment. In interaction with matter, an electron may undergo either elastic or inelastic COLLISIONS. The former results in deflection of the electron and dissipation of some of its energy in the form of heat: this is the basis of the localized heating of substances in a vacuum. In the case of inelastic collisions the electron loses discrete amounts of energy, accompanied by various phenomena according to the state of the matter: for gas atoms, ionization or excitation accompanied by light emission occurs; for solids and liquids, emission of X RAYS, FLUORESCENCE and secondary electron emission are among the phenomena which may result; for crystalline matter, ELECTRON DIFFRACTION can occur.

electron density. (1) In general, the number of electrons per unit mass.
(2) In astrophysics, the number of electrons per unit volume.

electron density map. A representation of the location of atoms in a crystal lattice as deduced from X RAY DIFFRACTION studies. The atoms do not appear as small point structures since the electrons are not localized; thermal vibrations of the atoms also have an effect. With the help of such representations, the structure of many compounds which defied chemical analysis has been determined.

electron diffraction. The DIFFRACTION of electrons by atoms or molecules. The fact that it occurs confirms the wave nature of the electron, the wavelength being $h/(mv)$ where h is the PLANCK CONSTANT and m and v the mass and speed of the electron. The penetrating power of an electron beam is low compared to that of an X ray beam and so electron diffraction is not much used to investigate crystal structure. It is widely used, however, for solid surface studies and for the measurement of bond lengths and angles of molecules in gases. The main techniques are low-energy electron diffraction, known as *LEED*, in which the diffracted electrons are reflected on to a fluorescent screen, and high-energy electron diffraction, known as *HEED*, in which electrons reflected from or transmitted through thin films are studied.

electronegative elements. Elements which tend to gain electrons and so form negative ions.

electron energy loss spectroscopy. Another name for ELECTRON IMPACT SPECTROSCOPY.

electron gas. Free electrons in a substance, regarded as analogous to a real gas dissolved in the substance. The model has proved useful in the theoretical treatment of for example CONDUCTIVITY and THERMIONIC EMISSION.

electron gun. A device, illustrated in fig. E2, for producing a narrow beam of high-velocity electrons. The intensity of the beam can be varied by adjusting the control-grid potential.

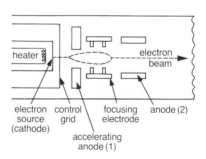

E2 Electron gun

electronic charge. The charge of the ELECTRON.

electronic mass. The mass of the ELECTRON.

electronics. The study, design and use of devices depending on electron movement in a gas, vacuum or semiconductor. Apart from a few specialized uses, gas and vacuum devices are now obsolescent.

electronic spectrum. A spectrum resulting from changes of energy of electrons in atoms.

electron impact spectroscopy. An inclusive term for forms of ELECTRON SPECTROSCOPY in which excited molecular states are studied by using electron beams to induce transitions between electronic energy levels.

electron lens. A device for focusing an electron beam using either an electrostatic or a magnetic field. The process is analogous to the focusing of a light beam with an optical lens. An example of the use of electron lenses is in the ELECTRON MICROSCOPE.

electron microscope. A device which produces a magnified image of a sample by means of high-energy electrons rather than by light. The electron wavelength is much shorter than that of light and so the resolution of the electron microscope is much better than that of its optical counterpart. Objects as small as 0.2 nanometre can be resolved and a magnification of 200 000 times obtained.

Three types of vacuum electron microscope are available: transmission, scanning and scanning-transmission. In the *transmission electron microscope* a combination of electrostatic and magnetic lenses focuses the electron beam on to a thin sample and the transmitted electrons form an image on a fluourescent screen. In the *scanning electron microscope* the incident beam of electrons is subjected to a varying field and thus made to scan the sample, which can be of any convenient shape and thickness. The incident electrons cause the emission of SECONDARY ELECTRONS from the sample surface and the secondary electrons form the image on a fluorescent screen. The scanning instrument has lower resolution than the transmission type but has the advantage of yielding a three-dimensional image. The hybrid microscope has the advantages of both the other types.

electron miltiplier. Another name for PHOTOMULTIPLIER.

electron optics. The study of the behaviour and control of an electron beam in magnetic and electric fields.

electron probe microanalysis. A technique for analysing amounts of a substance as small as 10^{-13} gramme. A very fine beam of electrons of diameter about 10^{-6} metre is focused on the sample and the resulting CHARACTERISTIC X RADIATION emitted by the sample is examined. From the intensity of this radiation the quantity of substance present can be deduced.

electron shell. The electrons having the same total QUANTUM NUMBER in a given atom. Each shell is denoted by a letter of the alphabet, starting with K for the innermost shell. For the K shell each of two electrons can have total quantum number 1; for the next shell, i.e. the L shell, each of eight electrons can have total quantum number 2, while for the following shell, i.e. the M shell, each of 16 electrons can have total quantum number 3. A shell containing the number of electrons mentioned is said to be complete; if it contains a smaller number it is said to be incomplete. Electrons in a shell are arranged in SUBSHELL groups. *See* ATOMIC ORBITAL.

electron spectroscopy. The measurement of the distribution of kinetic energies of the electrons in an electron beam. The information is used to determine energy levels in the electron source.

electron spin resonance. A technique similar to NUCLEAR MAGNETIC RESONANCE but applied to the investigation of the SPIN of electrons in atoms. The absorbed radiation lies in the microwave region.

electron stains. Substances such as osmic acid with high scattering power for electrons. They are used in conjunction with the electron microscope in an analagous way to staining media in optical MICROSCOPY.

electron synchrotron. *See* SYNCHROTRON.

electron tube. A vacuum tube, or a tube containing gas, in which electrons move between two electrodes. Examples are the THERMIONIC VALVE and GAS DISCHARGE TUBE.

electronvolt. Symbol eV. The energy acquired by an electron in moving through a potential difference of one volt; it equals

$$1.602\ 192 \times 10^{-19}\ \text{joule}$$

electro-optics. Another name for OPTO-ELECTRONICS (def. 1).

electro-osmosis. The movement of liquid through a porous diaphragm or other permeable solid as a result of an applied electric field.

electrophoresis. A technique for the analysis and separation of colloids by the application of an electric field. Generally a small amount of sample is applied to a paper moistened with salt solution and in contact with two electrodes. The components in the sample are identified by their different migration rates.

electrophorus. An early form of ELECTROSTATIC GENERATOR consisting of a metal plate with insulating handle. The device is placed on top of a charged dielectric plate. The metal plate is then momentarily earthed and withdrawn and thus acquires a charge of opposite sign to that of the dielectric plate.

electroplating. The process of coating a metal object with a thin layer of metal using ELECTRODEPOSITION.

electropositive elements. Elements which tend to lose electrons and so form positive ions.

electroscope. An electrostatic instrument for the detection of differences in electric potential. The best-known type is the *gold leaf electroscope* in which two gold leaves hang side by side from a metal rod inside a draught-proof insulated case. When the supporting rod is charged, the leaves separate due to the repulsion of the like charges on them.

electrostatic deflection. The deflection of an electron beam by applying an electrostatic field perpendicular to the direction of the beam. It is used in some CATHODE RAY TUBE types.

electrostatic field. Another name for ELECTRIC FIELD.

electrostatic focusing. The use of an electrostatic field to produce convergence of an electron beam. The field is usually obtained by applying different potentials to two or more hollow cylinders coaxial with the beam. *See also* CATHODE RAY TUBE.

electrostatic generator. A device for building up electric charge to a high potential, as required for many experimental purposes. Early examples are the ELECTROPHORUS and WIMSHURST MACHINE. A more sophisticated development is the VAN DE GRAAFF GENERATOR.

electrostatic induction. The separation of charge in a neutral body due to the effect of an electrostatic field. For example when an uncharged insulated conductor is placed near a negatively charged body, the portion of the conductor nearest to the body becomes positively charged and the remoter regions negatively charged; the algebraic sum of the charges is zero for the conductor.

electrostatic polarization. *See* POLARIZATION (def. 3).

electrostatic potential. *See* POTENTIAL.

electrostatic precipitation. The process whereby solid particles or liquid droplets are separated from suspension in a gas by applying an electrostatic field, i.e. by the use of ELECTROSTATIC INDUCTION. The process has wide industrial use in recovery and purification of materials.

electrostatics. The study of stationary electric charges and their effects.

electrostatic shielding. The enclosure of the volume to be shielded by an insulated wire mesh cage. The electric field inside the volume due to charges outside it is then zero. To prevent any external effects of charges inside such a cage, it is only necessary to earth the cage.

electrostatic units. *See* CGS UNITS.

electrostatic voltmeter. A type of ELECTROMETER used for measuring potential diffrences in the kilovolt range. *See also* QUADRANT ELECTROMETER.

electrostriction. The change in dimensions of a body when subjected to an electric field. If the field is not uniform, the body will tend to move into the region of higher (lower) field strength if the relative PERMITTIVITY of the body is higher (lower) than that of its surroundings. *Compare* MAGNETOSTRICTION;   PIEZOELECTRIC EFFECT.

electrovalent bond. A type of bond in which atoms or groups of atoms are held together by electrostatic forces between ions.

electroweak theory. A theory which combines ELECTROMAGNETIC INTERACTION and WEAK INTERACTION.

element. A substance composed of atoms all of which have the same ATOMIC NUMBER.

elementary charge. The smallest amount of charge which can be observed. It is the charge on the ELECTRON.

elementary particles. The fundamental constituents of all the matter in the universe. Their various properties are indicated in the table. Particles not marked stable eventually decay into stable ones by the emission of other elementary particles. The mean life depends on the type of interaction: it is longest, in the range 10^{-6} to 10^{-10} second, for WEAK INTERACTION, shortest, about 10^{-23} second, for STRONG INTERACTION, and intermediate, around 10^{-16} second, for ELECTROMAGNETIC INTERACTION. It is to be noted that according to modern theory only leptons are truly fundamental, the hadrons having constituents known as QUARKS. *See* HADRON; LEPTON; MESON; BARYON.

elevation of boiling point. The increase in the boiling point of a pure solvent which occurs when impurities are dissolved in it.

ellipse. A CONIC of eccentricity less than

Particle group	Particle name	Particle symbol	Charge (proton as unit)	Mass (proton as unit)	Isospin I	Spin J	Parity P	G parity	Charge conjugation parity	Strangeness	Mean Life/s
photon	photon	γ	0	0	1		−1		−1		stable
leptons	neutrino	ν	0	0		½					stable
	electron	e	−1	$^1/_{1800}$		½					stable
	muon	μ	−1	$^1/_9$		½					2.2×10^{-6}
	tau	τ	−1	2		½					
	positive tau	$\bar{\tau}^+$	1^+								
mesons	pion	π^-	±1	$^1/_7$	1	0	−1	−1		0	2.6×10^{-8}
	pion	π^0	0	$^1/_7$	1	0	1	−1	1	0	8.4×10^{-15}
	kaon	K	±1	½	½	0	−1			±1	1.2×10^{-8}
	kaon	K^0	0	½	½	0	−1			±1	
	kaon	K^0	0	½	½	0	−1			±1	8.6×10^{-10}
	kaon	K^0_2	0	½	½	0	−1			±1	5.2×10^{-8}
	eta	η	0	½	0	0	−1	1	1	0	
baryons	proton	p	1	1	½	½	1			0	stable
	neutron	n	0	1	½	½	1			0	932
	lambda	λ	0	1.1	0	½	1			−1	2.5×10^{-10}
	sigma	Σ^-	1	1.2	1	½	1			−1	8.0×10^{-10}
	sigma	Σ^0	0	1.2	1	½	1			−1	1.0×10^{-14}
	sigma	Σ	−1	1.2	1	½	1			−1	1.5×10^{-10}
	xi	Ξ^0	0	1.3	½	½	1^*			−2	3.0×10^{-10}
	xi	Ξ	−1	1.3	½	½	1^*			−2	1.7×10^{-10}
	omega	Ω	−1	1.8	0	$^3/_2{}^*$	1^*			−3	1.3×10^{-10}

*Predicted by theory

Very short life duration (10^{-23}s) particles such as delta, sigma star and Xi star particles are not included.

1. It is the locus of a point which moves so that the sum of its distances from two fixed points, i.e. from each *focus*, is constant. Taking axes along the axes of symmetry of the ellipse, as illustrated in fig. E3, its equation is

$$x^2/a^2 + y^2/b^2 = 1$$

where *a* and *b* are known as the *semimajor* and *semiminor axis* respectively. The area of the ellipse is πab.

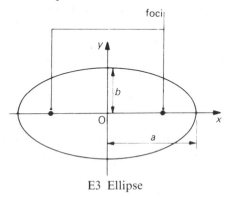

E3 Ellipse

ellipsoid. The three-dimensional surface or solid obtained by rotating an ELLIPSE about its major or minor axis.

elliptical. Having the shape of an ELLIPSE or an ELLIPSOID.

elliptical galaxy. An ellipsoidal-shaped GALAXY with no clearly defined internal structure. The observed outline shape varies from almost circular to narrow elliptical.

elliptical polarization. *See* POLARIZATION (electromagnetic).

elongation. The angular distance of a celestial object from the Sun, as measured from Earth.

e/m. The ratio of charge to mass for the electron. For electron speeds much less than the speed of light, the value of *e/m* is

$1.758\ 796 \times 10^{11}$ coulomb per kilogramme. At speeds approaching that of light, *e/m*

decreases with increase in speed due to the pronounced relativistic increase in electron mass. *See* RELATIVISTIC MASS.

emanation. An obsolete term for the radioactive gases given off in the radioactive decay of radium, actinium and thorium.

emf. Abbrev. for ELECTROMOTIVE FORCE.

emission spectrum. The SPECTRUM of the radiation emitted by a substance due to its constituent atoms or molecules returning from a higher to a lower energy state. The atoms or molecules can be excited to a higher energy state in many ways, for example by heat or by bombardment with electrons or with X rays.

emissivity. Symbol ε. The ratio of power per unit area radiated from the surface of a body to that radiated from a BLACK BODY at the same temperature.

emitter. The electrode in a TRANSISTOR through which CARRIERS enter the interelectrode region.

empirical. Based on the results of observation rather than on theory.

emu. Abbrev. for electromagnetic units. *See* CGS UNITS.

emulsion detector. A special photographic emulsion designed for investigating nuclear reactions. The emulsion is much thicker and has a much greater silver halide concentration than do emulsions for ordinary photography. Alpha particles, protons and neutrons are detected by the track of silver granules they produce, generally observed under a high-power microscope. The emulsions have been used in cosmic-ray measurements at different altitudes.

end correction. The distance beyond the end of a pipe containing a vibrating air column at which an antinode occurs. It has been shown to equal $0.6\ r$ where *r* is the pipe radius, but also depends on the wavelength of the vibration, tending to zero for very short wavelengths. *See also* VIBRATIONS IN PIPES.

endoergic. *See* endothermic process.

endothermic process. A process accompanied by heat absorption. Nuclear processes of this nature are known as *endoergic.* *Compare* EXOTHERMIC PROCESS.

energy. Symbol E. The capacity of a body or system to perform WORK.
Potential energy, symbol U, is the energy possessed by a body or system due to its position. A body of mass m situated at height h above a reference plane in a gravitational field has potential energy of mgh, where g is the acceleration due to gravity.
Kinetic energy, symbol T, is the energy stored in a system due to the movement of masses within the system, and is measured by the work necessary to bring the system to rest. A body of mass m and speed v has a kinetic energy of $mv^2/2$. A body whose moment of inertia and angular speed about the axis of rotation are respectively I and ω has a kinetic energy of $I\omega^2/2$.
See also INTERNAL ENERGY; MECHANICAL ENERGY; ROTATIONAL ENERGY; VIBRATIONAL ENERGY.

energy band. A band of allowed energies for an electron in a solid. *See* BAND THEORY.

energy density. The amount of energy per unit volume.

energy density of radiation. The radiant energy density per unit volume. For a BLACK BODY it is proportional to the fourth power of the thermodynamic temperature. *See* STEFAN'S LAW.

energy exchanges. The conversion of one type of energy to another. Thus a pendulum bob at its highest point has POTENTIAL ENERGY but no KINETIC ENERGY, while at its lowest point the reverse is true.

energy level. One of a number of discrete energies that may be possessed by a nucleus, atom, molecule or other system. For example the electron orbiting the nucleus in a hydrogen atom can only, according to the BOHR THEORY, occupy certain orbits of different radius, each corresponding to a certain energy. Likewise the VIBRATIONAL and ROTATIONAL ENERGIES of molecules can only have discrete values. The differences between successive vibrational energy levels are smaller than those between successive electronic energy levels, but are larger than the differences between successive rotational energy levels. Changes in energy level are associated with the emission or absorption of radiation whose frequency is proportional to the change. The lowest energy level of a system is its *ground state*; acquisition of energy may raise the system to an *excited state. See also* EXCITATION.

energy momentum tensor. A set of quantities for an electromagnetic or material field. The set, which transforms as a tensor, specifies the energy density, momentum density and stress.

energy of dissociation. The minimum amount of energy required to separate a molecule into its constituent atoms.

energy state. The energy associated with an allowed ENERGY LEVEL of a system.

enrichment. The process of increasing the concentration of one particular ISOTOPE in a mixture of isotopes. An important application is fuel preparation for some NUCLEAR REACTOR TYPES.

enthalpy. Symbol H. A quantity defined by the equation

$$H = U + pV$$

where U is the INTERNAL ENERGY of a system, p its pressure and V its volume. By convention, the change in H due to an exothermic reaction is taken to be negative.

entrance pupil. The APERTURE first encountered by light entering an optical system.

entropy. Symbol *S*. A quantity characteristic of the thermodynamic state of a system. For a REVERSIBLE PROCESS, the change Δ*S* in entropy is given by Δ*Q*/*T* where Δ*Q* is the amount of heat absorbed by the system at thermodynamic temperature *T*; when the process is reversed, the entropy returns to its initial value. For an IRREVERSIBLE PROCESS, i.e. for all actual processes, the entropy of a closed system is permanently increased (*see* HEAT DEATH). A value of *S* can be related to the probability of that value by the equation

$$S = k \ln W^C$$

where *k* is the BOLTZMANN CONSTANT, *W* the probability and *C* another constant; again the principle of permanent increase of entropy applies. In information theory, the concept of entropy has been used as a measure of the uncertainty of knowledge.

ephemeris. A table of calculated astronomical data giving positions and movements of various celestial bodies.

epicycloid. The locus of a point on the circumference of a circle as that circle rolls on the outside of another circle. *Compare* HYPOCYCLOID.

epidiascope. An apparatus that can function both as a slide projector and an EPISCOPE.

episcope. An optical apparatus for projecting an image of a flat opaque surface on a screen. The object is illuminated by a high-intensity beam and lenses of large aperture are used in order to obtain as bright a final image as possible.

epitaxy. A method of growing a thin layer of material on a single crystal substrate so that the lattice structure of layer and substrate are identical; the most usual method is to condense the vapour of the layer material on the substrate. The technique is extensively used in SEMICONDUCTOR manufacture, when a layer of different conductivity to the substrate is required.

EPM. Abbrev. for ELECTRON PROBE MICROANALYSIS.

equation. A mathematical statement that two expressions have the same value. An equation may be true for only certain values of a variable:

$$y^2 + 3y + 2 = 0$$

holds for *y* = −1 or −2 only. Alternatively it may be true for all variable values:

$$y^2 + 3y + 2 \equiv (y + 1)(y + 2)$$

is true for all values of *y*. Such an equation is known as an *identity* and is identified by the symbol ≡. Scientific equations are mathematical statements of relationships between physical quantities.

equation of state. An equation stating the relationship between the pressure, *p*, volume, *V*, and thermodynamic temperature, *T*, of a substance. The simplest example of an equation of state is the IDEAL GAS law

$$pV = nRT$$

where *n* is the number of moles of substance present and *R* is the UNIVERSAL GAS CONSTANT. An equation of state describing reasonably accurately the behaviour of a real gas is the VAN DER WAALS EQUATION.

equilateral. Denoting a plane figure with all its sides of equal length.

equilibrium. The state of a system with respect to a given observable quantity during the time for which there is no change in that quantity.
  Dynamic equilibrium exists in a system when an activity in one sense or direction is annulled by a similar reverse activity.
  Static equilibrium exists in a system when the RESULTANT of all the vectors acting is zero and when the sum of the moments of the vectors about three mutually perpendicular axes is also zero.
  Thermal equilibrium exists in a system when no net heat exchange occurs between the system and its surroundings.

equilibrium stability. A state whose nature is determined by the effect of the vectors brought into play when an equilibrium is slightly disturbed. If these vectors tend to increase the disturbance, then the system was in *unstable equilibrium*; if they tend to decrease the disturbance, the system was in *stable equilibrium*; when there is no effect, the system was in a state of *neutral equilibrium*. For a system of forces, the POTENTIAL ENERGY is a minimum for stable equilibrium and a maximum for unstable equilibrium.

equinox. Either of the points at which the ECLIPTIC crosses the CELESTIAL EQUATOR. These are thus the points at which the Sun in its apparent annual motion crosses the celestial equator. The Sun crosses from south to north at the *vernal equinox* and from north to south at the *autumnal equinox*. In the northern hemisphere, the vernal equinox crossing is around March 21st and the autumnal equinox crossing is around September 23rd; at these times (also referred to as solstices) night and day are of equal duration.

equipartition of energy. The principle that in a multi-particle system in equilibrium, the mean kinetic energy per particle is the same for each DEGREE OF FREEDOM and equal to $kT/2$ where k is the BOLTZMANN CONSTANT and T the thermodynamic temperature. A monatomic gas therefore has an energy per atom of $3kT/2$ since there are three degrees of freedom. The proposition is not generally true when quantum considerations are important, but it is frequently a good approximation.

equipotential. Denoting a line or surface connecting points with the same POTENTIAL.

equivalent circuit. A circuit arrangement with the same electrical characteristics, for specified conditions, as a much more complicated circuit.

equivalent resistance. The value of the resistance of a single resistor which, when used to replace all the resistances in a given circuit, would dissipate the same power as them.

equivalent simple pendulum. The simple pendulum of length such that it has the same periodic time as a given compound pendulum. *See* PENDULUM.

equivalent sine wave. A sine wave having the same ROOT MEAN SQUARE magnitude and the same fundamental frequency as a given wave.

erecting lens. A lens used in an optical instrument to produce an erect image from an inverted image.

erecting prism. A prism used instead of an ERECTING LENS in an optical instrument.

E region. Another name for E layer. *See* IONOSPHERE.

erg. The CGS UNIT of energy, equivalent to 10^{-7} joule.

error. (1) *random error.* A small measurement error due to instrumental imperfections and/or inaccurate human judgement. Whenever possible the random errors for each part of an experiment should be estimated and taken into account in the final result. When possible it is helpful to use a graphical representation of the results since anomalous seeming observations can then be easily identified and rechecked. The HEISENBERG UNCERTAINTY PRINCIPLE is considered to show that there is a small amount of uncertainty in all physical measurements, no matter how good the instrument and the observer.
(2) *systematic error.* An error other than a random error. An error caused by faulty calibration is an example.

erythema. The reddening of the skin due to exposure to radiation, for example ultraviolet radiation.

Esaki diode. Another name for TUNNEL DIODE.

escape velocity. The minimum speed required by an object to escape from a given gravitational field. The escape velocity from Earth is 11 200 metre per second; from the Moon it is 2370 metre per second and from the Sun it is 618 kilometre per second.

ESR. Abbrev. for ELECTRON SPIN RESONANCE.

esu. Abbrev. for electrostatic units. *See* CGS UNITS.

etalon. An interferometer comprising two semi-silvered optically flat and accurately parallel glass plates separated by an air gap of a few millimetre. Since the device gives sharp fringes and has high resolving power, it is used for the study of the HYPERFINE STRUCTURE of spectral lines. *See also* FABRY-PEROT INTERFEROMETER.

eta meson. Symbol η^0. *See* ELEMENTARY PARTICLES, table; MESON.

ether. A hypothetical nonmaterial fluid previously supposed to permeate all space. *See also* MICHELSON-MORLEY EXPERIMENT.

Euclidean geometry. The familiar two- and three-dimensional geometry in which many theorems concerning the properties of figures are deduced from a number of definitions and concepts. It is the geometry usually applied to physical measurements involving distance and angle. It embodies the *parallel postulate* that if a point lies outside a line, it is possible to draw only one line through the point which is parallel to the first line. *Compare* NON-EUCLIDEAN GEOMETRY.

Eulerian angles. Three angles of rotation describing the position of a body relative to Cartesian axes, origin O, fixed in space when the body rotates about an axis through O, such that this axis is also rotating. The angles are useful for the analysis of the dynamics of PRECESSION and NUTATION.

Euler's equations of motion. Equations which apply to a mechanical system consisting of connected particles. The equations are forms of the two vector equations

$$\mathrm{d}p/\mathrm{d}t = R \text{ and } \mathrm{d}h/\mathrm{d}t = G$$

where the vectors p and h are respectively the linear and angular momenta of the system referred to an INERTIAL FRAME OF REFERENCE with a stationary fixed origin; R is the vector sum of the forces applied to the system and G is the sum of the vector moments of these forces about the origin.

Euler's theorem. *See* POLYHEDRON.

Eustachian tube. The channel connecting the middle ear to the upper part of the throat. It opens when swallowing occurs, thus maintaining atmospheric pressure inside the middle ear despite the absorption of oxygen at the surface.

evaporation. The escape from the surface of a liquid or solid, at a temperature below the boiling point, of molecules whose energy of thermal agitation is sufficiently great. The substance cools as a result of the process since the escape of the most energetic molecules reduces the mean energy of the molecules left behind. Evaporation from a solid surface is employed for producing thin metal films to serve as TRANSISTOR connections.

evaporator. The part of a refrigerating plant in which the liquid refrigerant is evaporated, thus taking heat from the surroundings.

even-even nucleus. A nucleus containing an even number of protons and an even number of neutrons.

even-odd nucleus. A nucleus containing an even number of protons and an odd number of neutrons.

event. A point in space-time.

evolute. A curve that is the locus of the CENTRE OF CURVATURE of another specified curve, known as the *involute*.

Ewing's theory of magnetism. A theory based on the assumption that the individual atoms or molecules of ferromagnetic substances act as small magnets. When such a substance is unmagnetized, these elementary magnets are arranged in closed chains so that there is no external magnetic effect. When the substance is magnetized, the elementary magnets are considered to align themselves in the direction of the magnetizing field, saturation being reached when they are all so aligned. The material will obviously be left with all elementary south poles at one end and all elementary north poles at the other, i.e. it will be a magnet. Hysteresis occurs because of the force necessary to break up the molecular chains. The theory has been partially confirmed by modern investigations. *See* FERROMAGNETISM.

exa-. Symbol E. A prefix meaning 10^{18}.

excess pressure in bubble. An expression for this excess pressure may be obtained by considering the equilibrium of half a bubble, H, of radius r, as illustrated in fig. E4. The force exerted on H due to the pressure p_1 outside is $p_1 \pi r^2$, and to pressure p_2 inside is $p_2 \pi r^2$; the difference between these is balanced by the force due to SURFACE TENSION, γ. Hence

$$\pi r^2 (p_2 - p_1) = 2\pi r \gamma$$

and so

$$p_2 - p_1 = 2\gamma/r$$

If the bubble is a soap bubble then it has two faces in contact with air. The surface tension force on half of it is thus $4\pi r\gamma$, and hence the excess pressure is $4\gamma/r$.

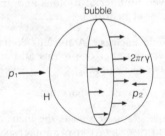

E4 Excess pressure in bubble

exchange. The hypothetical continuous exchange of charged particles between two similar quantum mechanical systems. The simple wave functions describing the two systems are combined with a function expressing the exchange of place of the particles and yielding an energy different from that of the no-exchange state. For example electron exchange can explain ferromagnetism and antiferromagnetism; the exchange of pions has been postulated to account for nuclear forces.

excitation. (1) A process whereby a particle acquires *excitation energy*, which increases its energy to a value greater than that in its ground state. The particle is then said to be in an *excited state*.
(2) The passage of current through the winding of an electromagnet.
(3) The application of a signal to the base of a transistor or to the control electrode of a thermionic valve.

excitation energy. *See* EXCITATION.

excited state. The state of an atom or nucleus with greater energy than that associated with the lowest ENERGY LEVEL. *See* EXCITATION.

exciton. An electron-hole pair in a crystalline solid. The electron is in an EXCITED STATE and is bound to the positive HOLE by electrostatic attraction. The exciton may migrate through the solid and then the pair may recombine with the emission of a photon.

exclusion principle. *See* PAULI EXCLUSION PRINCIPLE.

exoergic. *See* EXOTHERMIC PROCESS.

exosphere. The region of the Earth's atmosphere above an altitude of 400 kilometre. *See* ATMOSPHERIC LAYERS, fig. A9.

exothermic process. A process accompanied by heat emission. Nuclear processes of this nature are called *exoergic*. *Compare* ENDOTHERMIC PROCESS.

expanding universe. The hypothesis, based on observations of RED SHIFT, that the distance between galaxies is continuously increasing as required by the BIG BANG THEORY.

expansion. The increase of a given mass of substance in length, area or volume as a result of change in physical conditions, usually a change in temperature. *See also* COEFFICIENT OF EXPANSION.

expansion of scale. A factor to be taken into account in accurate measurement. If a scale reads correctly at one temperature, its readings at higher (lower) temperatures will be too low (high), since the distance between any two of its divisions will have increased (decreased), assuming its expansion is not anomalous.

expansion of water. *See* ANOMALOUS EXPANSION OF WATER.

expectation value. The average of many measurements of a quantity made on a system whose state is the same prior to each measurement.

exploring coil. A small coil of wire used for measuring magnetic flux. It is generally used in conjunction with a BALLISTIC GALVANOMETER. *See* FLUXMETER.

exponent. A number indicating the power to which a variable is raised; thus in y^n, n is the exponent. Exponents satisfy the following laws:

$$y^n \times y^m = y^{n+m}$$

$$(y^n)^m = y^{nm}$$

$$(yz)^n = y^n \times z^n$$

$$1/y^n = y^{-n}$$

exponential function. A mathematical function of a variable, y, of the form

$$A\, e^{ay}$$

where A and a are constants and e is the base of natural logarithms and is an irrational number, 2.718 28 ... The function is sometimes written exp y and can be expanded as the convergent infinite series

$$1 + y + y^2/2! + y^3/3! + ... + y^n/n! + ...$$

See also FACTORIAL.

exponential decay. The decrease of some physical quantity, usually with time, according to a negative exponential law. The law is represented by an equation of the type

$$y = y_0\, e^{-\lambda t}$$

where y_0 is the initial value of the quantity and y_t its value after time t and λ is the DECAY CONSTANT. Physical examples include the following: the fall in activity with time of a pure radioactive substance yielding a stable daughter product; the fall in amplitude with time for damped harmonic oscillations; the fall in voltage with time of a charged capacitor leaking through a high resistance.

exposure dose. *See* DOSE (def. 2).

exposure meter. An instrument used in photography. It consists of a photocell connected to a meter which indicates, for various types of film, the F NUMBER for a given shutter speed, or vice versa, to give the correct exposure. In some cameras the exposure meter is built-in and automatically controls the camera aperture.

external pacemaker. A pacemaker for the heart in which a coil and its power supply are strapped to the outside of the chest. A current in this coil induces a current in a corresponding coil attached to the heart inside the chest. The power-supply cells are thus free from attack by body chemicals and can easily be replaced; these advantages greatly outweigh the disadvantage that an external pacemaker requires more power than one completely implanted.

external work. The work done by a substance in expanding against an external resistance; it is given by

$$\int_{v_1}^{v_2} p\, dV$$

where p is the pressure and V_1 and V_2 are respectively the initial and final values of volume V. For a cyclic process, the work done per cycle is given by the area enclosed by the cycle on the pressure-volume plot.

extinction coefficient.   Another name for LINEAR ATTENUATION COEFFICIENT.

extracorporeal circulation.   The circulation of the blood in the body by means of an external agency: venous blood is removed from the body, oxygenated and pumped back as arterial blood.

extraordinary ray.   *See* DOUBLE REFRACTION.

extrapolation.   The estimation of the value of a function of a variable for a value of the variable lying outside the measurement range.

extremely high frequency.   A frequency in the range 30 gigahertz to 300 gigahertz.

extrinsic semiconductor.   *See* SEMICONDUCTOR.

extrusion.   A process whereby a substance is forced by compression through a suitably shaped aperture to yield a product of uniform cross sectional area.

eye.   A section through the human eye is illustrated in fig. E5. Light enters through the transparent *cornea* and is focused by the combination of cornea and *crystalline lens* on to the *retina*, which is a light-sensitive layer of nerve cells; the *fovea centralis* or YELLOW SPOT is the most efficient part of the retina. At the point where the optic nerve enters the eye is a small area with no light sensitivity and therefore known as the *blind spot*. The *sclera* is a tough membrane which protects the eye. The *iris* is an arrangement of muscle tissue in front of the lens; it controls the aperture of the PUPIL through which light passes. The *aqueous humour* is a watery substance whereas the *vitreous humour* is gelatinous. When the eye is focused for infinity, the *ciliary body* is relaxed, causing tightening of the zonule attaching it to the crystalline lens, which is therefore at its flattest. When the eye is focused for the NEAR POINT, the ciliary body tenses, reducing the tension in the zonule and thus permitting the crystalline lens to thicken at its centre and thereby increase its power. This process of lens adjustment is known as *accommodation*. Defects of the eye include

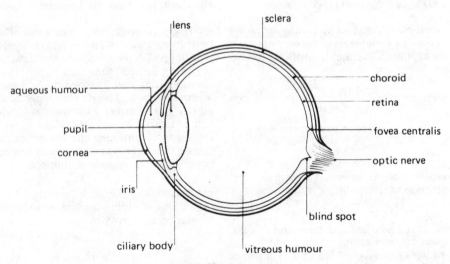

E5 Human eye

ASTIGMATISM, HYPERMETROPIA, MYOPIA and PRESBYOPIA. *See also* COLOUR VISION; VISUAL ACUITY.

eye lens.   The crystalline lens of the EYE.

eyepiece.   A single lens, doublet or lens combination forming a magnified image of the image formed by the OBJECTIVE of an optical instrument. Examples are the HUYGENS' EYEPIECE and the RAMSDEN EYEPIECE.

eye ring.   The position of the eye, when using an optical instrument, for which most light enters the eye. It coincides with the image of the objective formed by the EYEPIECE.

F

Fabry-Perot interferometer. An ETALON of variable plate separation.

faced-centred. Having or involving a crystal structure in which the UNIT CELL has eight atoms at corners and a further six atoms situated one at the centre of each face. If the unit cell is a cube, the lattice is described as *face-centred cubic*; this is illustrated in fig. F1.

F

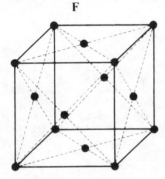

F1 Face-centred cubic unit cell

factor. Any of two or more integers that give a specified number when multiplied together; thus 3 and 5 are factors of 15.

factorial. The product of the first *n* positive integers, i.e.

$$1 \times 2 \times 3 \times 4 \times ... \times n$$

is known as factorial *n* and is written *n*!. By definition 0! is equal to one.

Fahrenheit scale. A temperature scale on which the ice point is defined as 32° F and the steam point as 212° F. The scale is no longer used for scientific purposes.

fallout. Radioactive material settling on the Earth's surface as a result of a nuclear explosion. *Local fallout* occurs within 250 kilometre of an explosion and happens within a few hours of it. *Tropospheric fallout* occurs all over the Earth in the approximate latitude of the explosion and within a week of it. *Stratospheric fallout* may occur anywhere on Earth over a period of years after the explosion. Iodine-131, which accumulates in the thyroid gland, and strontium-90, which accumulates in bones, are the most dangerous FISSION fragments since they may be taken up by grazing animals and so passed to humans.

family. A set of curves, surfaces etc. described, apart from a constant, by a single equation. Thus

$$y^2 = 4ax$$

represents a family of parabolas, each member corresponding to a different value of *a*.

farad. Symbol F. The SI unit of CAPACITANCE, equal to the capacitance of a parallel plate capacitor holding a charge of one coulomb when there is a potential difference of one volt across its plates. The microfarad is a more useful practical unit.

faraday. See FARADAY'S LAWS OF ELECTROLYSIS.

Faraday constant. See FARADAY'S LAWS OF ELECTROLYSIS.

Faraday dark space. See GAS DISCHARGE TUBE.

Faraday disc. Another name for HOMOPOLAR GENERATOR.

Faraday effect. The rotation of the plane of polarization of plane-polarized electromagnetic radiation (*see* POLARIZATION, electromagnetic) on passing through certain isotropic substances subject to a magnetic field, strength H, parallel to the direction of travel of the radiation. The angle of rotation is proportional to Hl, where l is the path length of the radiation in the medium. If the radiation retraverses its path in the opposite direction the rotation effect is doubled, i.e. the direction of rotation is independent of the sense in which the radiation traverses the magnetic field.

Faraday-Neumann law. *See* ELECTROMAGNETIC INDUCTION.

Faraday's ice pail experiment. An experiment concerned with electrostatic induction. By connecting the insulated metal ice pail to a gold leaf ELECTROSCOPE and lowering a suspended charged body into the pail, Faraday showed that a charged body enclosed in a *hollow conductor* induces a charge similar to its own on the outside of the conductor and a charge opposite to its own on the inside of the conductor. The total charge inside a hollow conductor is therefore always zero.

Faraday's laws of electrolysis. Two laws given as follows.
1. The mass of any substance liberated from an electrolyte by the passage of current is proportional to the current and the time for which it flows.
2. The masses of different substances liberated in electrolysis by the same quantity, Q, of electric charge are proportional to the relative atomic masses of the substances, each divided by the charge, Z, on its ion, i.e.

$$Q = FmZ/A_r$$

m is the mass of substance of relative atomic mass A_r liberated and F is a constant known as the *Faraday constant*, which has the value

9.648 670 × 10⁴ coulomb per mole

this quantity is sometimes known as a *Faraday*.

far infrared. *See* INFRARED RADIATION.

far point of the eye. The most distant point an EYE can focus clearly. For a normal eye it is at infinity.

far sight. Another name for PRESBYOPIA.

far ultraviolet. *See* ULTRAVIOLET RADIATION.

fast breeder reactor. *See* NUCLEAR REACTOR TYPES.

fast neutron. A neutron whose kinetic energy exceeds 0.1 mega-electronvolt. However, the term is also used to indicate a neutron capable of causing fission of $^{238}_{92}U$, i.e. a neutron of energy above 1.5 mega-electronvolt.

fast reactor. *See* NUCLEAR REACTOR TYPES.

fathom. A unit of length equal to 6 feet. Sea depths are usually expressed in fathom.

fatigue. The progressive decrease of the BREAKING STRESS of a material subjected to repeated applications of a stress which, had it been maintained at the same steady value, would not have caused failure.

feedback. Any process in which the output controls the input in some way, for example the coupling of a portion of the output of an electronic amplifier to the input. If the input is enhanced by the process it is known as *positive feedback*, and as *negative feedback* if the input is reduced. Negative feedback is the type mainly used in amplifiers since it tends to stabilize the amplifier, reducing noise and distortion; such feedback may be accomplished by using either a capacitor or an inductor.

femto-. Symbol f. A prefix meaning 10^{-15}.

Fermat's principle of least time. The path of a ray of light from one point to another, through one or more media, is such that the time taken is a minimum, i.e. less than the time required for other possible paths. The principle also applies to other electromagnetic radiations.

fermi. A unit of length equal to 10^{-15} metre, mainly used in nuclear and atomic physics.

Fermi-Dirac statistics. A form of quantum statistics concerned with the distribution of indistinguishable particles among various allowed energy levels, no level being occupied by more than one particle.

Fermi level. The highest occupied energy level in a solid. *See* BAND THEORY.

fermions. Any elementary particle obeying FERMI-DIRAC STATISTICS, such as BARYONS and LEPTONS. Fermions have a spin of ½. *Compare* BOSON.

Fermi surface. The surface in momentum space formed by electrons occupying the FERMI LEVEL.

ferrimagnetism. A phenomenon resembling ANTIFERROMAGNETISM but associated with unequal neighbouring antiparallel magnetic moments.

ferrite. An insulating low-density ceramic oxide of iron with another added oxide, which determines whether the ferrite exhibits FERRIMAGNETISM or FERROMAGNETISM.

ferroelectricity. The property, exhibited by some dielectrics such as barium titanate, of retaining electric POLARIZATION in the absence of a polarizing field. The effect is due to the lining up in domains of electric dipoles in the dielectric in an analagous way to magnetic dipoles in ferromagnetic material (*see* FERROMAGNETISM). The graph of electric displacement against applied electric field exhibits a HYSTERESIS loop analogous to that of ferromagnetic hysteresis.

ferromagnetism. A type of magnetism exhibited by some solids such as iron, cobalt and nickel which become magnetized in weak magnetic fields and which have large positive MAGNETIC SUSCEPTIBILITY. Such substances exhibit HYSTERESIS and have high relative PERMEABILITY. The magnetic behaviour of ferro-

magnetic substances is temperature dependent (*see* CURIE TEMPERATURE). Some ferromagnetic materials such as steel retain much of their magnetism when the magnetizing field is removed, and are said to be *hard*; others, such as soft iron, lose most of their magnetism on removal of the field and are said to be *soft*.

Ferromagnetism is explicable on the *domain theory*. This postulates the existence of groups of atoms, known as *domains*, volume 10^{-12} cubic metre to 10^{-8} cubic metre; the magnetic moment of each atom is aligned in the same direction. A domain therefore behaves like a saturated magnet and is held together by strong interatomic forces. The magnetic moment of each atom is thought to be due to the spin of electrons in an unfilled inner shell. In the unmagnetized state of a ferromagnetic material, it is considered that the domains are randomly arranged so that there is no resultant magnetic moment. Under the influence of a magnetic field, domains aligned in the field's direction increase in size at the expense of domains not so aligned; eventually, for a sufficiently strong field, all the domains are aligned and the substance is saturated. Permanent magnetism occurs when the domains remain aligned in the absence of the field.

fertile Denoting an ISOTOPE which can be transformed into FISSILE material in a NUCLEAR REACTOR. Uranium-238 is an example of a fertile isotope.

Féry total radiation pyrometer. A PYROMETER suitable for the direct determination of temperature up to 1400° C by measuring the total energy of radiation of all wavelengths emitted by a source.

FET. Abbrev. for FIELD EFFECT TRANSISTOR.

Feynman diagram. A graphical scheme for representing the interactions between elementary particles and fields.

Fibonacci numbers. A sequence of numbers each of which is the sum of the two preceding numbers:

$$1, 1, 2, 3, 5, 8, 13, ...$$

fibre optics. An optical technique for transmitting images along flexible transparent fibres. Light falling on the end of a high refractive index glass fibre, of less than 1 millimetre diameter, travels along it by repeated TOTAL INTERNAL REFLECTION and therefore suffers little or no absorption. By using a bundle of such fibres (each of diameter 0.01 millimetre to 0.5 millimetre) in a fixed array, complete images can be transmitted, each fibre being responsible for a small area of the whole. Since total internal reflection can occur even if the fibres are appreciably curved, the system is of great practical significance for viewing or photographing inaccessible objects such as machine parts and internal organs of the body.

Fick's law. The rate of diffusion of a solute per unit area in a direction perpendicular to the area is proportional to the gradient of concentration of solute in that direction.

field. A region under the influence of some physical agency, for example electrical, magnetic, gravitational, thermal; each of the first three is a VECTOR while the fourth is a SCALAR. A vector field is frequently represented by a set of curves, each known as a *line of force*, whose density at any point gives the strength of the field at that point and whose direction is the direction of the field at the point.

field effect transistor. A TRANSISTOR in which current flow depends on the movement of majority carriers only. There are two main types: *junction field effect transistors*, written *JUGFET* or *JFET*, and *insulated gate field effect transistors*, written *IGFET*; the latter are also known by the abbreviations *MISFET* or *MIST* or *MOSFET* or *MOST*. For both types the electrodes are SOURCE, GATE and DRAIN. The basic structures are illustrated in fig. F2. For the n-type channel JFET, making the drain positive relative to the source causes electrons to flow from source to drain. If the gate is then made negative relative to the source, the n-channel width available for electron flow is restricted and therefore drain current is reduced; hence changes in voltage applied to the gate cause changes

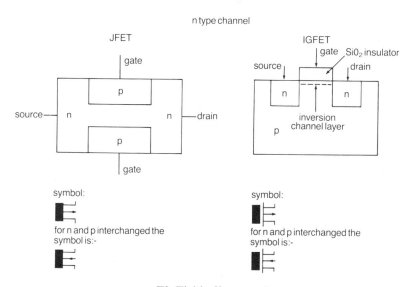

F2 Field effect transistors

in the drain current, i.e. the operation is similar to that of of a bipolar transistor.

In the n-type channel *enhancement IGFET*, the gate is insulated from the rest of the device by a thin film of silicon dioxide and the source and drain are connected with highly doped n-type regions which have been previously introduced into the p-material by diffusion. When the drain and gate are both positive with respect to the source, a narrow n-channel is induced just below the gate and so current will flow between source and drain. *Depletion IGFETs* are also available. They differ from the enhancement variety by having a built-in narrow strip of doped material in the space below the gate; the depletion IGFET with no signal applied to the gate thus looks like the enhancement variety with the gate connected to make it conduct. For all p-type channel field effect transistors, the polarities described for the n-types are reversed.

One of the attractions of field effect transistors is their very small size: they are therefore extensively used in integrated circuits. They are mostly low-power devices.

field emission. The emission of electrons from the surface of a solid due to a high electric field at the surface. Use is made of the phenomenon in the *field emission microscope* to study the surface structure of a metal point which is situated at the centre of a hemispherical fluorescent screen; the screen is at a high positive potential with respect to the point. The whole is enclosed in gas at low pressure. An image of the point is formed on the fluorescent screen by the ejected electrons.

field emission microscope. *See* FIELD EMISSION.

field glasses. Another name for BINOCULARS.

field ionization. The ionization of gaseous atoms or molecules at the surface of a solid under the influence of a high electric field: electron transfer occurs from a gaseous atom or molecule to the solid so producing a positive gaseous ion, which is accelerated away by the field. The effect is used in the *field ion microscope*, which is similar to the FIELD EMISSION microscope except that the fluorescent screen is negative with respect to the metal point and the image is due to the impact of gas ions on the screen. The device has high resolution.

field ion microscope. *See* FIELD IONIZATION.

field magnet The magnet providing the magnetic field in an electrical machine.

field of view. The area of an object which can be seen by using an optical instrument. It is usually expressed in terms of the angle subtended at the eye.

field theory. The theory of the manner in which field potentials, for example electromagnetic, gravitational and nuclear potentials, account for the propagation of a field.

Quantum field theory is a field theory in which physical observables are represented by suitable operators which obey certain rules.

Unified field theory is any theory attempting to unify Maxwell's electromagnetic theory and general relativity theory.

film badge. A piece of photographic film in a plastic holder, worn by those exposed to radiation: the blackening of the film indicates the extent of exposure. If the received amount of a particular radiation is of interest, filters to remove other types of radiation are placed in front of the film.

filter. (1) A device for absorbing radiation of particular energies while transmitting radiation of other energies. For example light may be filtered by pieces of coloured glass or film.

(2) An electric circuit which allows passage of alternating currents within a certain range or ranges of frequency, while stopping currents which have frequencies outside the allowed range(s).

filter pump. A type of vacuum pump, illustrated in Fig. F3, in which a jet of water forced through a narrow nozzle traps air

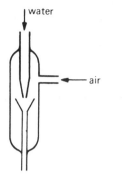

F3 Filter pump

molecules and removes them from the system. It cannot reduce the pressure below the vapour pressure of water.

finder. A small telescope attached to the side of a large astronomical telescope to facilitate the directing of the large telescope towards the object to be observed.

fine beam tube. A low-pressure hydrogen GAS DISCHARGE TUBE in which the path of a fine beam of electrons issuing through a hole in the anode is rendered visible due to ionization of the hydrogen molecules. By using a pair of HELMHOLTZ COILS to apply a uniform magnetic field to the tube, and measuring the resulting electron deflection, the tube may also be used for the measurement of e/m for the electron.

fine structure. The structure observed in a spectral line or band when it is viewed at high resolution; line splitting due to electron spin can therefore be observed. To detect HYPERFINE STRUCTURE even higher resolution is required. Hyperfine structure results from energy-level changes in atoms of different isotopes of a substance and also from interaction of nuclear spin with electron spin.

fine structure constant. Symbol α. A measure of the strength of ELECTROMAGNETIC INTERACTION, equal to

$$e^2/(2hc\varepsilon_0)$$

where e is the charge on the electron, h the PLANCK CONSTANT, c the speed of light and

ε_0 the ELECTRIC CONSTANT. The approximate value is 1/137.

fissile. Capable of undergoing nuclear FISSION. Sometimes the description is restricted to materials that undergo fission by the impact of slow neutrons.

fission. The process in which an atomic nucleus splits into fragment nuclei of comparable size. It is usually accompanied by emission of neutrons and of gamma rays and the release of large amounts of energy. Fission may be either spontaneous or induced by bombardment. *See also* NUCLEAR REACTOR; NUCLEAR WEAPON.

fission bomb. *See* NUCLEAR WEAPON.

fission reactor. Another name for NUCLEAR REACTOR.

fixed point. A reference point of defined value on a TEMPERATURE SCALE.

fixed star. A star which appears not to alter its position on the CELESTIAL SPHERE. A fixed star was so named to distinguish it from a planet, which was known as a *wandering star.*

flash barrier. *See* FLASHOVER.

flashover An unwanted arc discharge between two electrical conductors. Damage can be minimized by using a screen of fireproof material, i.e. a *flash barrier.*

flash point. The lowest temperature to which a substance must be heated before it can be ignited.

flavour. The generic name for the qualities which distinguish the various types of QUARK, and in addition the various types of LEPTON.

F layer. *See* IONOSPHERE.

Fleming's rules. Mnemonics for the relationship between direction of electric current, motion of conductor carrying it, and magnetic field, taken as mutually perpendicular. The *left-hand rule*, also known as the *dynamo rule*, gives the direction of induced current. The *right-hand rule*, also known as the *motor rule*, gives the direction of motion. For either hand, the first and second fingers and thumb are held mutually perpendicular; the First finger indicates Field direction, the second finger indicates current direction and the thuMb indicates the direction of Motion. To apply the left-hand rule, the first finger and thumb of the left hand are suitably aligned and so the second finger gives the current direction. To apply the right-hand rule, the first and second fingers of the right hand are suitably aligned and so the thumb gives the direction of motion.

flicker photometer. A PHOTOMETER in which a screen is illuminated alternately by a standard source and by the test source in rapid succession. The distances of the sources are adjusted until the screen does not appear to flicker, i.e. until the intensities of illumination of the screen due to the two sources are equal. The instrument is especially useful for the comparison of sources of different colours.

flint glass. Another name for OPTICAL FLINT.

flip-flop. An electronic circuit which has two stable states and is switched from one state to the other by a triggering pulse. Such a circuit is widely used as a LOGIC CIRCUIT.

floppy disc. A flexible plastic DISC with a magnetic coating, held in a stiff envelope. It is used for information storage in a small COMPUTER system.

flow. (1) of an ideal liquid. Steady streamline flow along a tube of an ideal liquid is governed by the BERNOULLI EQUATION. Moreover, since the liquid is friction-free, its speed is the same at all points on a given cross section and, since the liquid is incompressible, the product of speed of flow and cross-sectional area at any point is constant, i.e. the narrower the tube the faster the liquid flows.

(2) of a real liquid. Since a real liquid has internal friction there is a fall of pressure along a tube in which it flows, showing that energy is being dissipated. Furthermore, the speed of a particle of liquid in a tube depends on the distance of the particle from the tube axis. A typical graph of speed against distance from axis is shown in fig. F4. *See also* POISEUILLE'S EQUATION; STOKES' LAW.

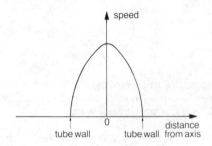

F4 Liquid flow speed profile across tube

fluid. Any substance that can flow, such as a liquid or gas.

fluidics The use of fluid flow through narrow pipes in order to perform tasks normally or previously carried out by electronic equipment. The fluid circuits are very much slower than electronic ones but are unaffected by magnetic fields or radiation and are less temperature sensitive. They are therefore finding increasing use in NUCLEAR REACTORS and SPACECRAFT.

fluidity. The reciprocal of the VISCOSITY of a fluid.

fluid upthrust. *See* UPTHRUST.

fluorescence. LUMINESCENCE which ceases when the exciting source is cut off (*Compare* PHOSPHORESCENCE). Fluorescence is a term used scientifically to describe luminescence lasting less than 10^{-8} second after removal of excitation, or luminescence which appears to the eye to cease with excitation. In general usage the term is synonymous with luminescence.

fluorescence assay. A method used extensively for identification and location studies. It is often an alternative to radioactive techniques since it is hazard-free and capable of detecting impurities at a concentration of about 1 in 10^{10}. Monochromatic radiation is used to excite the FLUORESCENCE and allowance is made for any resulting fluorescence of containers, solvent etc. For some work it is sufficient to establish the presence of fluorescence; in more elaborate investigations the fluorescent light is investigated by SPECTROMETER. Important applications are the measurement of the protein content of milk, vitamin assay of food subsequent to preparation and processing and the study of excretion rates of substances in urine, for example small quantities of LSD.

fluorescent lamp. A lamp in which the light is mainly produced by LUMINESCENCE. Generally the lamp consists of a long glass GAS DISCHARGE TUBE whose inner surface is coated with a PHOSPHOR; the gas is usually mercury vapour which, when current passes through it, emits ultraviolet radiation which in turn excites visible radiation from the phosphor coating. Such lamps convert more of the energy supplied into light than do tungsten filament lamps.

fluorescent screen. A screen coated with luminescent material, which emits visible radiation when struck by electrons, X rays etc.

flux. (1) The flow of physical entities, such as particles, charge, energy or radiation, across a given area in a given direction.
(2) A substance used in soldering to keep the surface clean and free from oxide.

flux density. The FLUX per unit area.

fluxmeter. An instrument for measuring MAGNETIC FLUX. In one type an EXPLORING COIL connected to a BALLISTIC GALVANOMETER is quickly introduced into the flux with its plane perpendicular to it. The galvanometer reading is proportional to the MAGNETIC INDUCTION.

flying spot microscope. A microscope in which a small spot of light scans the object. The image is received on a photocell, and the resulting output is amplified and fed to a cathode ray oscillograph whose time base is synchronized with the light spot. The contrast of the display can be varied and the resolution is better than that obtained by PHOTOMICROGRAPHY.

flywheel. A wheel with an appreciable MOMENT OF INERTIA, mounted in a system with a fluctuating drive (i.e. most engines) in order to smooth out the fluctuations.

f number. The ratio of the focal length of a lens to its diameter. In a photographic camera, different f numbers for the same lens are obtained by using a variable diaphragm to change the effective lens diameter. The image brightness decreases with increasing f number and so longer exposure times are required at high f numbers; the f number of a lens is sometimes called the *speed* of the lens.

focal length. The distance between a principal point of an optical system and the corresponding focal point (*see* CENTRED OPTICAL SYSTEM). In general a system has two focal lengths, f_1 and f_2, related by the equation

$$n_1/f_1 = -n_2/f_2$$

where n_1 and n_2 are the REFRACTIVE INDICES of the media on the two sides of the system. For a thin lens and paraxial rays, the principal points coincide at the centre of the lens; for a spherical mirror and paraxial rays, they coincide at the midpoint of the mirror.

focal plane. A plane perpendicular to the axis of an optical system and passing through a focal point.

focal points. *See* CENTRED OPTICAL SYSTEM.

focus. (1) A point to which rays of light are converged, or appear to be converged, by an optical system.
(2) *See* CONIC.

foot. An imperial unit of length. *See* Table 6A.

foot candle. An obsolete unit of illumination equal to 1 lumen per square foot.

foot, pound, second units. A scientifically obsolescent system of units employing the pound for mass, the foot for length and the second for time. *See* Table 6.

forbidden band. A band of forbidden energy for an electron in a solid. *See* BAND THEORY.

forbidden transition. A transition by an atom or molecule between two energy levels which would result in a QUANTUM NUMBER change not allowed by the SELECTION RULES.

force. Symbol *F*. The agency responsible for altering the state of rest or of uniform motion of a body (*see* NEWTON'S LAWS OF MOTION). Some examples of forces are BUOYANCY, CENTRIFUGAL FORCE, CENTRIPETAL FORCE, ELECTROMOTIVE FORCE, INTERMOLECULAR FORCE, SATURATION FORCE, and VISCOUS FORCE. The SI unit of force is the NEWTON.

forced convection. *See* CONVECTION.

forced oscillation. The motion produced when a vibrating system is acted upon by an external vibrating force. The system oscillates with a frequency equal to that of the external force. The amplitude is a maximum when this frequency is the same as the natural frequency of the system; RESONANCE is then said to occur.

force pump. A pump somewhat similar to the LIFT PUMP but with a valve in the cylinder side rather than in the piston. The force pump is capable of raising water from a depth of not more than 100 metre.

Fortin's barometer. A type of mercury BAROMETER.

Foucault pendulum. A simple pendulum consisting of a heavy metal ball suspended on a long wire. When the pendulum is set in vibration the plane of its swing slowly rotates, making a complete rotation in 24/sin λ hour, where λ is the latitude of the plane of swing. The pendulum provides direct evidence of the Earth's rotation.

four-dimensional continuum. A reference system comprising the three dimensions of space and one of time. The idea of the four-dimensional continuum followed from the theory of RELATIVITY.

Fourier series. A series of the form

$$a_0 + a_1 \cos \omega t + b_1 \sin \omega t + a_2 \cos 2\omega t + b_2 \sin 2\omega t + ...$$

where ω and the *a*s and *b*s are constants and *t* is the time. Any periodic function of time can be expressed in this form, i.e. as a sum of harmonic terms.

four vector. A set of four quantities which behave under a LORENTZ TRANSFORMATION in the same way as the space-time coordinates of an EVENT.

fovea centralis. *See* EYE.

FPS units. Short for FOOT, POUND, SECOND UNITS.

fractals. Geometric shapes which imitate objects which appear self-similar under different magnifications. Examples are shapes of coast lines, of polymers and mass distributions in random alloys. The concept is being increasingly employed in statistical physics.

fraction. An expression in the form of one quantity, the *numerator*, divided by another, the *denominator*. If both numerator and denominator are integers, the fraction is said to be a *common fraction*; if both numerator and denominator are themselves fractions, the fraction is said to be a *complex fraction*. A fraction if known as a *proper fraction* if the numerator is smaller than the denominator; if the numerator is larger than the denominator, the fraction is said to be an *improper fraction*.

frame of reference. A particular set of co-ordinate axes used as a reference system for making physical measurements.

Franck-Condon principle. A quantum mechanical criterion for the probability of an electronic transition in a molecule or crystal. It corresponds to the classical concept that there is only a chance of a transition provided the time required for it is sufficiently small for the position and velocity of the atom or molecule to be sensibly unchanged during it.

Fraunhofer diffraction. DIFFRACTION associated with plane wave fronts.

Fraunhofer lines. Fine dark lines in the solar spectrum, mainly caused by the absorption of certain wavelengths in the Sun's PHOTOSPHERE. A few lines are due to absorption in the Earth's atmosphere.

free convection. *See* CONVECTION.

free electron. An electron that is not permanently attached to a specific atom or molecule and so is free to move under the influence of an applied electric field. *See also* BAND THEORY.

free energy. The portion of a system's energy which is available for conversion to work. *See* GIBBS FREE ENERGY; HELMHOLTZ FREE ENERGY.

free fall. Motion due to gravitational attraction only, i.e. unimpeded by atmospheric buoyancy and viscous retardation.

free oscillations. Oscillations that occur when a system is displaced from its equilibrium position and then left to itself. The system performs vibrations about its equilibrium position with its *natural frequency*, i.e. the characteristic frequency of the system. The amplitude of oscillation diminishes gradually until all the energy supply in the initial displacement has been expended.

free space. A region in which there is no matter, no electromagnetic fields and no gravitational fields. The refractive index is 1, the temperature absolute zero and the speed of light has its maximum value.

free surface energy. *See* SURFACE TENSION.

freeze drying. The rapid freezing and drying of a substance in a high-vacuum system. The process is widely used in food preservation.

freezing point. The temperature at which the solid and liquid phases of a substance can exist in equilibrium together at a defined pressure, usually standard atmospheric pressure.

F region. Another name for F layer. *See* IONOSPHERE.

Frenkel defect. *See* DEFECT.

frequency. Symbol f or v. The number of complete cycles of a periodic process occurring in unit time. It is measured in hertz. *See also* ANGULAR FREQUENCY.

frequency band. A particular range of frequencies within the radio frequency spectrum. The internationally agreed frequency bands are shown in the table.

Internationally agreed frequency bands

Wavelength	band	Frequency
1 mm – 1 cm	Extremely high frequency; EHF	300 – 30 GHz
1 cm – 10 cm	Super-high frequency; SHF	30 – 3 GHz
10 cm – 1 mm	Ultra-high frequency; UHF	3 – 0.3 GHz
1 m – 10 m	Very high frequency; VHF	300 – 30 MHz
10 m – 100 m	High frequency; HF	30 – 3 MHz
100 m – 1000 m	Medium frequency; MF	3 – 0.3 MHz
1 km – 10 km	Low frequency; LF	300 – 30 kHz
10 km – 100 km	Very low frequency; VLF	30 – 3 kHz

frequency distribution. A table, graph or equation describing how a particular attribute is distributed among the members of a group. An example is a plot of the number of people within a group who have heights lying in various sepcified intervals into which the height range is divided. The number in each interval is known as the frequency appropriate to that interval. *See also* HISTOGRAM.

frequency modulation. MODULATION in which the carrier frequency is increased or diminished as the signal amplitude increases or diminishes, and at the frequency of the signal, the carrier amplitude remaining constant.

frequency modulated cyclotron. Another name for SYNCHROCYCLOTRON.

fresnel. A unit of frequency equal to 10^{12} hertz.

Fresnel biprism. A prism of isosceles triangular section having one very obtuse angle. It will thus produce two very close coherent images from the same source and can therefore be used to demonstrate the INTERFERENCE of light.

Fresnel diffraction. DIFFRACTION associated with curved wave fronts.

Fresnel lens. A lens, illustrated in fig. F5, with one surface convex and the other

stepped. It is used in headlights, searchlights etc. since it has a smaller mass than a double convex lens of the same aperature and focal length.

friction. (1) *dynamic sliding friction.* The force necessary to keep one solid surface just sliding over another horizontal surface. It is equal to $\mu_d R$ where R is the normal reaction between the surfaces and μ_d is a constant known as the *coefficient of dynamic friction.*

(2) *static sliding friction.* The force necessary to start one solid surface just sliding over another horizontal surface. It is equal to $\mu_s R$ where R is the normal reaction between the surfaces and μ_s is a constant known as the *coeficient of static friction;* μ_s is somewhat greater than μ_d, the coefficient of dynamic friction.

(3) *rolling friction.* The force resisting the rolling of a body on a horizontal plane. It is much smaller than dynamic sliding friction, especially for hard surfaces, a fact which explains the use of wheels and ball bearings.

fringes. Circular or rectilinear or other shaped patterns of colour, or of alternate monochromatic brightness and darkness, produced by INTERFERENCE or DIFFRACTION of light.

fuel cell. An electrochemical cell in which the energy of the reaction between a conventional fuel and oxygen, usually from the air, is directly and continuously converted into low-voltage direct-current electric energy. It does not require charging.

fulcrum. The pivot about which a LEVER turns.

full wave rectification. Rectification of an alternating current or voltage such that the

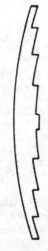

F5 Fresnel lens

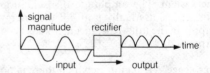

F6 Full wave rectification

negative half waves are made positive, as illustrated (fig F6), so that the whole of the input signal is available to the load. *Compare* HALF WAVE RECTIFICATION.

function. A mathematical expression involving variables.

fundamental. The component with the lowest frequency in a complex wave.

fusion. (1) Melting.
(2) A nuclear reaction in which two light nuclei join to form a heavier nucleus, for example the formation of helium from hydrogen:

$$\ce{^2_1H + ^3_1H -> ^4_2He + ^1_0n}$$

Such reactions are accompanied by the evolution of large amounts of energy. High temperature is necessary for the reaction since it is only then that the kinetic energy of the nuclei is sufficiently great to overcome their repulsion. The process is the basis of the production of energy in stars and was first demonstrated on Earth by the explosion of the hydrogen bomb. Efforts are being made to harness the process as a controlled energy source. *See also* FUSION REACTOR.

fusion bomb. *See* NUCLEAR WEAPON.

fusion reactor. A device for obtaining energy by nuclear FUSION. Usually gas is initially heated by passing a very high current through it, resulting in the formation of a PLASMA. The plasma is contained by suitable magnetic fields. Problems of plasma stability and the extraction of its energy are still under investigation. *See* ZETA; JOINT EUROPEAN TORUS.

G

g. Symbol for the ACCELERATION DUE TO GRAVITY.

Gaede molecular pump. A vacuum pump in which a grooved cylinder rotates with very small clearance in a casing. On either side of a fixed comb which projects into the grooves, are the gas inlet and outlet. Rotation of the cylinder drags the gas from inlet to outlet, giving a pressure of about 0.1 pascal for a rotation speed between 8000 and 12 000 revolutions per minute.

gain. A measure of the efficiency of an electronic system. For an amplifier it is the ratio of output to input power.

galaxy. A giant assembly of stars, gas and dust, containing on average about 10^{11} stars. The mean intergalactic distance is millions of light years, but galaxies usually occur in *clusters*. They contain most of the matter of the universe. *See also* ELLIPTICAL GALAXY; IRREGULAR GALAXY; SPIRAL GALAXY.

Galaxy. The spiral catherine wheel shaped galaxy containing the Sun. The Galaxy diameter is about 10^5 light years and its thickness 20 000 light years. It belongs to a local group of around 20 galaxies.

Galilean telescope. A simple terrestrial TELESCOPE with a convex objective lens and a concave eyepiece separated by the difference between the focal lengths of the two lenses. Since the final image is erect, the system is used in low-power *opera glasses* and simple binoculars.

Galilean transformation. A set of equations relating space and time co-ordinates, i.e. x', y', z' and t', in a frame of reference moving with constant speed v along the x axis of another frame, to the co-ordinates x, y, z and t in this other frame. Each observer is assumed to be at the origin of co-ordinates of the appropriate frame and to coincide at $t' = 0 = t$. The equations are

$$x' = x - vt, y' = y, z' = z, t' = t$$

They conform to NEWTONIAN MECHANICS. *Compare* LORENTZ TRANSFORMATION.

gallon. *See* Table 6B.

Galton whistle. A short cylindrical pipe blown from an annular adjustable nozzle. For a suitable air-blast pressure and nozzle setting, resonant vibrations of frequency above the audible limit are obtained.

galvanic cell. Another name for primary cell. *See* CELL (def. 1).

galvanometer. Any instrument for detecting or measuring small electric currents. The most common type is a MOVING COIL INSTRUMENT, which is, however, unsuitable for alternating current. *See also* ASTATIC GALVANOMETER; EINTHOVEN GALVANOMETER; TANGENT GALVANOMETER.

galvanometer sensitivity. The galvanometer deflection produced by unit current.

games theory. A mathematical theory used to predict the optimum strategy in situations of conflict.

γ. (1) The symbol used to denote the ratio of the specific heat of a gas at constant volume to that at constant pressure.
(2) The gradient of the linear part of the graph of density of processed photographic material against logarithm of exposure.

gamma camera. An instrument for demonstrating the distribution of radioactive isotope in a part of the human body. A large thin SCINTILLATION crystal is

116

located above the part under investigation and below an array of PHOTOMULTIPLIER tubes. When gamma rays from the isotope strike the crystal at point A say, the size of the resulting pulse from each photomultiplier depends on its position relative to A. By feeding the outputs to a cathode ray OSCILLOGRAPH via an amplifier, a picture of the isotope distribution is obtained and can be photographed.

γ ray. *See* GAMMA RAY.

gamma ray. Very penetrating electromagnetic radiation of wavelength in the range 10^{-10} metre to 10^{-13} metre, i.e. having the smallest wavelengths in the known electromagnetic spectrum and hence the largest photon energies. Gamma rays are emitted spontaneously by some radioactive substances during nuclear decay and are also formed in particle ANNIHILATION. Unlike alpha and beta rays, gamma rays are uncharged and are therefore not deflected by electric and magnetic fields.

gamma ray astronomy. The study of gamma rays from space. Observations have been made from several satellites launched from Earth and the origins of various gamma ray sources located.

gamma ray heating. A method of supplying a controlled amount of heat to a thermally isolated sample at a temperature below 1 K by gamma ray irradiation, usually from cobalt-60. The technique is used in low-temperature calorimetry.

gamma ray spectrometer. An instrument for measuring the distribution of energy in gamma ray spectra.

gamma ray spectrum. The emission spectrum of sharp lines, corresponding to well-defined quantum energies, which results from nuclear transitions between different energy levels.

gamma ray transformation. A radioactive disintegration with the emission of gamma rays.

Gamow barrier. Another name for NUCLEAR BARRIER.

ganged circuits. Two or more circuits each containing variable elements which can be simultaneously adjusted using a single control.

gas. A state of matter characterized by ease of flow, compressibility and spontaneous expansion in order to fill any container in which it is placed. *See also* IDEAL GAS; KINETIC THEORY OF MATTER.

gas amplification. Another name for GAS MULTIPLICATION.

gas breakdown. A phenomenon occurring when the voltage across a gas-filled tube exceeds a certain value. Ions in the gas then become sufficiently energetic to produce more ions by collision with netural atoms and to avoid recombination. Hence multiplication occurs, causing breakdown of the gas insulation property. Gas breakdown is somewhat analogous to AVALANCHE BREAKDOWN in a semiconductor.

gas constant. *See* UNIVERSAL GAS CONSTANT.

gas cooled reactor. A thermal nuclear reactor using a gaseous coolant (*see* NUCLEAR REACTOR TYPES). Carbon dioxide emerging at 350° C is used in the mark I MAGNOX reactor; in the advanced gas cooled reactor it emerges at 600° C.

gas discharge tube. An ELECTRON TUBE whose behaviour is markedly influenced by the presence of a gas. For a sufficiently large electric field between the electrodes, ionization of gas atoms and molecules near the electrodes occurs. As an ion moves towards the electrode of opposite sign to itself, collision with other atoms and molecules may occur resulting in their excitation or ionization and so the process spreads. Radiation accompanies both return to the unexcited state and ion recombination.

The gas pressure also influences the appearance of the tube. At relatively high pressure, emanation of radiation is confined to regions close to the electrodes and is known as *positive glow* near the anode and as *negative glow* near the cathode; the glows spread throughout the tube as the pressure is reduced. Further pressure reduction results in the appearance of dark regions, for example the *Crookes dark space* and *Faraday dark space*, which fill the tube as the pressure is still further reduced. These phenomena are explicable in terms of collision probabilities.

gaseous ions. Ions formed in gases by the action of ionizing radiation. They readily recombine to form neutral atoms and molecules and so few remain shortly after cutting off the ionizing radiation.

gas filled relay. Another name for THYRATRON.

gas filled tube. An electron tube containing a sufficient quantity of gas vapour to ensure that, once ionized, the gas is entirely responsible for the electrical characteristics of the tube.

gas laser. *See* LASER.

gas laws. Laws governing the relationship between the pressure, volume and temperature of a gas. *See* EQUATION OF STATE.

gas multiplication. (1) The process whereby, in a sufficiently strong electric field, ions produced in a gas by ionizing radiation can produce additional ions.
(2) The ratio of final ionization produced by a process of gas multiplication to initial ionization.

gas thermometer. *See* CONSTANT PRESSURE GAS THERMOMETER; CONSTANT VOLUME GAS THERMOMETER.

gate. (1) The electrode(s) in a FIELD EFFECT TRANSISTOR whose voltage controls the conductivity of the channel.
(2) An electronic circuit which gives an output signal for only certain combinations of two or more input signals.

(3) An electronic device that passes signals for only a specified fraction of the input signal.

gauge symmetry. A symmetry in which no measurable property of the world changes if protons and neutrons can be substituted for each other at each point in space independently.

gauge theory. Any theory incorporating GAUGE SYMMETRY. Examples are QUANTUM CHROMODYNAMICS and ELECTROWEAK THEORY.

gauss. Symbol G. The unit of magnetic flux density in CGS UNITS, equal to 10^{-4} tesla.

Gaussian distribution. A plot associated with a set of values of y and described by the equation

$$f = [(2\pi)^{1/2}\sigma]^{-1} \exp \ [-(y - \mu)^2/(2\sigma^2)]$$

f is the frequency of occurrence of the experimentally determined variable y, μ is the mean of the values of y and σ is the standard DEVIATION.

Gaussian eyepiece. An eyepiece frequently used in telescopes, illustrated in fig. G1. The hole in the side of the eyepiece tube permits illumination of the crosswires; when they are in focus and seen alongside the image of them formed by reflection in the plane mirror M, the telescope is focused for parallel light and its axis is perpendicular to M. Such an adjustment is essential for the correct use of a spectrometer.

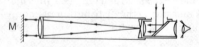

G1 Telescope with Gaussian eyepiece

Gaussian points. Another name for cardinal points. *See* CENTRED OPTICAL SYSTEM.

Gaussian units. A type of CGS UNITS employing both electrostratic and electromagnetic units.

gaussmeter. Another name for FLUX-METER.

Gauss' theorem. For any closed surface in an electric field, the integral over the surface of the electric flux component in a direction perpendicular to the surface is equal to the total electric charge enclosed by the surface. Analogous statements apply to gravitational, magnetostatic and fluid velocity fields.

Gay-Lussac's law. The volumes at the same temperature and pressure in which perfect gases combine chemically bear a simple whole number relation to each other and to that of the resulting product, if gaseous.

gegenschein. A patch of light infrequently seen in the night sky in a position opposite to that of the Sun. It results from the scattering of sunlight by meteor particles.

Geiger counter. An instrument for detecting ionizing radiation and for counting particles. As shown in fig. G2, it consists of a cylindrical metal cathode containing either halogen or rare gas at low pressure, and with a thin mica, glass or metal window at one end; a thin wire anode is mounted along the cylinder axis. The potential difference between cathode and anode is slightly lower than that required to produce a discharge in the gas, i.e. between 300 and 400 volt for halogen filled tubes and around 1000 volt for ones filled with rare gas.

Radiation passing through the window produces gas ions and electrons, which are accelerated by the electric field and so produce further ionization. This results in a pulse of current lasting until the discharge is quenched by the gas. The momentary current passes through a high resistance producing a potential difference which is amplified and registered by a detecting or counting device. The disadvantage of the tube is that it has a 'dead' time, i.e. it is necessary for one pulse to die away before another can be counted.

Geiger and Marsden's experiment. An experiment concerned with the scattering of α particles by thin metal films of high atomic mass. The principle of the vacuum-mounted apparatus and the results are indicated in fig. G3. Rutherford deduced that the results are consistent with the existence of a small massive positively charged nucleus of radius about 10^{-15} metre, around which revolve electrons of path radius about 10^{-10} metre so that the atom as a whole is electrically neutral. The experiment is thus one of great significance.

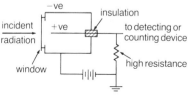

G2 Geiger counter

Geiger-Muller tube. The tube employed in a GEIGER-COUNTER.

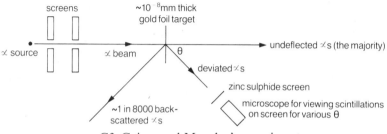

G3 Geiger and Marsden's experiment

Geiger-Nuttal law. The range, R, of a particle emitted in alpha decay is given by

$$\log \lambda = C \log R + B$$

where λ is the DECAY CONSTANT and B and C are constants; C has the same value for all four RADIOACTIVE SERIES but B has a different value for each series. This empirical law is only approximately true.

Geissler tube. A GAS DISCHARGE TUBE comprising a glass or quartz capillary tube with an electrode-holding bulb attached at each end. The discharge is intensified in the capillary tube, making the device a useful spectroscopic light source.

generalized co-ordinates. (1) Co-ordinates which describe the motion of a mechanical system without their exact nature being specified.
(2) The minimum number of co-ordinates necessary to specify the state of a mechanical system subject to constraint.

generalized momenta. Momenta related to GENERALIZED CO-ORDINATES by the expression

$$p_i = L/\partial \dot{q}_i$$

where $p_1, p_2, \dots p_n$ are the generalized co-ordinates and $q_1, q_2 \dots q_n$ respectively the associated momenta; L is the LAGRANGIAN FUNCTION for the system.

general theory of relativity. *See* RELATIVITY.

generation. A group of two LEPTONS and two QUARKS. The first generation comprises electron, neutrino, up quark, down quark; the second generation comprises muon, neutrino, charmed quark and strange quark; the third generation comprises tau lepton, neutrino, hypothetical top quark and bottom quark.

generator. (1) A machine for generating electric current by ELECTROMAGNETIC INDUCTION. In the simple example illustrated in fig. G4, a coil of insulated wire C, i.e. the armature, is rotated about a fixed axis in a constant magnetic field produced by a field magnet. C is connected to conducting cylindrical slip rings SR on the driving shaft; the rings make continuous contact with fixed carbon brushes B. So an alternating current flows in any circuit connected to BB; the direction of the current reverses each time it passes through zero, which corresponds to C vertical. To obtain direct current SRSR are replaced by a split ring commutator as shown; the new

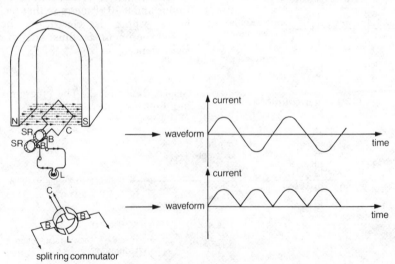

G4 Principle of generator

wave form is as illustrated. In more sophisticated machines, several coils are used.

In very large generators, the magnetic field is produced by rotating field coils and the current is induced in a stationary armature; the advantage is that brushes are not needed to collect the large generated currents but are used to supply the relatively small direct current to the field coils.

(2) Any of various devices for producing electrostatic charge; examples are the VAN DE GRAAFF GENERATOR and the WIMSHURST MACHINE.

(3) A line on the surface of a cone or cylinder, lying in the same plane as the axis and such that the surface would be generated by rotating the line about the axis.

geocentric. (1) Having the Earth at the centre.

(2) Measured with reference to Earth.

geodesic. The shortest line joining two points on a surface. For points lying on a sphere it is part of a great circle.

geomagnetism. Another name for TERRESTRIAL MAGNETISM.

geometric mean. The nth root of the product of n numbers; thus the geometric mean of α, β, γ and δ is $(\alpha\beta\gamma\delta)^{\frac{1}{4}}$.

geometric optics. *See* OPTICS.

geometric progression. A sequence of numbers each of which is the product of the preceding term and a fixed number. A geometric progression thus has the form

$$a, \ ar, \ ar^2, \ ..., \ ar^{n-1}$$

where r is known as the *common ratio* and n is the number of terms. The sum of these terms is

$$a(1 - r^n)/(1 - r)$$

If $-1 \langle r \langle 1$, the sum becomes

$$a/(1 - r)$$

as n tends to infinity.

geometry. The mathematical study of shape. If confined to two or to three dimensions it is known respectively as *plane geometry* and *solid geometry. See also* EUCLIDEAN GEOMETRY; NON-EUCLIDEAN GEOMETRY; ANALYTIC GEOMETRY.

geophysics. The study of the behaviour and physical properties of Earth. It includes geophysical prospecting, seismology, oceanography, and the study of atmospheric electricity, gravitational properties and terrestrial magnetism.

geosynchronous orbit. A SYNCHRONOUS ORBIT of a satellite about the Earth.

germanium. Symbol Ge. A metalloid element used in a pure form for semiconductor manufacture.

getter. A substance, usually a pure metal, used for removing unwanted atoms or molecules from an environment. For example heated magnesium will combine with unwanted residual oxygen and nitrogen in a sealed vacuum system, and phosphorous is used in the manufacture of metal oxide semiconductors to remove unwanted sodium.

GeV. Symbol for giga-electronvolt, i.e. 10^9 electronvolt; it is thus equal to one thousand million (one US billion) volt.

g factor. Another name for LANDÉ FACTOR.

giant planet. A planet whose mass is greater than that of Earth, i.e. JUPITER, SATURN, URANUS and NEPTUNE.

giant star. A star large in size and brightness but small in density compared with the Sun. *See also* RED GIANT.

gibbous. The phase of the Moon between half and full.

Gibbs free energy. Symbol G. A quantity equal to

$$H - TS$$

where H is the ENTHALPY, S the ENTROPY and T the thermodynamic temperature of a system. In an isothermal reversible change at constant pressure, the work done on a system equals its change in G; G is a minimum when the system is in equilibrium.

Gibbs function. Another name for GIBBS FREE ENERGY.

Gibbs-Helmholtz equation. The equation

$$U = F - T(\partial F/\partial T)_V$$

where U is the INTERNAL ENERGY, F the FREE ENERGY and T the thermodynamic temperature of a system. Subscript V indicates differentiation at constant volume.

giga-. Symbol G. A prefix meaning 10^9, i.e. one thousand million. In the USA the symbol B is sometimes used.

gilbert. Symbol Gb. The electromagnet unit of magnetomotive force in CGS UNITS: a magnetomotive force of 0.4π gilbert is produced by one turn of wire carrying a current of one ampere.

Giorgi units. A system of units based on the metre, kilogram, second and (up to 1950) the ohm and thereafter the ampere. SI UNITS superseded the system.

Gladstone-Dale law. The ratio of $n - 1$ to ρ is constant, where n and ρ are respectively the refractive index and density of a substance; thus if ρ is altered by change in temperature etc., n also varies.

glancing angle. The difference of 90° and the angle of incidence.

glide. The movement of one atomic plane over another in a crystal, i.e. the process whereby a solid undergoes PLASTIC DEFORMATION.

glide plane. A plane upon which GLIDE occurs due to a suitable shearing stress. The glide may be in a particular direction in the plane.

globar. A carborundum rod which when heated is a good source of infrared radiation.

globular cluster. See CLUSTER.

glow discharge. The emanation of radiation along the length of a GAS DISCHARGE TUBE at a suitable gas pressure. Characteristic features of the glow discharge are nonuniform variation of voltage with distance along the tube, relatively high current of a few milliampere, independence of voltage drop on current. Because of the latter property the glow discharge is a useful voltage stabilizer. Other applications are in luminous signs, lighting purposes and as a source for spectral analysis of the gas or vapour in which the discharge occurs.

gluon. A hypothetical particle exchanged between a pair of QUARKS so binding the pair together. The gluon plays a role in QUANTUM CHROMODYNAMICS analagous to that played by the photon in QUANTUM ELECTRODYNAMICS.

golden section. The division of a line or area so that the ratio of the whole to the larger part equals the ratio of the larger part to the smaller one. Proportions based on golden sections are considered aesthetically pleasing and so appear in many paintings.

gold leaf electroscope. See ELECTROSCOPE.

gold point. See INTERNATIONAL TEMPERATURE SCALE (table).

goniometer. An instrument for measuring the angles between the faces of crystals.

G-parity. A quantum number associated with elementary particles of zero BARYON NUMBER and STRANGENESS. It is conserved in STRONG INTERACTION only.

grad. See GRADIENT.

grade. A unit of plane angle equal 0.9 degree and written 1^g.

gradient. A vector derived from a scalar function, φ say, of co-ordinates x, y and z. It is defined by

$$\text{grad } \phi = \nabla\phi = i\, \partial\phi/\partial x + j\, \partial\phi/\partial y + k\, \partial\phi/\partial z$$

where i, j and k are UNIT VECTORS along the x, y and z axes respectively and ∇ is the operator DEL. Electric field, for example, is the negative gradient of electric potential.

Graham's law. At constant temperature and pressure, the diffusion rate of a gas is inversely proportional to the square root of the density.

grain. *See* Table 6D.

gram-force. Symbol gf. An obsolescent unit defined as the force that would give a mass of 1 gramme an acceleration equal to the standard ACCELERATION DUE TO GRAVITY. *Compare* GRAM-WEIGHT.

gramme. *See* Table 6D.

gramme-atom. Former name for MOLE.

gramme-molecule. Former name for MOLE.

gram-weight. Symbol gwt. An obsolescent unit defined as the force that would give a mass of 1 gramme an acceleration equal to the local ACCELERATION DUE TO GRAVITY. *Compare* GRAM-FORCE.

grand unified theories. Attempts to unite the theories of the strong, electromagnetic, weak and gravitational forces of nature. The starting point is the general theory of RELATIVITY.

Grashof number. A dimensionless number defined as

$$(l^3 g\alpha\rho^2\theta)/\eta^2$$

where l is a typical dimension of a hot body which produces natural convection in a fluid of density ρ, coefficient of viscosity η and temperature coefficient α of fluid density; g is the acceleration due to gravity and θ the temperature difference between the hot body and the fluid.

graticule. A network of fine lines introduced into an optical instrument so that it may be seen simultaneously with the image viewed. It is used as a reference system and also for making measurments. The lines may be either fine wires or threads, or engraved on transparent material.

gravimeter. An instrument for the determination of fluid density by using a total immersion float attached by a spiral spring to the bottom of the containing vessel. The extension of the spring gives a measure of the liquid density.

gravitation. The mutual attraction of bodies due to their mass. It is responsible for objects falling to Earth. *See also* RELATIVITY.

gravitational acceleration. Another name for ACCELERATION DUE TO GRAVITY.

gravitational constant. (1) The constant G appearing in NEWTON'S LAW OF GRAVITATION. It has the value 6.6732×10^{-11} newton metre squared per kilogramme squared. *See also* CAVENDISH'S EXPERIMENT.
(2) The Gaussian constant of gravitation, equal to $G^{\frac{1}{2}}$.

gravitational field. The region in which gravitational forces are operative. Its magnitude at any point due to mass m is Gm/x^2 where x is the distance of the point from m; its direction is along the line joining the point to m.

gravitational force. One of the fundamental forces believed to account for all the observed interactions of matter.

gravitational interaction. The interaction occurring between bodies or particles as a result of their masses. It is about 10^{40} times weaker than ELECTROMAGNETIC INTERACTION.

gravitational potential. Symbol V. The work done in moving unit mass from infinity to the point under consideration. For a mass m, concentrated at a point, the potential equals

$$\int_{\infty}^{x} (Gm/x^2)\mathrm{d}x = -Gm/x$$

where G is the GRAVITATIONAL CONSTANT and x the distance of the point under consideration from m. For a hollow homogeneous spherical shell of mass m, the potential at a point outside the shell is $-Gm/x$ where x is the distance of the point from the shell centre; the potential of the shell is therefore the same as if its mass were concentrated at its centre. For a point lying inside the shell, the potential is $-Gm/r$ where r is the shell radius; the inside of the shell is thus an equipotential region. For a solid homogeneous sphere of mass m and radius r, the potential at a point outside the sphere is $-Gm/x$ where x is the distance from the centre of the sphere to the point; for a point inside the sphere, the potential is

$$-Gm(3r^2 - x^2)/(2r^3)$$

gravitational unit. A unit involving the ACCELERATION DUE TO GRAVITY.

gravitational waves. Waves propagated as the result of the acceleration or deformation of a mass. The general theory of RELATIVITY predicts the existence of gravitational radiation travelling at the speed of light. Coincidences between the signals from widely separated detectors, consisting of solid bars which should be set into oscillation by the waves, have been sought. The results are claimed to indicate that gravitational pulses from a point near the centre of the Galaxy are reaching Earth; however, the evidence is disputed.

graviton. A hypothetical elementary particle, the quantum of gravitation, responsible for the effects of gravity. It is assumed to be its own antiparticle, to have zero charge and zero rest mass and a spin of 2.

gravity. The attraction of one body for another as the result of gravitation.

gravity balance. A sensitive spring balance: the variations of the reading for a fixed mass at different places enable the variation of the ACCELERATION DUE TO GRAVITY to be studied.

gravity cell. A primary electric cell in which two electroytes are kept apart by their different densities.

gravity wave. A fluid surface wave of large wavelength whose motion is controlled by gravity rather than by SURFACE TENSION. Thus the magnitude of the velocity v of shallow sea waves is given by

$$v = (g\lambda)^{1/2}$$

where g is the acceleration due to gravity and λ the wavelength.

gray. *See* RADIATION UNITS.

grease spot photometer. A visual photometer in which the photometer head consists of thin white opaque paper with a translucent spot at the centre. The two sources under comparison, A and B, of power P_A and P_B respectively, are located on either side of the spot so that a line joining them passes normally through the spot centre. The distances y_A and y_B of A and B respectively from the spot when it disappears as viewed from one side are found, and also the corresponding distances y'_A and y'_B for disappearance from the other side. Then

$$P_A/P_B = y_A y'_A/(y_B y'_B)$$

great circle. A plane section of a sphere through its centre. *Compare* SMALL CIRCLE.

Green's theorem. The vector form of GAUSS' THEOREM.

Gregorian telescope. An astronimical telescope, illustrated in fig. G5. It is similar to the CASSEGRAINIAN TELESCOPE but uses two concave mirrors rather than one concave and one convex.

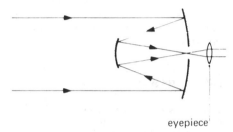

eyepiece

G5 Gregorian telescope

Grenz rays. Long wavelength X rays.

grid. (1) A CONTROL ELECTRODE.
(2) The high-voltage transmission line system interconnecting many large generating stations. Voltages range from 275 kilovolt to 735 kilovolt.

grid bias. A potential difference applied between the cathode and control grid of a thermionic valve so that operation on any chosen part of the characteristic curve may be obtained.

grid leak. A high resistance between the cathode and grid of a thermionic valve. It prevents charge accumulating on the grid.

ground. Another name (mainly US) for EARTH.

ground plane. A conducting sheet at earth potential, incorporated in an electric circuit to provide a low-impedance path to earth throughout the circuit.

ground state. *See* ENERGY LEVEL.

ground wave. A radio wave transmitted so that it travels over the surface of the Earth to the receiver, rather than reaching it by IONOSPHERE reflection.

group. A set of entities with certain mathematical properties. Members of the set can be combined by addition, multiplication etc.: entities form a group if combination of any two of them yields a third and if the combination is an ASSOCIATIVE OPERATION. Moreover there must be one entity, known as the *identity*, such that com-

bination with it produces no change; for example one is the identity associated with multiplication. Furthermore for every entity there must be another that combines with it to yield the identity. A group member may be a number, a VECTOR or a MATRIX collection. Group theory is used in spectroscopy, crystallography, and particle physics.

group velocity. The velocity of propagation of a pulse. The pulse may be considered as resulting from the superposition of sinusoidal waves of slightly different wavelength. For example with two such waves BEATS are obtained whose velocity of propagation is

$$c - \lambda\, \delta c / \delta\, \lambda$$

where c is the velocity (i.e. the PHASE VELOCITY) of one sinusoidal wave of wavelength λ and $c - \delta c$ is the velocity and $\lambda - \delta\lambda$ the wavelength of the other sinusoidal wave. In general the group velocity is

$$c - \lambda\, dc/d\lambda$$

and is the wave velocity as experimentally determined.

Grove cell. A two-fluid primary CELL, the fluids being dilute sulphuric and fuming nitric acids separated by a porous partition. The negative electrode is a zinc rod dipping into the sulphuric acid and the positive electrode is a platinum plate immersed in the nitric acid. The electromotive force is 1.93 volt.

grown junction. A semiconductor junction formed by adding extra impurities to the melt while the crystal is growing in it.

Grüneisen's law. The ratio of the linear COEFFICIENT OF EXPANSION of a metal to its SPECIFIC HEAT CAPACITY is a constant, independent of the temperature at which measurements are made.

guard ring. A large metal plate coplanar with and surrounding a small metal plate but separated from it by an air gap. An applied electric field is then uniform over the small plate, which can be treated as an infinite plane since edge effects are eliminated. An extra electrode which performs a similar function is often used in semiconductor devices and vacuum tubes.

A guard ring is also used in accurate experiments on heat flow: the central plate and ring are independently heated to the same temperature so that edge effects are eliminated and heat evenly distributed over the central plate.

guard wires. Earthed conductors placed beneath overhead line conductors so that, should the latter snap, they will be earthed without striking the ground. Where high-voltage lines cross telephone wires or a thoroughfare, a protective net of ground wires is used.

Guillemin effect. A type of MAGNETO-STRICTION in which a magnetic field, applied along the length of a bent bar of ferromagnetic material, tends to straighten the bar.

Gunn effect. The production of a coherent MICROWAVE current in certain homogeneous n-type semiconductors under the influence of an electric field of several thousand volt per centimetre. Such devices provide simple solid-state sources of microwave power.

gyrator. A device which reverses the phase of signals transmitted in one direction, but has no effect on the phase of signals transmitted in the opposite direction. It is mainly used at microwave frequencies.

gyrocompass. A GYROSCOPE free to swing in any plane, rotated by an electric motor about a north-south axis. The axis thus maintains this direction irrespective of any movements of the mounting. The instrument is unaffected by stray magnetic fields and is widely used on ships, aircraft, missiles etc.

gyrodynamics. The study of rotating bodies, particularly when subject to PRECESSION.

gyromagnetic effects. The relationships between the magnetization of a body and its rotation. An iron cylinder rotated at high speed about its axis develops weak magnetization and conversely an iron cylinder free to rotate will do so slightly when suddenly magnetized.

gyromagnetic ratio. Symbol γ. The ratio of the magnetic dipole moment of a dipole to its angular momentum. Thus for an orbiting electron $\gamma = e/(2m)$, and for a spinning electron $\gamma = e/m$, where e and m are the electronic charge and mass respectively.

gyroscope. A symmetrical well-balanced wheel which has its mass distributed as far as possible from its axis and is mounted in a double gimbal so that it can spin freely about its axis, which can adopt any orientation in space. When the wheel is spinning, the direction of the axis of the wheel resists change. When subjected to a sufficiently large torque tending to alter the axis direction, the gyroscope turns about an axis perpendicular to both its initial axis of spin and to the axis of the torque, i.e. the gyroscope manifests PRECESSION.

gyrostabilizer. A device used on ships to prevent excessive rolling. A large and heavy GYROSCOPE may be used to counteract the roll. Alternatively a small gyroscope is employed: this senses the roll and through a servo-mechanism actuates stabilizing fins below water level.

gyrostat. A gyroscope used to indicate or use the constancy of axis direction of a fast-moving gyroscope.

H

h. The symbol used for the PLANCK CONSTANT.

habit. *See* CRYSTAL HABIT.

hadron. Any particle that can participate in STRONG INTERACTION. Hadrons fall into two groups: MESONS and BARYONS.

haemorrhage location. A process most easily accomplished by injecting the patient with blood labelled with the radioactive isotope chromium-51. This isotope is taken up by red cells and so, for normal blood circulation, the radioactivity is distributed evenly throughout the circulatory system. A build-up of radioactivity at some site therefore indicates a haemorrhage at that site; the rate of build-up indicates the rate of blood loss.

hair hygrometer. A HYGROMETER in which a hair attached to a pointer is kept taut by a spring. Increase in the relative humidity of the surrounding air causes the hair to increase in length, and to decrease in length for a decrease of relative humidity. The pointer therefore moves, and the relative humidity can be read off from a precalibrated scale.

halation. The luminous patch surrounding the spot on the screen of a CATHODE RAY TUBE. It is usually caused by internal reflection in the glass of the screen.

half cell. One electrode of an electrolytic CELL and the electrolyte with which it is in contact.

half life. Symbol $T_{\frac{1}{2}}$ or $t_{\frac{1}{2}}$. The time in which the amount of a radioactive nuclide decays to half of its original value. The half life is equal to $(\ln 2)/\lambda$ where λ is the DECAY CONSTANT.

half period zones. Zones into which a wave front is conveniently divided when determining the intensity at a given point resulting from FRESNEL DIFFRACTION. The zones are such that the radiation from one zone reaching the point is half a period out of phase with that arriving from adjacent zones.

half silvered mirror. A MIRROR, the silvering on whose boundary surface is such that equal amounts of light are transmitted and reflected by it.

half thickness. The thickness of a uniform sheet of material which when inserted in the path of a beam will transmit only half of it.

half wave dipole. A straight AERIAL approximately half a wavelength long. When excited, the dipole has voltage ANTINODE and current NODE at each end, while at the centre there is a voltage node and current antinode. The feeder is usually connected across a small gap in the dipole centre.

half wave plate. A plate of quartz or mica cut parallel to the OPTICAL AXIS and of thickness such that the ordinary and extraordinary rays of transmitted light differ in phase by 180°. Plane polarized light incident normally on the plate undergoes an angle of rotation of its plane of polarization, equal to twice the angle between the axis and the direction of the incident vibrations. The plates are used in polarimeters.

half wave rectification. The conversion of alternating to direct current in such a way that current passes only in alternate half cycles. Compare FULL WAVE RECTIFICATION.

half width. Either the width or half the width of a spectrum line as measured at half its height; it is thus a confusing term.

Hall coefficient. The coefficient R_H in the expression

$$R_H BJT$$

for the potential difference produced in the HALL EFFECT; B is the magnitude of the magnetic induction, J the current density and T the thickness of the specimen in the direction of the potential difference.

Hall effect. The production of a potential difference between opposite surfaces of a current-carrying conductor, or semiconductor, placed in a magnetic field whose direction is perpendicular to both the current and the potential difference. The potential difference is caused by deflection of the charge carriers and is in opposite directions for positive holes and electrons; its magnitude depends on the concentration of the charge carriers. *See also* HALL COEFFICIENT.

Hall mobility. Symbol μ_H. The product of the HALL COEFFICIENT and the ELECTRIC CONDUCTIVITY.

Hall probe. A small conductor or semiconductor of known HALL COEFFICIENT used to determine magnetic induction by measuring the other variables in the expression for the potential difference.

Halley's comet. One of the most spectacular and best known comets. It has a period of 76 year and last appeared in 1985/86.

halo. A luminous ring sometimes observed around the Sun or Moon. It is due to the refraction of light by small ice particles high in the Earth's atmosphere.

Hamiltonian. Symbol H. An expression for the energy of a system. In simple cases it is the sum of the kinetic and potential energies; thus for a particle of momentum p and mass m,

$$H = p^2/(2m) + V$$

where V is the potential energy.

Hamiltonian function. Another name for HAMILTONIAN.

Hamilton principle. For a conservative system,

$$\int_0^t (T - V)\,dt$$

has a stationary value; T and V are the kinetic and potential energies respectively and t is the time. This principle is a more general form of the LEAST ACTION PRINCIPLE and of FERMAT'S PRINCIPLE OF LEAST TIME.

Hamilton's equations. A restatement of LAGRANGE'S EQUATIONS but emphasizing momentum rather than force. They are much used in advanced mechanics, including quantum mechanics.

Harcourt pentane lamp. A lamp burning a mixture of pentane and air under specified conditions. Previously it was used as a standard for the international CANDLE.

hardness. The resistance of a crystal face to scratching. It usually varies with direction and may be measured by the BRINELL TEST. Arranged in order of diminishing hardness, some frequently used substances are: diamond, corundum, topaz, quartz, felspar, apatite, fluorspar, calcspar, rocksalt and talc.

hard radiation. Ionizing radiation with a high degree of penetration.

hard vacuum tube. A vacuum tube such that ionization of the residual gas has a negligible effect on the electrical characteristics of the tube.

hardware. Computer equipment other than SOFTWARE.

Hare's apparatus. A device for measuring the RELATIVE DENSITY of a liquid, illustrated in fig. H1. With the clip open the liquid L and pure water are sucked up and their respective levels recorded after closing the clip. Since the pressures at X and Y are the same, i.e. both atmospheric, then

$$\rho_w h_w g = \rho_L h_L g$$

where h_w and h_L are as shown and ρ_L and ρ_w are the densities of liquid and water respectively. Hence the relative density of L is given by

$$\rho_L/\rho_w = h_w/h_L$$

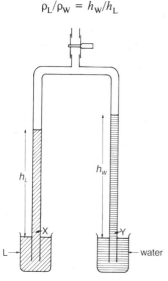

H1 Hare's apparatus

harmonic. (1) A vibration whose frequency is an integral number of times that of the FUNDAMENTAL, which is the first harmonic. *See also* VIBRATIONS IN PIPES; VIBRATIONS IN STRINGS.

(2) A continuous function which satisfies the LAPLACE EQUATION and whose first derivative is also a continuous function.

harmonic analyser. A device for evaluating the coefficients of the FOURIER SERIES for a particular function.

harmonic distortion. DISTORTION arising from the presence in the output of harmonics not present in the input.

harmonic oscillator. An oscillator for which the restoring force varies linearly with the displacement of the system from its equilibrium position.

harmonic series. The series

$$1 + 1/2 + 1/3 + 1/4 \ldots 1/n + \ldots$$

hartley. A unit of information equal to

$$\log_2 10 \text{ bits}$$

It is used in a digital COMPUTER.

Hartmann dispersion formula. For a prism:

$$dn/d\lambda = -C(\lambda - \lambda_0)^2$$

where n and λ are refractive index and wavelength respectively and C and λ_0 are constants.

Hartmann generator. A device for generating high-power ultrasound waves in fluids. It operates on the same principle as the GALTON WHISTLE but has a higher top frequency of 100 000 hertz and higher power output.

hartree. An energy unit equivalent to

$$4.8505 \times 10^{-18} \text{ joule}$$

It is used in atomic physics and spectroscopy.

H bomb. *See* NUCLEAR WEAPON.

HCF. Abbrev. for HIGHEST COMMON FACTOR.

health physics. The investigation of health and safety at work in medicine, science and industry. In particular it deals with protection from RADIOACTIVITY; this involves detection and measurement of the radioactivity, decontamination, disposal of radioactive waste, laboratory and shielding design and the monitoring of radioactivity tolerance doses received by workers.

heart. The organ which maintains the circulation of the blood by pulsating rhythmically at about 76 pulses per minute normally. The electrical activity of the heart, i.e. the spike potentials associated with the heart's action, are responsible for opening and closing the heart's valves in the correct sequence. A device which takes over the heart's function is called an *artificial heart*.

heat. Symbol Q. The energy transferred from a higher temperature body to a lower temperature one as a result of the temperature difference between them. A body's heat is the KINETIC ENERGY of translation, vibration and rotation in its molecules. For a body remaining in the same state, the change in its heat content due to a temperature change θ is given by $mc\theta$ where m and c are respectively the mass and SPECIFIC HEAT CAPACITY of the body. Change of state of a body occurs at constant temperature and is accompanied by the evolution or absorption of an amount ml of heat where l is the SPECIFIC LATENT HEAT. Transfer of heat can occur by CONDUCTION, CONVECTION and RADIATION.

heat capacity. The amount of heat required to raise the temperature of a body by 1 K. *See also* SPECIFIC HEAT CAPACITY.

heat content. Another name for ENTHALPY.

heat death. The state of a closed system when its total ENTROPY has increased to its maximum value. There is then no available energy since the temperature is uniform and all the matter is disordered. If the universe is a closed system it will eventually reach this state.

heat engine. A device in which heat energy is converted to mechanical energy. The engine absorbs heat from a reservoir at one temperature and gives out heat to a HEAT SINK at a lower temperature; the cycle is then repeated. *See* CARNOT CYCLE; OTTO CYCLE; EFFICIENCY.

heat exchanger. A device for transferring heat between fluids. The fluids may flow through different channels separated by a

wall, or flow either alternately or continuously through the same channel. An exchanger may be used for cooling, as in a car radiator, or for extracting waste heat from outlet gases, furnaces, chemical plants etc. Heat exchangers may be provided with devices such as metal fins, coiled tubes etc. to prolong the time and area of contact of the fluids and so improve the efficiency of the exchange.

heating effect of current. *See* JOULE HEATING.

heating of atomization. The heat required to decompose one MOLE of a substance into its atoms.

heat of combustion. The heat evolved when one MOLE of substance is burnt in oxygen.

heat pump. A device for extracting heat from large quantities of relatively low temperature material, such as air, and supplying the heat at a higher temperature. A volatile liquid vaporizes in tubes in the low temperature material, the necessary heat coming from that material. The vapour is then compressed and reliquefies, giving up heat.

heat sink. (1) A device for conducting and radiating heat away from electronic components to prevent damage to them.
(2) A system considered to absorb heat at a constant temperature, as in the operation of a heat engine.

heat transfer. A process which may occur in three ways: CONDUCTION; CONVECTION and RADIATION.

heat transfer coefficient. The amount of heat emitted per second per unit area by a surface at a temperature of 1 K above its surroundings.

Heaviside layer. *See* IONOSPHERE.

heavy atom technique. A method used in the investigation of molecular structure by X RAY DIFFRACTION. It involves the introduction into the substance under investigation of atoms which scatter X rays well, i.e. atoms of high atomic number, and mathematical treatment of the results obtained; by successive approximations a reasonably accurate structure is arrived at.

heavy hydrogen. Another name for DEUTERIUM.

heavy lepton. A particle having properties similar to the electron but more massive. *See* LEPTON.

heavy water. A liquid with the same chemical properties as water but with different physical properties. The term is mainly used for *deuterium oxide*, which contains the hydrogen isotope deuterium with a relative atomic mass of 2. Heavy water is extracted from natural water either by fractional distillation or by electrolysis or exchange processes.

heavy water reactor. A type of thermal NUCLEAR REACTOR using HEAVY WATER as the moderator and sometimes also as the coolant. *See also* NUCLEAR REACTOR TYPES.

hectare. A unit of area equal to 10 000 square metre.

hecto-. Symbol h. A prefix meaning 100.

HEED. *See* ELECTRON DIFFRACTION.

Heisenberg picture. The representation of dynamic states of a quantum mechanical system by stationary vectors, and of physical quantities by operators which evolve with time.

Heisenberg representation. Another name for HEISENBERG PICTURE.

Heisenberg uncertainty principle. The principle that it is impossible to measure both the momentum and position of a particle to any desired degree of accuracy. Errors in measuring instruments and human errors are not responsible for this *indeterminancy*, which arises from the wavelike behaviour of particles and from the observation itself disturbing the system in an unpredictable way. If Δp and Δx are the uncertainties in momentum and position respectively, then

$$\Delta p \, \Delta x \simeq h$$

where h is the PLANCK CONSTANT. A similar relationship holds for uncertainties in the measurement of energy and time, represented by ΔE and Δt respectively, i.e.

$$\Delta E \, \Delta t \simeq h$$

heliocentric. (1) Having the Sun at the centre of the universe.
(2) Measured with reference to the Sun.

helium. Symbol He. An element present in some natural gas, in some radioactive ores and in the atmosphere, from which it is extracted as a byproduct of the liquefaction of air. It is used for filling air ships and some types of fluorescent lamps.

Liquid helium exists in two forms, I and II, respectively stable above and below 2.19 K. Form II exhibits superfluidity and has high thermal conductivity and low viscosity.

helix. A curve drawn on a cylindrical or conical surface so that all the generators of the surface are cut at the same angle.

Helmert's formula. The empirical formula

$$g = 9.806\ 16 - 0.025\ 928 \cos 2\phi$$
$$+ 6.9 \times 10^{-5} \cos^2 2\phi - 3.086 \times 10^{-6}\, h$$

where g is the ACCELERATION DUE TO GRAVITY at latitude ϕ and altitude h metre.

helmholz. A unit of DIPOLE moment per unit area, equal to

$$3.335 \times 10^{-10} \text{ coulomb per metre}$$

Helmholtz coils. A pair of identical flat coils of wire mounted parallel to each other and separated by a distance equal to the radius of either coil. When a current is passed through the coils connected in series, a uniform magnetic field is produced on either side of the mid-point between the coils.

Helmholtz electric double layer. A mono-molecular layer of dipoles on the surface of any solid or liquid, the negative charges being outermost. The attraction between the positive and negative dipole charges is smaller the larger the dielectric constant of the substance. Hence if materials of different relative PERMITTIVITY are brought into contact, the one with the smaller value will capture negative charge from the other. When the substances are separated, the one of higher relative permittivity is therefore left positively charged and the other is negatively charged. This is in accordance with observation.

Helmholtz free energy. Symbol A. A quantity defined as

$$U - TS$$

where U is the INTERNAL ENERGY, S the ENTROPY and T the thermodynamic temperature of the system. The difference in A for two different states of the system gives the maximum work that could be done on or by the system.

Helmholtz function. Another name for HELMHOLTZ FREE ENERGY.

Helmholtz resonator. An acoustic resonator in the form of a spherical or cylindrical bulb containing air which is connected to the atmosphere via a small neck in the bulb. The air resonates at a single frequency given by

$$c(S/lV)^{1/2}/(2\pi)$$

where c is the speed of sound, l and S respectively the length and cross sectional area of the neck and V the volume of the cavity.

henry. Symbol H. The SI unit of self and mutual inductance, equal to the inductance of a closed circuit with a magnetic flux of 1 weber per ampere.

Henry's law. The amount of gas dissolved in a given volume of solvent with which it does not react is directly proportional to the equilibrium pressure of gas above the liquid.

heptagon. A seven-sided POLYGON.

heptode. A thermionic valve with seven electrodes: a cathode, five grids and an anode.

Herschelian telescope. A type of astronomical telescope, illustrated in fig. H2, in which light is reflected by a concave mirror at a small angle to the direction of incidence.

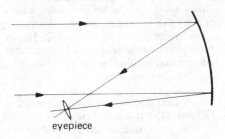

eyepiece

H2 Herschelian telescope

hertz. Symbol Hz. The SI unit of frequency, defined as the frequency of a periodic phenomenon with a period of 1 second.

Hertzian oscillator. A device, illustrated in fig. H3, first used for the production of radio waves. Whenever the potential across the gap is sufficiently high to render the air in the gap conducting, an oscillatory discharge occurs. This is accompanied by the emission of radio waves of the same frequency, usually around 10^8 hertz.

Hertzian waves. Former name for RADIO FREQUENCY RADIATION.

Hertzsprung-Russell diagram. A plot of stellar magnitude against spectral classification, shown in fig. H4. Various classes of stars are concentrated on different parts

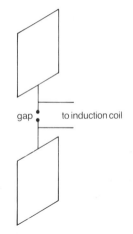

gap • to induction coil

H3 Hertzian oscillator

of the diagram. The greatest number of stars, including the sun, lies in the diagonal band, which is known as the *main sequence*. The radius *r* of a *main-sequence star* is

related to its temperature T and luminosity L by the formula

$$\log r = 0.5 \log L - 2 \log T + 1.57$$

The highest luminosity stars in the band are bluish white in colour.

heterodyne. Using or involving the radio frequency version of BEATS. If the radio signals are close in frequency the heterodyne frequency difference is in the audio-frequency range, an effect used in radio reception. *See also* SUPERHETERODYNE RECEIVER.

heterogenous radiation. Nonmonochromatic radiation, usually of a particular type such as visible, infrared etc.

heterogenous reactor. A NUCLEAR REACTOR whose core consists of thin fuel rods embedded in the moderator.

heterojunction. *See* P-N JUNCTION.

hexagon. A six-sided POLYGON.

hexagonal system. *See* CRYSTAL SYSTEM.

hexahedron. A POLYHEDRON with six faces. An example is a cube, which is a regular hexahedron.

hexode. A thermionic valve with six electrodes: a cathode, four grids and an anode.

HF. Abbrev. for HIGH FREQUENCY.

Higgs particle. A particle whose existence is predicted by ELECTROWEAK THEORY. It is believed that the massiveness of the W PARTICLE and of the Z PARTICLE is associated with it.

highest common factor. The largest number that exactly divides into each of a given set of numbers. For example the highest common factor of 9, 12 and 15 is 3.

high fidelity. Faithful sound reproduction. An appreciable increase in fidelity

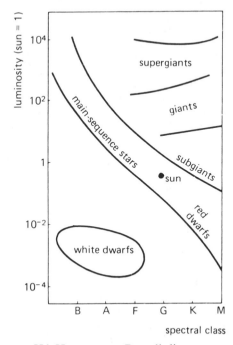

H4 Hertzsprung-Russell diagram

has been achieved by using LASER radiation in sound reproduction.

high frequency. A frequency in the range 3 megahertz to 30 megahertz.

high pressure technique. For the 1 gigapascal to 6 gigapascal range, pressure may be generated by forcing a piston into a cylinder. At higher pressure more sophisticated equipment is required; the subject is important for compressibility studies, phase equilibrium work and research on geological phenomena.

high tension. Usually, the voltage applied to the anode of a thermionic valve, generally between 60 volt and 250 volt.

high vacuum tube. Another name for HARD VACUUM TUBE.

high voltage. A voltage in excess of 650 volt. The term is generally used in connection with electrical power transmission and distribution.

Hilbert space. A multidimensional space in which each CHARACTERISTIC FUNCTION of wave mechanics is represented by an orthogonal UNIT VECTOR.

histogram. A graphical representation of data in which the range of values of the observations is divided into suitable, usually equal, intervals and plotted along the x axis. Each interval forms the base of a rectangle whose height, i.e. y co-ordinate, represents the number of observations whose values lie within that interval. Fig. H5

number of scripts

examination mark/(%)

H5 Histogram

shows the histogram for the examination marks of a group of 28 students; the interval here is 10%.

hole. An unfilled vacancy in an electronic energy level in a solid. Such a vacancy is mathematically equivalent to a POSITRON. In a p-type semiconductor holes are the majority carriers, i.e. they are responsible for most of the electric conductivity. *See also* HOLE CONDUCTION.

hole conduction. A mechanism of conduction in which HOLES appear to move through a solid under the influence of an electric field: when an electron fills a hole, another hole appears where the electron came from and so on. The apparent movement of the holes is equivalent to a movement of positive charges in the same direction.

hologram. *See* HOLOGRAPHY.

holography. A technique for obtaining a stereoscopic image of an object without the use of lenses. A LASER beam is divided in two by means of a half-silvered mirror, as shown in fig. H6. One part illuminates the object and the other is directed towards a high-resolution photographic plate where it meets the light reflected, diffracted or transmitted by the object; interference thus occurs and the phase information is manifest as an intensity variation. The image on the plate therefore carries information about both intensity and phase. This photograph, known as a *hologram*, consists of a very large number of closely spaced dots visible under a microscope. Direct illumination of the hologram by a beam of the same laser light results in the production of diffracted waves. These are of the same amplitude and phase distribution as the original waves from the object. As shown, one of the diffracted beams forms a real stereoscopic image; another gives rise to a virtual stereoscopic image which is seen on looking into the hologram. By moving the head while so doing, more of the object is made visible. Moreover, if a hologram is cut into pieces, the whole of

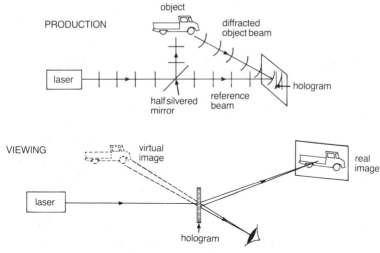

PRODUCTION

object

diffracted
object beam

laser

half silvered
mirror

reference
beam

hologram

VIEWING

virtual
image

laser

real
image

hologram

H6 Hologram production and viewing

the image can be seen in any piece, although there is a loss of quality.

It is now possible to reconstruct coloured images by hologram. Such holograms are made from thick photographic emulsions, information being recorded in planes throughout the thickness; the plane separation depends on the wavelength used. The scene is first illuminated with helium-neon laser light, which is red, and then with argon-ion laser light, which is green and blue. When the resulting hologram is illuminated with light from both these lasers, i.e. with white light, a coloured three-dimensional image of the original object results. For accurate reproduction of colour it is necessary to swell the hologram in order to compensate for emulsion shrinkage on development.

hollow conductor. *See* FARADAY'S ICE PAIL EXPERIMENT.

homocentric. Converging to or diverging from a common point.

homogeneous radiation. Radiation of uniform wavelength.

homogeneous reactor. A type of NUCLEAR REACTOR whose core is a uniform medium consisting, for example, of fuel dissolved in the moderator.

homogeneous solid. A solid for which the chemical and physical properties do not vary from point to point.

homogeneous strain. A strain which satisfies the following relationships:

$$x' = a_1 x + b_1 y + c_1 z$$
$$y' = a_2 x + b_2 y + c_2 z$$
$$z' = a_3 x + b_3 y + c_3 z$$

x, y, z and x', y', z' are respectively the co-ordinates relative to the same rectangular Cartesian axes outside the body, before and after straining of the body, of a point in it; all the coefficients are constants. In such a strain, a plane in the body remains plane and a sphere becomes an ellipsoid.

homopolar generator. A copper disc of radius r rotating at f revolutions per second between the poles of a magnet and having brush connection to axle and circumference. If the axle radius is negligible compared to r, then the direct electromotive force obtained across the brushes is $\pi r^2 B f$ where B is the magnitude of the magnetic induction. Such generators are useful for supplying small direct voltages, for example as sometimes required in electroplating.

Hooke's law. Up to a certain stress, the strain produced in a body is proportional to the stress applied and is independent of the time. The strain disappears completely on removal of the stress. *See also* ELASTIC LIMIT; YIELD POINT.

Hope's experiment. An experiment designed to demonstrate the ANOMALOUS EXPANSION OF WATER; the apparatus used is shown in fig. H7. Water at about 10° C is poured into the tall cylinder and the apparatus left to stand for several hours. It is then found that the top thermometer records 0° C and the surface layer of water may even freeze. The lower thermometer however records 4° C, showing that 4° C is the temperature of maximum density of water; water heated between 0° C and 4° C therefore contracts.

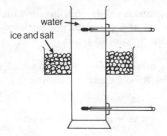

H7 Hope's experiment

horizontal component. The component of the Earth's magnetic field along the MAGNETIC MERIDIAN at a point on the Earth's surface.

horizontal pendulum. *See* PENDULUM.

horsepower. Symbol hp. The unit of power is IMPERIAL UNITS, equivalent to 745.7 watt.

hot. Highly radioactive.

hot atom. An atom in an excited state, or one with kinetic energy above the thermal level of its surroundings, usually produced as a result of nuclear processes.

hot cathode tube. A discharge tube in which thermionic emission from a heated element provides the electron beam for carrying the discharge.

hot wire ammeter. An ammeter in which the current to be measured is passed through a wire, so raising its temperature and causing it to expand. The expansion produces rotation of a pointer over a graduated scale. The measurement of either alternating or direct current is possible since the heating effect is proportional to the square of the current. Like any other ammeter it can be used as a voltmeter by incorporating a suitable series resistance.

hot wire anemometer. An instrument for measuring fluid velocity by the cooling effect produced in a heated wire due to the fluid movement over it. The change in temperature of the wire is obtained from the change in its resistance, which is measured by a WHEATSTONE BRIDGE; if the fluid velocity varies, a cathode ray OSCILLOGRAPH detector, rather than a galvanometer, is used with the bridge. The advantage of this type of anemometer is that the part of the instrument in contact with the fluid can be made much smaller than in other types.

hot wire gauge. A gas pressure depending on the cooling of a hot filament by the gas. *See* PIRANI GAUGE.

hot wire microphone. A device for measuring the intensity and amplitude of sound waves by recording the decrease in resistance of an electrically heated fine wire due to the impact of sound waves on the wire.

Hubble constant. Symbol H. The ratio of the recessional velocity of a galaxy, as measured by its RED SHIFT, to its distance. It is not generally accepted that H is a true constant; its value is between 1.6×10^{-18} per second and 3.2×10^{-18} per second. Provided the rate of expansion of the universe is constant, $1/H$ gives the age of the universe; the upper limit of the age is thus 20×10^9 years. Furthermore c/H, where c is the speed of light, gives the radius of the observable universe; the upper limit of the radius is thus 1.85×10^{26} metre.

hue. *See* COLOUR HUE.

hum. Extraneous alternating currents appearing in an amplifier output. They originate in an associated or nearby electric power circuit.

humidity. (1) *absolute humidity*. Symbol d. The mass of water vapour in unit volume of air. It is given by

$$0.002\ 17e/T$$

where e is the water vapour pressure and T the thermodynamic temperature.
  (2) *relative humidity*. Symbol U. The ratio of the actual absolute humidity to the absolute humidity of air saturated with water vapour at the same temperature. It is often expressed as a percentage. *See also* HYGROMETER; HYGROSCOPE.
  (3) *specific humidity*. The mass of water vapour present in air per unit mass of the moist air.

Hund's rules. Empirical rules for the sequence of the energy states of atomic terms due to a particular electronic configuration. With some reformulation they are also relevant to molecules.

hunting. The variation of a controlled quantity above and below the desired value.

Huygens' eyepiece. An eyepiece, illustrated in fig. H8, which is commonly used with telescopes and microscopes. The focal length of the planoconvex field lens A, the lens separation and the focal length of the planoconvex eye lens B are in the ratio $2:3:4$. The eyepiece is corrected for lateral but not longitudinal CHROMATIC ABERRATION; it also shows pin cushion DISTORTION and other aberrations.

Huygens' principle. Each point on a wave front can be considered as the origin of *secondary waves*, propagated only in the direction of transmission. The new wave front is the envelope of the secondary waves. The principle leads to correct prediction of the results of reflection, refraction, interference and diffraction of waves.

HWR. Abbrev. for HEAVY WATER REACTOR.

hybrid computer. *See* COMPUTER.

hybridization. A combination of two or more constituent orbitals of an atom to form *hybrids*, i.e. equivalent and directed orbitals. *See* ATOMIC ORBITAL.

hydrated electron. An electron trapped at the centre of a collection of water molecules. Such electrons may occur when an aqueous solution is irradiated with ionizing radiation; they are extremely reactive.

hydraulic press. A device, illustrated in fig. H9, which is used for applying pressure. Since liquids are very difficult to compress, pressure applied to surface a of

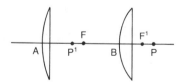

F and F¹ are focal points
P and P¹ are principal points

H8 Huygens' eyepiece

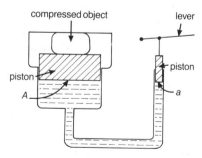

H9 Hydraulic press

the liquid is transmitted through it and so reappears over the considerably larger area A, giving a force there greater than that applied at a by a factor A/a.

hydrodynamics. The study of the motion of incompressible fluids.

hydroelectric power station. A power station in which hydraulic energy is converted to electric energy. Generally water acquires kinetic energy by falling from a higher to a lower level, then uses the energy to work a water turbine, which in turn drives a generator.

hydrogen bomb. *See* NUCLEAR WEAPON.

hydrogen electrode. An electrode system in which hydrogen is in contact with a solution of hydrogen ions. It comprises a HALF CELL in which platinum foil, immersed in a dilute acid, has hydrogen gas bubbled over it. The hydrogen electrode is used as a standard for measuring electrode potentials.

hydrogen spectrum. *See* BALMER SERIES; BRACKETT SERIES; LYMAN SERIES; PASCHEN SERIES. *See also* BOHR THEORY.

hydrometer. An instrument for measuring the RELATIVE DENSITY of a liquid. In a constant mass hydrometer, a weighted bulb with a long calibrated stem is used so that it always floats vertically when immersed in a liquid; the relative density is read off from the length of calibrated stem immersed. In a constant displacement hydrometer, such as *Nicholson's hydrometer*, scale pans are located at the top and bottom of a constant mass hydrometer and masses placed in them such that the instrument is always immersed to the same mark on the stem; the relative density is calculated from the values of the masses required. *See also* HARE'S APPARATUS.

hydrophone. An instrument for detecting underwater sounds.

hydrostatic balance. A beam BALANCE used to measure the RELATIVE DENSITY of a

substance. The balance is provided with a wooden bridge which is placed across but not touching one of the pans so that the substance is suspended directly from the balance arm above this pan. The substance is weighed first in air and then in water contained in a beaker standing on the bridge. The relative density of the substance can be calculated from the observations.

hydrostatics. The study of fluid equilibrium.

hygrometer. An instrument for measuring the relative HUMIDITY of air. *See* CHEMICAL HYGROMETER; HAIR HYGROMETER; WET AND DRY BULB HYGROMETER.

hygroscope. A device for roughly indicating the relative HUMIDITY of air. In one form a card impregnated with cobalt chloride is used: its colour changes from blue to pink when the moisture content of the atmosphere is sufficiently high.

hyperbola. A CONIC of eccentricity greater than 1. It is the locus of a point which moves so that the difference between its distances from the foci is constant. Its equation in rectangular Cartesian coordinates is

$$x^2/a^2 - y^2/b = 1$$

where a and b are constants; the plot is shown in fig. H10. The curves approach the asymptotes

$$y = \pm bx/a$$

H10 Hyperbola

hyperbolic. Having the shape of a HYPER-BOLA or HYPERBOLOID.

hyperbolic functions. The functions sinh, cosh, tanh, cosech, sech and coth of a variable. They are analogous to the TRIGONO-METRIC FUNCTIONS, thus

$$\tanh y = \sinh y/\cosh y \text{ etc.,}$$

where

$$\sinh y = (e^y - e^{-y}/2$$
$$\cosh y = (e^y + e^{-y})/2$$

hyperbolic logarithms. See LOGARITHM.

hyperboloid. A solid of revolution obtained by rotating a HYPERBOLA of equation

$$x^2/a^2 - y^2/b^2 = 1$$

about either the x or y axis. As shown in fig. H11, rotation about the y axis yields the one sheet form and rotation about the x axis yields the two sheet form. The term hyperboloid also covers solids whose appropriate cross sections are elliptical rather than circular.

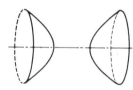

two sheets

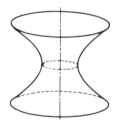

one sheet

H11 Hyperboloids

hypercharge. Symbol Y. A quantum number associated with an elementary particle and equal to the sum of BARYON NUMBER and STRANGENESS. Hypercharge is conser-

ved in STRONG INTERACTION and in ELEC-TROMAGNETIC INTERACTION but not in WEAK INTERACTION.

hyperfine structure. The various wavelengths differing by about 10^{-12} metre which appear when high-resolution apparatus, such as the FABRY-PEROT INTER-FEROMETER, is used to examine some radiation apparently monochromatic at low resolution. See also FINE STRUCTURE.

hypermetropia. An EYE defect in which parallel rays of light are focused to a point behind the retina when the eye is at rest, i.e. the eyeball is too short. Distant objects can be clearly seen by using the accommodating power of the eye, but for near objects a convex lens is required for distinct vision.

hypernuclens. See LAMBDA PARTICLE.

hyperon. Any elementary particle with half-integral spin, participating in STRONG INTERACTION and with mass greater than that of a proton, i.e. any BARYON which is not a nucleon. Lambda, sigma, xi and omega minus particles are hyperons. Any hyperon decays into a nucleon.

hypersonic velocity. A velocity greater or equal to five times that of sound.

hypocycloid. The locus of a point on the circumference of a circle which rolls on the inside of another circle.

hypothalamus. A region of the brain which regulates sweating. When the blood temperature increases, sweating is increased so more body heat is lost as the sweat evaporates. When a fall in blood temperature is detected, sweating is reduced. The system is thus an example of negative FEEDBACK.

hypothesis. A proposition for explaining observed facts. It may be either tentative or highly probable; if the former it is used as a guide to further investigations.

hypsometer. An apparatus used to determine the upper fixed point of a MERCURY

IN GLASS THERMOMETER M. As shown in fig. H12, M is held in steam above water boiling at atmospheric pressure. The outer jacket ensures that condensed steam flows back into the boiler. The thermometer is read when the mercury level becomes steady; the exact temperature at which this occurs is found from tables of pressure versus boiling point and so any necessary correction to the thermometer reading can be applied.

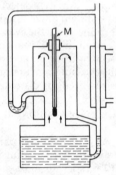

H12 Hypsometer

hysteresis. The phenomenon manifested by a system whose response to some stimulus depends on the previous history of the system.

Magnetic hysteresis is a phenomenon found in substances exhibiting FERROMAGNETISM: the magnetic flux produced by a given magnetizing field depends on the previous magnetization of the material. It can be demonstrated by means of a HYSTERESIS LOOP.

Electric hysteresis is shown by materials exhibiting FERROELECTRICITY. It is analogous to magnetic hysteresis; thus the curve displaying the magnitude of the electric displacement against the magnitude of the electric field strength has a similar form to the curve of the magnitude of magnetic induction against that of magnetic field strength. *See* HYSTERIS LOOP.

Torsional hysteresis is a phenomenon whereby the angle of twist of a wire for the same applied TORQUE depends on whether that torque is reached by increase or by decrease in torque.

hysteresis loop. The closed curve obtained by plotting MAGNETIC INDUCTION; magnitude B, of a ferromagnetic material against MAGNETIC FIELD STRENGTH, magnitude H, for numerically equal maximum and minimum values of H. It is shown in fig. H13 as a solid line; the dotted line OA is obtained only for initially unmagnetized material. APQB is obtained for H decreasing from A to B; BP'Q'A results from H increasing from B to A.

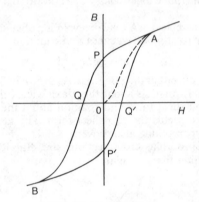

H13 Hysteresis loop

The area enclosed by the loop divided by 4π is the energy loss per unit volume in taking the specimen through the prescribed magnetizing cycle; it depends on the nature and heat treatment of the magnetic substance. On the plot, OP and OQ are respectively known as the REMANENCE and COERCIVITY of the material. For permanent magnets these quantities need to be large; however for magnetic materials used in generators transformer cores and electromagnets, low coercivity and energy loss, i.e. small area loops, are desirable.

hysteresis loss. The dissipation of energy which occurs in any type of HYSTERESIS.

I

ice. Water in the solid state. It is crystalline and birefringent and has several allotropic forms mostly only stable under high pressure.

iceland spar. Another name for CALCITE.

ice point. The temperature at which ice and water are in equilibrium at standard atmospheric pressure. It was used to define 0° C on the INTERNATIONAL TEMPERATURE SCALE but has now been replaced as a fixed point by the TRIPLE POINT of water, which is slightly higher than the ice point.

icosahedron. A POLYHEDRON having 20 faces. If these are congruent equilateral triangles, the icosahedron is said to be *regular.*

ideal crystal. A single crystal whose lattice has a perfectly regular structure, free from defects and impurities.

ideal gas. An idealized gas composed of atoms of negligible volume and all of whose collisions are perfectly elastic. An ideal gas conforms exactly to AVOGADRO'S HYPOTHESIS, BOYLE'S LAW, CHARLES' LAW, JOULE'S LAW and DALTON'S LAW OF PARTIAL PRESSURES. Real gases behave as ideal gases only at low pressure. *See also* EQUATION OF STATE.

identity. (1) *See* EQUATION.
(2) *See* GROUP.

IGFET. *See* FIELD EFFECT TRANSISTOR.

illuminance. Symbol E_v. The energy of light striking a surface per unit area per unit time, i.e. $d\Phi_v/dA$ where Φ_v is the LUMINOUS FLUX and A the area. *Compare* IRRADIANCE.

illumination. Another name for ILLUMINANCE.

image. A representation of an object. The image is said to be a *real image* if the rays forming it actually pass through it, i.e. are convergent; if they do not, but appear to diverge from the image, it is said to be a *virtual image.* A real image can therefore be focused on a screen whereas a virtual one cannot. Image formation by light can be investigated by the methods of both geometric and physical optics.
Image in a plane mirror. The image is the same size as the object, virtual, erect, situated as far behind the mirror as the object is in front and laterally inverted.
Image in a spherical mirror. For a convex mirror the image is always diminished, erect and virtual. For a concave mirror the nature of the image depends on the distance of the object from the mirror. For object distances greater than the focal length the image is real and inverted; for an object distance intermediate between the focal length and the radius of curvature the image is magnified, while for an object distance greater than the radius of curvature it is diminished. For object distances less than the focal length the image is virtual, erect and magnified.
To locate an image graphically it is only necessary to construct two rays: one ray is drawn from the object tip directed towards the centre of curvature of the mirror, which reflects it back along the same path; the other ray is drawn from the object tip parallel to the axis of the mirror which, if concave, reflects it to pass through focal point, and, if convex, reflects it so that it appears to come from the focal point. The constructions for the various cases are illustrated in fig. I1. The diagrams are only valid for rays close to the axis and making

141

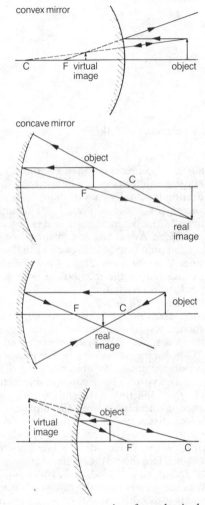

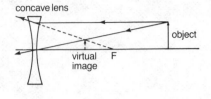

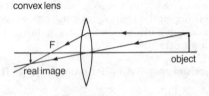

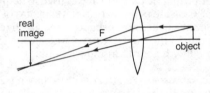

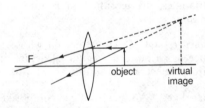

I1 Image construction for spherical
mirrors

I2 Image construction for thin spherical
lenses

small angles with it; they are not so drawn in the interests of clarity. The image position may also be found by calculation. *See* MIRROR FORMULA.

Image formed by a thin spherical lens. For a concave lens the image is always diminished, erect and virtual. For a convex lens, the nature of the image depends on the distance of the object from the lens. For object distances greater than the focal length the image is real and inverted; it is diminished for object distances greater than twice the focal length and magnified for object distances lying between the focal length and twice the focal length. For object distances less than the focal length the image is virtual, erect and magnified.

As shown in fig. I2, the image can be located by construction in a similar way to that described for an image in a spherical mirror and with the same reservations. The construction rays are the undeviated ray from the object tip through the centre of the lens, and the ray from the object tip parallel to the axis of the lens. The image position may also be found by calculation. *See* LENS FORMULA.

image converter. An electronic device for obtaining a visible image from an image formed by invisible radiation such as infrared. The radiation is focused on a photocathode so causing the release of electrons. The electrons are then attracted to a positively charged fluorescent screen anode, thus producing a visible image.

image intensifier. An apparatus for intensifying the brightness of an image. It works in a similar way to the IMAGE CONVERTER.

image orthicon. A type of camera tube. *See* TELEVISION.

imaginary number. A number that is the square root of a negative number. An imaginary number is written as i*a* where i is the square root of -1 and *a* is a real number.

immersion objective. A type of objective used in an optical MICROSCOPE. The part of the lens nearest the object is immersed in a liquid placed on top of the cover glass of the microscope specimen slide. The cover glass has a refractive index equal to that of the liquid, which is generally cedar wood oil or sugar solution. The effect of the liquid is to increase the NUMERICAL APERTURE of the objective and hence the resolution of the microscope.

impedance. (1) *electric impedance.* Symbol Z. A complex quantity which measures the ability of an alternating current circuit to resist the flow of current. Its magnitude equals the ratio of root mean square voltage to root mean square current. Z may be written as $R + iX$ where R is the resistance and X the reactance of the circuit and i is the square root of -1. The impedance magnitude is

$$(R^2 + X^2)^{1/2}$$

The phase angle θ between current and voltage is given by

$$\tan \theta = X/R$$

(2) *acoustic impedance.* A measure of the ability of a device carrying sound waves to resist the passage of the sound. It equals the product of the density of a medium and the speed of sound in it.

(3) *mechanical impedance.* A measure of the ability of a device supporting mechanical motion, such as a channel through which fluid flows, to resist the motion.

impedance matching. The selection of suitable impedance values for connected devices, such as circuits. For maximum efficiency in transferring power from one device to another, it is essential that their impedances should be carefully chosen rather than left to chance.

impedance plethysmography. A technique used to detect volume changes in conducting regions of the body by measuring the associated changes in electric IMPEDANCE. The main applications are the monitoring of breathing and of arterial blood flow. For breathing investigations leads are usually attached to the chest, and for blood flow measurements to the legs or arms. A low voltage of frequency in the range 15 kilohertz to 20 kilohertz is applied to one pair of electrodes, since the impedance change, which is anyway small, is then a maximum. To keep the current sufficiently low to be unnoticed by the patient, the signal from another pair of electrodes is amplified for observation.

imperial units. A system of weights and measures based on the pound as unit of mass and the yard as unit of length (*see* Table 6). The system is gradually being replaced by SI UNITS.

improper fraction. *See* FRACTION.

impulse. The time integral of the magnitude F of a force, i.e.

$$\text{impulse} = \int_0^t F \, dt$$

where t is the time for which the force acts. If F is very large and t is very short, the force is said to be *impulsive.* An impulse also equals the change in MOMENTUM which it produces.

impulse noise. *See* NOISE.

impulsive force. *See* IMPULSE.

impurity semiconductor. A SEMICONDUC-TOR containing foreign atoms, either naturally occurring or introduced, which have an overriding effect on the amount and type of conductivity. An impurity semiconductor is a type of extrinsic semiconductor.

incandescence. The emission of radiation, whose wavelengths lie in the visible region of the spectrum, by a substance at a high temperature.

inch. An imperial unit of length. *See* Table 6A.

inclination. Another name for DIP.

inclinometer. Another name for DIP CIRCLE.

inclined mirrors. Two plane mirrors set up as illustrated in fig. I3. By applying the laws of reflection it can be shown that, no matter what the value of the angle of incidence on the first mirror, a ray of light is deviated by twice the angle between the mirrors (i.e. by 2θ) after reflection at both of them in turn. On looking into the mirrors the number of images seen is $360/\theta - 1$; thus for mirrors inclined at 90° three images are seen. All the images lie on a circle centred on the meet of the mirrors.

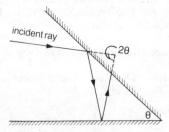

I3 Light deviation by inclined mirrors

incoherent radiation. Radiation that is not COHERENT. Radiation from any source other than a LASER is incoherent.

incoherent scattering. *See* SCATTERING.

independent variable. *See* DEPENDENT VARIABLE.

indeterminancy. *See* HEISENBERG UNCERTAINTY PRINCIPLE.

index. A number indicating the power to which a number of expression is raised. For example in y^7 the index is 7.

induced charge. Charge produced by the process of CHARGING BY INDUCTION.

induced electromotive force. Electromotive force produced in a conductor whenever the magnetic flux linked with it changes, either by actual change in the flux or by suitable movement of the conductor or both. *See* ELECTROMAGNETIC INDUCTION.

inductance. The measure of the ability of an electric circuit to resist the flow of a changing current as a result of an induced electromotive force opposing the current flow. *See* ELECTROMAGNETIC INDUCTION.

induction. Any change in a body due to the action of a field. *See* ELECTROMAGNETIC INDUCTION; ELECTROSTATIC INDUCTION; MAGNETIC INDUCTION.

induction coil. A device, illustrated in fig. I4, for producing a high voltage by ELECTROMAGNETIC INDUCTION. A primary coil of a few turns of insulated wire is wound on a laminated iron core; it is surrounded by a secondary coil of many turns of insulated wire. A low current, I_p, continuously interrupted by a contact breaker B, is supplied to the primary coil. This produces a voltage output V_s from the secondary coil. V_s consists of a succession of sharp pulses corresponding to breaks in the primary circuit, with much smaller inverse pulses in between corresponding to makes in the primary circuit. The value of capacitor C is chosen so as to just prevent sparking at the contacts and thus give the greatest possible secondary voltage. An important application of the induction coil is to produce the sparks in the sparking plugs of internal combustion engines.

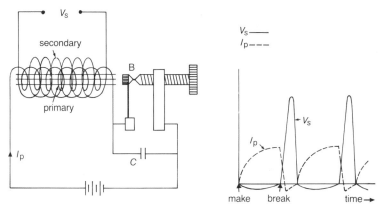

I4 Induction coil

induction heating. A method of heating metals by inducing EDDY CURRENTS in them using an alternating magnetic field.

induction motor. An alternating-current type of ELECTRIC MOTOR with an effectively rotating magnetic field. Such motors are nonsynchronous. *Compare* SYNCHRONOUS MOTOR.

inductive. Involving ELECTROMAGNETIC INDUCTION or inductance.

inductor. A choke or other component used to introduce inductance into a circuit.

inelastic collision. *See* COLLISION.

inelastic scattering. *See* SCATTERING.

inequality. A mathematical statement that one quantity is not the same as another. For example $a > b$ is an inequality signifying that a is greater then b; $a < b$ is an inequality signifying that a is less than b.

inertia. The tendency of a body to resist changes in its motion. *See also* MASS.

inertial force. The FORCE which when compounded with the vector sum of the applied forces acting on an accelerating body yields a zero resultant. Thus in the equation $F = ma$, where F is the force (a

vector) acting on a body of mass m moving with acceleration a (a vector), the inertial force is $-ma$.

inertial frame of reference. A frame of reference that remains constant. It is usually a co-ordinate system based on the fixed stars.

inferior planet. A planet nearer the Sun than is Earth. *Compare* SUPERIOR PLANET.

information theory. An analytical technique for establishing the minimum but sufficient amount of information required to solve a specified problem in communication or control.

infrared astronomy. The study of INFRARED RADIATION from space. Some of the infrared radiation is absorbed in the atmosphere, but there are atmospheric windows in the near infrared at the following wavelengths, all expressed in micrometre: 1.15–1.3, 1.5–1.75, 2–2.4, 3.4–4.2, 4.6–4.8, 8–13 and 16–18. For longer-wavelength observations the instrumentation requires transport above the atmosphere by either balloon, rocket or satellite. The infrared sources may be thermal or electronic.

infrared radiation. Radiation of wavelength range 0.73 micrometre to 1 millimetre. It is roughly classified into the *near* *infrared*, wavelength range 0.73 micrometre to 75 micrometre, and the *far infrared*, wavelength range 75 micrometre to 1 millimetre. Emission of infrared radiation involves changes in the VIBRATIONAL ENERGY of atoms and molecules. When infrared radiation is absorbed, it causes an increase in vibrational energy of the absorbing body, i.e. the body becomes hotter; for this reason infrared radiation is sometimes known as *heat radiation* or as *radiant heat*.

infrared spectrometer.   An instrument used for producing an infrared spectrum. For wavelengths up to about 2 micrometre, an optical spectrometer with a thermopile, bolometer, semiconductor or photographic detector is suitable. At longer wavelengths, glass is too absorbent and is replaced by prisms and lenses of quartz (up to 4 micrometre), or of fluorite (up to 10 micrometre), or of rocksalt (up to 15 micrometre) or of sylvin (up to 23 micrometre). For work at even longer wavelengths, a reflecting concave DIFFRACTION GRATING mounted in vacuo is required.

infrared spectrum.   An emission or absorption spectrum in the infrared region. The spectra are widely studied in the determination of molecular vibrational frquencies.

infrasonic.   Denoting or using frequencies of sound below the lowest audible frequency, which is about 20 hertz.

infrasound.   Sound of INFRASONIC frequencies.

injector.   A device for mixing fluids, using the principle that the pressure of a fluid flowing through a constricted region in a pipe is less than the pressure in the unconstricted part. A simple form is shown in fig. I5. Fluid X flows through a nozzle into a constriction in main tube T at such a rate that the pressure at Z falls below atmospheric; the side tube opening into a region at atmospheric pressure thus permits a second fluid Y to be pulled into the tube. An important application is in the

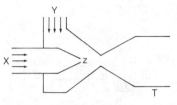

I5   Injector principle

administration of for example an air-oxygen mixture to a patient.

input.   (1) The signal, current, voltage etc. fed into an electric circuit or device.
(2) The information fed into a computer: this may be by punched card, paper tape, magnetic tape magnetic disc or conventional printed characters.
See also OUTPUT.

instantaneous value.   The value of a changing physical quantity at a particular instant of time.

instrumental error.   A random ERROR due to instrumental imperfection.

insulator.   A substance which is a poor conductor of electricity or of heat.

integral calculus.   The branch of calculus concerned with INTEGRATION.

integral equation.   An equation containing an integral which involves an unknown function. The equation is said to be linear only if it is linear in the unknown function. Solution of the equation requires the determination of the function.

integrated circuit.   An electronic circuit made in a single small unit. For example a network of junction or field effect transistors interconnected by thin metal films all formed on the surface of a SILICON CHIP constitutes an integrated circuit.

integration.   The determination of the sum of a number of entities, one of whose measurements is tending to zero. Thus in the plot of y against x illustrated in fig. I6,

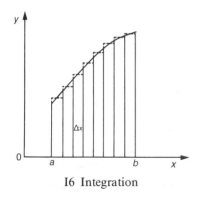

I6 Integration

the area under the curve between $x = a$ and $x = b$ is approximately the sum of the areas of rectangles of equal width Δx into which the curve may be divided as indicated. When Δx tends to zero the sum is written

$$\int_a^b y\mathrm{d}x$$

known as the *integral* of y with respect to x over the range $x = a$ to $x = b$. Integration is the inverse operation of DIFFERENTIATION.

intensity. A quantitative measure of the strength of a physical effect. For a beam of radiation, such as heat, light, sound, subatomic particles etc., it is the amount of radiant energy crossing unit area per second, the area being perpendicular to the direction of propagation. *See also* ELECTRIC INTENSITY; MAGNETIC INTENSITY.

interaction. *See* ELECTROMAGNETIC INTERACTION; GRAVITATIONAL INTERACTION; STRONG INTERACTION; WEAK INTERACTION.

interest formulae. An amount of money P invested for n year at $r\%$ per annum interest becomes

$$P(1 + r/100)^n$$

if the interest is compound and

$$P(1 + (r/100)n)$$

if the interest is simple.

interference. (1) A phenomenon which is exhibited by PROGRESSIVE WAVES and results from the interaction of wave trains of the same phase (i.e. coherent wave trains) and of the same frequency.

Interference of sound waves may be demonstrated as shown in fig. I7. The sound waves are generated by two loudspeakers connected in parallel to an audio frequency amplifier. An observer moving along the line AB experiences alternate zones of comparative silence S and of loud sound L. The S regions occur when the path difference to them of the sound waves from the two loudspeakers is an odd number of half wavelengths; the waves are thus out of phase and therefore cancel each other. For the L regions, the path difference is a whole number of wavelengths and so the waves are in phase and therefore reinforce each other. *See also* QUINCKE'S TUBE.

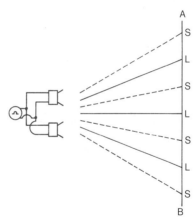

I7 Sound wave interference

For interference of light waves, demonstrations are provided by the FRESNEL BIPRISM, NEWTON'S RINGS. YOUNG'S FRINGES. colours of thin films etc.

Interference of water waves may be demonstrated using a RIPPLE TANK.
(2) The disturbance caused by undesired signals, atmospherics, hum, whistle etc. in a communication system. It often arises due to the presence of nearby electrical apparatus.

interference filter. An optical filter which depends on the INTERFERENCE of light in a thin film to give selective transmission of a narrow wavelength band.

interference microscope. *See* MICROSCOPE.

interferometer. Any instrument or apparatus designed to produce interference fringes, for example the FABRY-PEROT INTERFEROMETER. The instruments are used to measure wavelengths and small distances, to test the flatness of surfaces etc.

intermediate vector boson. A generic name for the two W PARTICLES and the Z PARTICLE.

intermolecular energy. A form of energy of two main types: *thermal*, which depends on the kinetic energy of the molecules, i.e. on the temperature, and *potential*, which is due mainly to electrical interaction between molecules. The latter is approximately of the form

$$a/r^p - b/r^q$$

for two molecules distance r apart; a and b are constants and p and q are integers depending on the nature of the substance. For an ionic solid such as sodium chloride, $p = 9$ and $q = 1$. The relative magnitudes of the two types of intermolecular energy determine the PHASE of a substance.

intermolecular force. For a pair of molecules, separation r, a force equal to $-dV/dr$, where V is the molecular potential energy. For the two ions referred to in the entry INTERMOLECULAR ENERGY, the force is

$$9a/r^{10} - b/r^2$$

which is illustrated in fig. I8. The equilibrium separation r_0 occurs at the absolute zero of temperature; at higher temperatures the ions oscillate about a position corresponding to r somewhat greater than r_0, the actual position depending on the amount of expansion which occurs. The approximate straight line character of the graph in the region of r_0 is in agreement with HOOKE'S LAW: the point Z beyond which the magnitude of the force decreases with increasing separation corresponds to the BREAKING STRAIN region.

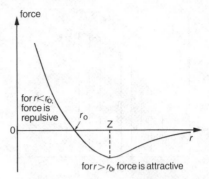

for $r < r_0$, force is repulsive

for $r > r_0$, force is attractive

I8 Intermolecular force variation

internal conversion. A process whereby an excited atomic nucleus returns to its ground state by transferring its excess energy to a bound electron, usually in a K, L or M shell, of its atom rather than by gamma ray emission. The excited electron is then ejected from the atom as a *conversion electron*, leaving behind an excited ion which usually returns to the ground state with the emission of an AUGER EFFECT electron or an X ray photon.

internal energy. Symbol U. The sum of the KINETIC and POTENTIAL ENERGIES of all the atoms and/or molecules of a system. The internal energy is the part of the energy of a system which is determined only by the state of the system. The absolute value of the internal energy of any system is unknown, but changes in its value can be determined: (*see* THERMODYNAMICS (first law)). A change in internal energy depends only on the initial and final states of a system, not on how the change took place.

internal resistance. The electric resistance within a source of electric current such as a cell or generator. It is given by

$$(E - V)/I$$

where E is the electromotive force and V the potential difference across the source when supplying a current I.

internal work. The work done in separating the molecules of a system against their intermolecular forces. For an IDEAL GAS the value is always zero.

international temperature scale. A practical temperature scale based on the meaning of THERMODYNAMIC TEMPERATURE and employing experimentally determined values of particular temperatures, i.e. primary fixed points, and specified experimental methods for measuring other temperatures. The 11 defined fixed points are shown in the table. Specified instruments are the platinum resistance thermometer below 630° C, a platinum-platinum/rhodium thermocouple in the range 630° C to 1064° C and a radiation pyrometer above 1064° C.

International Practical Temperature Scale

fixed point	TK	C
triple point of hydrogen	13·81	−259·34
temperature of hydrogen with vapour pressure 25 76 atmosphere	17·042	−256·108
b.pt. of hydrogen	20·28	−252·87
b.pt. of neon	27·102	−246·048
triple point of oxygen	54·361	−218·789
b.pt. of oxygen	90·188	−182·962
triple point of water	273·16	0·01
b.pt. of water	373·15	100
m.pt. of zinc	692·73	419·58
m.pt. of silver	1235·08	961·93
m.pt. of gold	1337·58	1064·43

interpolation. The estimation of the value of a function of a variable for a value of the variable lying between values for which the function value is known. The process may be done by calculation using an interpolation formula, or graphically. *See also* EXTRAPOLATION.

interrupter. A device for periodically stopping the flow of direct current so as to produce pulses.

interstellar dust. Dust existing in deep interstellar space and often seen as dark nebulae.

interstellar gas. Gas, mostly low density hydrogen but with traces of other elements, existing in deep interstellar space.

interstitial position. *See* DEFECT.

interval. In special RELATIVITY the quantity

$$c^2(\Delta t)^2 - (\Delta x)^2 - (\Delta y)^2 - (\Delta z)^2$$

where c is the speed of light and Δx, Δy, Δz and Δt are respectively the differences in the x, y, z and time co-ordinates of two specified events. In general relativity an interval is a generalization of the above.

intrinsic semiconductor. A pure SEMICONDUCTOR with equal electron and hole densities. In practice it does not exist, but the term is applied to nearly pure materials.

invar. A nickel-iron alloy containing 36% nickel. Its coefficient of thermal expansion is very small and so it is often used in clock manufacture.

invariant system. A system with no DEGREES OF FREEDOM.

inverse square law. Any law in which the magnitude of a physical quantity at any point is proportional to the reciprocal of the square of the distance of the point from the source of the physical quantity. Examples are NEWTON'S LAW OF GRAVITATION, LAMBERT'S LAW and COULOMB'S LAW.

inversion. (1) A mathematical transformation.
(2) A reversal in the usual direction of a process, for example the behaviour of water when heated from 0° C to 4° C (*see* ANOMALOUS EXPANSION OF WATER).
(3) The transformation of an optically active substance into one having the opposite rotatory effect without change of chemical composition.
(4) A form of speech scrambling.
(5) The production of a layer of opposite type to that of the main body in the surface of a semiconductor.

inversion symmetry. The theory that physical laws are invariant under the mathematical operation of INVERSION. WEAK INTERACTION violates the theory.

inversion temperature. (1) *See* JOULE-KELVIN EFFECT.

(2) The temperature of the hot junction of a thermocouple, whose cold junction is at 0° C, at which the direction of the thermal electromotive force reverses.

involute. *See* EVOLUTE.

iodopsin. A protein containing a retinene group found in human retinal receptors. Since its absorption spectrum does not match any visual response curve, its function is not at present understood. *See also* COLOUR VISION.

ion. An atom or goup of atoms with a net charge. Positive ions have a deficiency of electrons and negative ions have an excess.

ion engine. A rocket propulsion engine in whch a stream of ions is ejected from the exhaust jet by a high electric field.

ionic bond. Another name for ELECTROVALENT BOND.

ionic crystal *See* CRYSTAL.

ion implantation. The bombardment of a SEMICONDUCTOR with ions of sufficiently high energy to penetrate the surface. Controlled DOPING of the semiconductor may thus be achieved.

ionization. The process of forming ions from neutral atoms or molecules. Ionization may occur spontaneously when a substance is dissolved. Other methods of ion formation include electron capture, bombardment with ionizing radiation, photon bombardment and thermal ionization.

ionization chamber. An instrument for detecting IONIZING RADIATION. It consists of two electrodes of opposite charge mounted in a chamber containing a gas. When ionizing radiation enters the chamber, the gas atoms or molecules are ionized and the ions flow to the electrode of opposite sign. This causes an ionization current which, under suitable conditions, is proportional to the intensity of the radiation. *See also* GEIGER COUNTER.

ionization gauge. A vacuum pressure gauge consisting of a three-electrode thermionic tube fused to the gas system whose pressure is to be measured; the electrical connections are as illustrated in fig. I9. Electrons which pass through the grid ionize gas molecules positively so that they travel to the negatively charged plate. This produces a plate current which is a measure of the number of gas molecules present. Pressures as low as 10^{-3} pascal can be measured.

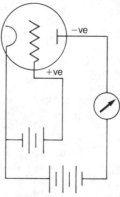

I9 Ionization gauge

ionization potential. Symbol I. The minimum energy required to remove the least strongly bound electron to an infinite distance from its parent atom and so produce a positive ion. The minimum energy required to remove the second least strongly bound electron from a neutral atom or molecule is called the *second ionization potential* and so on.

ionizing radiation. A stream of particles or a beam of electromagnetic radiation which can cause ionization in passing through a medium. For this to happen, the photon energy of the radiation or the kinetic energy of the particles must exceed the IONIZATION POTENTIAL of the substance under bombardment. Examples of ionizing radiation are gamma rays, X rays, electrons, protons and alpha particles.

ionosphere . A region of ionized air and free electrons extending from a height of 50 kilometre to 1000 kilometre above the Earth's surface. It is caused by X ray and short-wave ultraviolet ionizing radiation from the Sun. Since the free electrons in the ionosphere reflect radio waves, radio transmission is possible between points hidden from each other by the Earth's curvature.

The ionosphere is divided into three layers: the *D layer* between 50 kilometre and around 90 kilometre; the *E layer*, alternatively known as the *Heaviside layer*, extends between about 90 kilometre and about 150 kilometre; the *F layer*, also known as the *Appleton layer*, exists between around 150 kilometre and 1000 kilometre. The electron concentration is lowest in the D layer and highest in the F layer. Since no solar radiation arrives at night, the thickness of the layers varies from night to day. It also shows seasonal and latitude variations. *See also* ATMOSPHERIC LAYERS.

ionospheric wave. Another name for SKY WAVE.

ion pump. A pump for the production of a very high vacuum. The residual gas is ionized by an electron beam and the resulting positive ions are collected at a cathode where they remain trapped. The pump is only useful at a pressure of less than 1 micropascal. To overcome the problem of cathode saturation by ions, the *sputter ion pump* has been evolved. In this device electrodes of titanium, which are sputtered by a discharge, are used and so a film of titanium which acts as a GETTER is continuously produced.

IR. Abbrev. for INFRARED.

iris. *See* EYE.

iris diaphragm. An adjustable aperture for controlling the light entering an optical instrument.

irradiance. Symbol *E*. The energy of radiation striking a surface per unit area per unit time, i.e. $d\Phi/dA$ where Φ is the RADIANT FLUX and *A* the area. *Compare* ILLUMINANCE.

irradiation. Exposure to radiation, either by accident or intent.

irrational number. A number which cannot be expressed as the ratio of two integers.

irregular galaxy. A GALAXY with no discernible symmetry in shape or structure and of a size less than average.

irreversible process. A process such that, when concluded, the system involved cannot be returned to its original thermodynamic state. All natural physical processes are irreversible.

isobar. (1) A line on a weather map passing through places of equal atmospheric pressure.
(2) Any of a number of nuclides having the same MASS NUMBER but of different ATOMIC NUMBER.

isocline. A line on Earth connecting points having the same magnetic DIP.

isodiapheres. Nuclides having the same difference between total neutron and total proton numbers. For example $^{190}_{78}Pt$ and $^{186}_{76}Os$ both show a difference of 34.

isodynamic line. A line on Earth connecting points with the same HORIZONTAL COMPONENT.

isoelectronic. Having the same number of electrons.

isogonal line. A line on Earth connecting points with the same magnetic DECLINATION.

isomers. Atomic nuclei with the same MASS NUMBER and the same ATOMIC NUMBER but different energy states.

isometric change. A change taking place at constant volume.

isomorphous replacement technique. A method used in X ray diffraction analysis to establish the structure of complicated molecules for which the HEAVY ATOM TECHNIQUE is of little use. The method consists in making two compounds identical apart from the replacement of a particular heavy atom by another; by studying the X ray patterns from the two compounds it is possible to deduce the position of the replaceable atoms.

isosceles triangle. A triangle with two sides equal.

isospin. Symbol I A quantum number associated with a MULTIPLET. Allowed isospin values are

$$0, 1/2, 1, 3/2, ...$$

the number being the same for each member of a multiplet. Individual members are distinguished by different values of the *isospin quantum number*, I_3, which can have values

$$-I, -I + 1, ..., I - 1 \text{ and } I$$

Values of I and I_3 are shown in the table for some multiplets. In general,

$$I_3 = Q - Y/2$$

where Q and Y are respectively the charge and HYPERCHARGE of an elementary particle. I and I_3 are both conserved in STRONG INTERACTION; for ELECTROMAGNETIC INTERACTIONS, I_3 is conserved but I is not. The use of the word 'spin' implies only an analogy to angular momentum, which isotopic spin formally resembles.

isospin quantum number. *See* ISOSPIN.

isotherm. A line joining points on the Earth's surface at which the temperature is the same.

isothermal. A line joining all points on a graph that correspond to the same temperature.

isothermal bulk modulus. The BULK MODULUS of elasticity for constant temperature conditions.

isothermal process. A process occurring at constant temperature. *Compare* ADIABATIC PROCESS.

isotones. Nuclides with the same neutron number but different atomic numbers. For example $^{39}_{19}K$ and $^{40}_{20}Ca$ each have 20 neutrons.

isotopes. Nuclides having the same atomic number but different mass numbers, for example $^{235}_{92}U$ and $^{238}_{82}U$.

isotopic number. The difference between the number of neutrons and the number of protons in a nucleus.

isotopic spin. Another name for ISOSPIN.

isotopic weight. The relative atomic mass of an isotope.

isotropic. Possessing properties which are independent of direction. Thus an isotropic crystal has the same physical properties along all its axes.

Some multiplet I and I_3 values.

Particle group	name	symbol	I	I_3
nucleon	neutron	n	$\frac{1}{2}$	$-\frac{1}{2}$
	proton	p	$\frac{1}{2}$	$\frac{1}{2}$
meson	pion	π^-	1	-1
	pion	π^0	1	0
	pion	π^+	1	1
baryon	sigma	Σ^-	1	-1
	sigma	Σ^0	1	0
	sigma	Σ^+	1	1

J

J. The symbol for the MECHANICAL EQUIVALENT OF HEAT.

Jaeger's method. A practical method of measuring SURFACE TENSION of a liquid at various temperatures. An air bubble is formed inside the liquid on the end of a glass CAPILLARY tube of radius *r*. The excess pressure inside the bubble is measured and the surface tension γ determined from the formula

$$\gamma = rg(h\rho - h_1\rho_1)/2$$

h_1 is the depth of the end of the capillary below the surface of the liquid of density ρ_1 at temperature t_1; h is the pressure gauge reading, ρ the density of the pressure gauge liquid and g the acceleration due to gravity. By repeating the experiment for various values of t_1, the variation of surface tension with temperature for the liquid can be determined.

jamming. Radio reception interference such that the desired signal is unintelligible.

Jansky noise. A high-frequency static disturbance of cosmic origin.

JET. Abbrev. for JOINT EUROPEAN TORUS.

jet propulsion. Forward propulsion produced by one or more jets of high-velocity hot gas issuing from backwardly directed nozzles.

jet tone. The variable hiss accompanying the emergence of a stream of moving air into still air.

JFET. *See* FIELD EFFECT TRANSISTOR.

jj coupling. *See* COUPLING (def. 1).

Joint European Torus. A FUSION REACTOR under development at Culham, UK by several European countries for controlled fusion reaction research.

J meson. Early name for J/PSI PARTICLE.

Joly's steam calorimeter. An apparatus for measuring the SPECIFIC HEAT CAPACITY C_v of a gas at constant volume. Two copper spheres, as nearly identical as possible, are suspended from the opposite arms of a beam BALANCE. The spheres are surrounded by an enclosure through which steam can be passed. One sphere is evacuated and the other filled with a mass m of the sample gas at high pressure. When both spheres have reached steam temperature t_s, the excess mass of water m_s condensed on the gas-filled sphere is recorded. Then

$$C_v = m_s l/m(t_s - t_r)$$

where l is the SPECIFIC LATENT HEAT of steam and t_r is room temperature. To compensate for any differences in the spheres, a repeat control experiment is performed with both spheres evacuated; m_s is corrected by any difference in the masses of steam condensing on the two spheres.

The spheres are shielded from drops of water falling from the enclosure roof. Any drops falling from the spheres are trapped and so included in the mass measurement.

Josephson effect. The occurrence in two superconductors, separated by a thin dielectric, of an oscillatory current of frequency proportional to a steady potential difference applied between the conductors.

Joule. Symbol J. The SI unit of energy, equal to the work done when the point of application of a force of magnitude one newton is moved one metre along the line of action of the force. The joule is also the work done by a current of one ampere flowing through a resistance of one ohm for one second.

Joule heating. The amount of heat produced in a conductor by an electric current flowing through it. If a steady current of I ampere flows for t second through a resistance of R ohm, the heat produced is I^2Rt joule.

Joule-Kelvin coefficient. A quantity associated with the JOULE-KELVIN EFFECT and equal to the rate of change of temperature with pressure at constant enthalpy.

Joule-Kelvin effect. The change in temperature of a gas when undergoing a process of ADIABATIC expansion. It may be demonstrated experimentally by allowing a volume V of air at room temperature and high pressure p to flow through a porous plug into a vacuum; by the air temperature is observed to fall by about 0.1 K per atmospheric pressure difference.

The ENTHALPY of a gas is constant. Therefore at initial gas temperature if pV increases (decreases) as p decreases (increases), then the INTERNAL ENERGY U decreases (increases) with p. A decrease in U is associated with a fall, and an increase in U with a rise in gas temperature. Deviation from BOYLE'S LAW may thus produce either heating or cooling. In contrast, deviation from JOULE'S LAW produces only cooling because the POTENTIAL ENERGY of the attractive forces between the gas molecules is increased by increased volume or by decreased pressure.

The gas temperature at which the net effect is zero is known as the *inversion temperature*. It is pressure dependent and for a given gas can have two values for the same pressure. Below the lower (higher) inversion temperature there is a cooling (heating) effect, and above it a heating (cooling) effect. The inversion temperatures of nitrogen at 100 atmospheres pressure are about 100 K and 580 k.

Since an IDEAL GAS obeys both Boyle's law and Joule's law it would not exhibit the Joule-Kelvin effect.

Joule's equivalent. Another name for MECHANICAL EQUIVALENT OF HEAT.

Joule's law. At constant temperature the INTERNAL ENERGY of a gas is independent of volume.

Joule-Thomson effect. Another name for JOULE-KELVIN EFFECT.

J/psi particle. A MESON of mass 3 gigaelectronvolt, composed of a charmed QUARK and a charmed antiquark. Its discovery gave support to ELECTROWEAK THEORY and the CHARM concept, and impetus to the development of STRONG INTERACTION theory.

JUGFET.   See FIELD EFFECT TRANSISTOR.

junction diode.   Another name for semiconductor diode. *See* DIODE.

junction transistor.   Short for bipolar junction transistor. *See* TRANSISTOR.

Jupiter.   A GIANT PLANET of diameter 142 800 kilometre. Its mass is 318 times that of the Earth but its density only one fifth of Earth's. Its gravity is 2.5 times and its magnetic field about 17 000 times those of Earth. Jupiter is 778 million kilometre from the Sun and is the fifth planet out from it. Its orbital and average axial rotation periods are 11.86 year and 9 hour 51 minute respectively. Jupiter's atmosphere consists of hydrogen, helium, methane, ammonia and nitrogen. The southern hemisphere is characterized by what is known as the *Great Red Spot*. The planet and its satellites have been the subject of investigation by US space probes.

Juvin's rule.   The formula

$$h = 2\gamma \cos \alpha/(rg\rho)$$

where h is the difference in height between the level of liquid inside and outside a vertical open-ended capillary tube, internal radius r, standing in the liquid; ρ and γ are respectively the density and surface tension of the liquid, g is the acceleration due to gravity and α is the ANGLE OF CONTACT between the liquid surface and the capillary tube wall. For liquids which wet the tube, α is less than 90° and so cos α is positive; these liquids therefore rise in a capillary tube. For liquids which do not wet the tube, α is greater than 90° and so cos α is negative; such liquids therefore fall below the general liquid level outside the tube.

K

k. The symbol for the BOLTZMANN CONSTANT.

kaon. Another name for K MESON.

Kater's pendulum. A reversible compound PENDULUM used for the accurate measurement of the ACCELERATION DUE TO GRAVITY.

K capture. *See* CAPTURE.

keeper. A small piece of iron or steel used to complete a magnetic circuit and so prevent loss of magnetism while magnets are in store.

Kellner eyepiece. A variety of RAMSDEN EYEPIECE having a cemented eye lens which corrects for chromatic abberation and distortion more successfully than the original The eyepiece is mainly used in prism binoculars.

kelvin. Symbol K. The unit of THERMODYNAMIC TEMPERATURE in SI units. It is defined as 1/273.16 of the thermodynamic temperature of the triple point of water. The unit equals one degree on the Celsius scale of temperature, i.e.

$$1 \text{ K} = 1^\circ \text{ C.}$$

Kelvin balance. A type of AMPÈRE BALANCE.

Kelvin contacts. A method of testing electronic circuits and components whereby the effect of lead resistance on the measurements is eliminated.

Kelvin double bridge. A type of WHEATSTONE BRIDGE, illustrated in fig. K1, used for accurate measurement of a low resistance R. The resistance R' is of the same

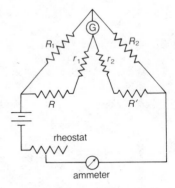

K1 Kelvin double bridge

order as R. The relationship

$$R_1/R_2 = r_1/r_2$$

is maintained throughout but its value is changed until the galvanometer G indicates zero current. The ratio value is then R/R'. The effects of errors due to contact and lead resistances are eliminated.

Kelvin effect. Another name for THOMSON EFFECT.

Kelvin's formula. The formula

$$T = 2\pi(LC)^{1/2}$$

where T is the period, L the inductance and C the capacitance in an electric circuit of negligible resistance.

Kennelly-Heaviside layer. Another name for Heaviside layer. *See* IONOSPHERE.

Kepler's laws. Three laws which describe *planetary motion.* They are as follows.
  1. The orbit of every planet is an ELLIPSE with the Sun at one focus of the ellipse.
  2. The radius vector, i.e. the line drawn

from Sun to planet, sweeps out equal areas of space in equal times.

3. The square of the time taken by a planet to complete an orbit is proportional to the cube of the semimajor axis of the orbit.

Kepler telescope.   A refracting astronomical TELESCOPE whose objective and eyepiece are both convex lens systems. When used in normal adjustment the focal points of the two systems coincide. To convert the instrument to a terrestrial telescope an erecting prism is inserted in the system.

kerma.   The ratio of the sum of the initial kinetic energies of charged particles indirectly produced by ionizing radiation in a small mass of substance, to that mass.

Kerr cell.   *See* KERR EFFECT.

Kerr effect.   (1) An electro-optical effect in which DOUBLE REFRACTION is induced in some liquids and gases when subjected to a strong electric field perpendicular to the direction of illumination. If n_1 and n_2 are respectively the refractive indices for the ordinary and extraordinary rays, then

$$n_1 - n_2 = k\lambda E^2$$

where λ is the wavelength of the radiation used, E, is the electric field strength magnitude and k is *Kerr's constant.* A practical application of the effect is in the *Kerr cell*, which is a transparent device containing a suitable liquid and two electrodes and which behaves as a shutter: a beam of plane polarized light incident on a Kerr cell can be stopped by applying a suitable voltage across the electrodes.

(2) A magneto-optical effect in which a beam of plane polarized light striking the polished pole face of an electromagnet shows slight elliptical polarization after reflection there.

Kerr's constant.   *See* KERR EFFECT (def. 1).

Kew magnetometer.   A MAGNETOMETER employing a steel tube as magnetic needle. A convex lens is mounted at one end of the tube and a graduated transparent scale, which lies in the focal plane of the lens, at the other end. The scale is observed through a telescope in normal adjustment. The instrument is used to make accurate measurements of the Earth's magnetic field.

kilo-.   (1) Symbol k. A prefix meaning 1000.

(2) A prefix used in computing to mean 2^{10}, i.e. 1024.

kilocycle.   A former unit of frequency equal to 1000 hertz.

kilogramme.   Symbol kg. The basic SI unit of mass. *See* Table 6D.

kilohertz.   Symbol kHz. A unit of frequency equal to 1000 hertz.

kilometre.   Symbol km. A unit of length equal to 1000 metre. *See* Table 6A.

kilowatt.   Symbol kW. A unit of power equal to 1000 watt.

kilowatt hour.   Symbol kWh. A unit of energy equal to a power of 1 kilowatt available for 1 hour.

kinematics.   A branch of mechanics dealing with motion without reference to mass or force.

kinematic viscosity.   The ratio of the coefficient of VISCOSITY of a fluid to its density.

kinetic energy.   Symbol T. The energy stored in a system due to the movement of masses within the system and measured by the work necessary to bring the system to rest. A body of mass m and speed v has a kinetic energy of $mv^2/2$. A body whose moment of inertia and angular speed about an axis of rotation are respectively I and ω has a kinetic energy of $I\omega^2/2$.

kinetic energy density.   The KINETIC ENERGY per unit volume.

kinetics. (1) The study of rates of chemical reactions. It yields information on reaction mechanisms.

(2) The DYNAMICS of material bodies.

kinetic theory of matter. A theory relating the motion of individual molecules to the macroscopic properties of a substance. For solids, intermolecular forces are so large that molecular motion is mainly confined to vibration about a fixed position. Application of kinetic theory leads to DULONG AND PETIT'S LAW. For liquids, attractive forces between molecules are smaller than for solids so that the molecules move around at random mainly inside the liquid (*see* EVAPORATION).

For gases, the intermolecular forces are even smaller so that the molecules occupy all the available space. Gas pressure is interpreted in terms of the incessant impacts of the gas molecules on the container walls. At each impact a reversal of momentum occurs, leading for an IDEAL GAS to the expression $\rho C^2/3$ for the gas pressure where ρ is the gas density and C the root mean square velocity of the molecules. The pressure expression is consistent with the experimentally established GAS LAWS if the thermodynamic temperature of the gas is proportional to C^2; this is a very reasonable assumption since heat is a form of energy and kinetic energy is proportional to the square of the speed. The distribution of velocities of gas molecules is given by the MAXWELL-BOLTZMANN DISTRIBUTION LAW. The kinetic theory predicts the relationship

$$\gamma = 1 + 2/n$$

where γ is the RATIO OF SPECIFIC HEATS of the gas and n the number of DEGREES OF FREEDOM for each gas molecule. Further evidence of molecular agitation is provided by DIFFUSION and BROWNIAN MOVEMENT.

Kirchoff's law. (1) The EMISSIVITY of a body equals its ABSORPTANCE at the same temperature.

(2) Either of two laws applying to the flow of current in a network. They are as follows.

1. The algebraic sum of the currents which meet at any point in a network is zero.

2. In any closed electric circuit, the algebraic sum of the products of current and resistance in each part of the network is equal to the electromotive force in the circuit.

Klein-Gordon equation. A relativistic form of the SCHRÖDINGER EQUATION, used in nuclear quantum theory:

$$\nabla^2\psi + [(E - V)^2 - m^2c^4]\psi/(h^2c^2) = 4\pi\rho$$

where ψ is the Schrödinger wave function, E the total particle energy, V the potential energy, m the rest mass of the particle, c the speed of light in vacuo, h the Planck constant and ρ a quantity proportional to the nucleon density.

klystron. *See* VELOCITY MODULATION.

K meson. A variety of strange MESON. *See* STRANGENESS.

knot. A unit of speed equal to 1.15 mile per hour. It is used for expressing the speed of ships and aircraft.

Knudsen flow. Another name for MOLECULAR FLOW.

Knudsen gauge. A device used to measure very low gas pressures for which the mean free path of the molecules is large compared with the apparatus dimensions. Gas molecules, after striking electrically heated stationary plates, temperature T_1, and then a cooler rotatable vane structure of temperature T_2, produce a resultant torque on the vane. The pressure can be calculated from the observed vane deflection θ and equals

$$k\theta/[(T_1/T_2)^{1/2} - 1]$$

where k depends on the torsional constant of the vane suspension.

Kramer's theorem. The lowest energy level in a paramagnetic material is at least two-fold degenerate if the magnetic ions have an odd number of electrons.

krypton. Symbol Kr. A gas used in
FLUORESCENT LAMP and LASER manufacture. The gas is obtained from the atmosphere as a byproduct of the liquefaction
of AIR.

Kundt's rule. The refractive index of a
medium does not vary continuously with
wavelength in the region of absorption
bands. *See* ANOMALOUS DISPERSION.

Kundt's tube. An aparatus, shown in fig.
K2, which is used in investigations on the
speed of sound. The inside of tube T is
sprinkled with a little dry powder such as
lycopodium. Rod AB is clamped at its midpoint so that the disc at B just clears the
sides of the tube. When the rod is stroked
with a cloth in the direction BA, it vibrates
longitudinally and a high-pitched note is
heard; disc B acts as a vibrating source of
the same frequency and so a sound wave
travels through the tube and is reflected
back from end C. Adjustment of the rod in

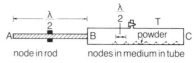

node in rod   nodes in medium in tube

K2 Kundt's dust tube

the direction BC will yield a position in
which the standing wave in the air in T
causes violent agitation of the lycopodium
powder, which then settles into small
definite heaps at the nodes; the average
distance between these is measured and
equals half a wavelength in air. It can be
shown that

$$lV_r = l_r V_a$$

where l_r and l are the length of the rod and
the internode distance respectively and V_r
and V_a are the sound speeds in rod and air
respectively. By using rods of various
materials and liquids or gases other than
air in the tube, and speeds of sound in
many media may be compared.

L

label. A radioactive atom in a molecule, used to monitor the behaviour of the molecule.

ladder filter. A series of identical four-terminal symmetrical networks connected together to yield a TRANSMISSION LINE with continuously repeated impedance sections.

laevorotatory. Denoting a substance which imparts an anticlockwise rotation, as seen by an observer facing the light source, to the plane of polarization of polarized light. *See* POLARIZATION (electromagnetic).

lag. (1) The interval of time, or the angle, by which a specific phase in one periodically varying quantity is delayed with respect to the same phase in a similar quantity. There may for example be a lag between an alternating current and the electromotive force producing it, or vice versa. *Compare* LEAD.
(2) The time interval between transmission and reception of a signal.

lagging load. A load (2) carrying a CURRENT lagging behind the electromotive force producing it. An example is an INDUCTOR.

Lagrange's equations. Second-order differential equations expressing the relationship between the LAGRANGIAN FUNCTION L of a system of particles, the generalized co-ordinates q_i, the generalized forces Q_i and the time t:

$$\mathrm{d}/\mathrm{d}t[\partial L/\partial \dot{q}_i] - \partial L/\partial q_i = Q_i$$

where $i = 1, 2, ..., n$ and n is the number of degrees of freedom of the system.

Lagrangian function. Symbol L. The kinetic energy of a system minus its potential energy.

Lalande cell. A primary CELL having zinc and iron electrodes, a caustic-soda solution electrolyte and using copper oxide for depolarization.

lambda particle. An uncharged elementary particle classified as a HYPERON of mass 2183 times the electron mass. It can replace a neutron in a nucleus yielding an extremely unstable *hypernucleus*.

lambda point. (1) The temperature 2.186 K at which HELIUM types I and II are in equilibrium.
(2) The temperature at which a sharp maximum occurs in the specific heat capacity of a substance; for example the temperature of the change from a ferromagnetic to a paramagnetic state.

lambert. An obsolete unit of luminance, equal to the luminance of a surface emitting one lumen per square centimetre.

Lambert's law. (1) The luminous intensity in any direction of a small element of a perfectly diffusing surface is proportional to the cosine of the angle between the normal and the direction.
(2) The temperature at which a sharp maximum occurs in the specific heat capacity of a substance; for example the temperature of the change from a ferromagnetic to a paramagnetic state.

Lamb shift. A small energy difference between the energy levels of the $^2P_{1/2}$ and $^2S_{1/2}$ states of hydrogen, which arises from interaction between the electron and the radiation field.

laminar flow. The steady flow of a fluid in parallel layers with little or no mixing between adjacent layers. *Compare* TURBULENT FLOW.

lamination. A form of construction used for the cores of transformers, transducers, relays, chokes and similar alternating current apparatus. The core is made of thin strips of surface-oxidized or varnished iron or steel so that it presents a high resistance to EDDY CURRENTS.

Lamy's theorem. For a particle in equilibrium under the action of three forces of magnitude A, B and C

$$A/\sin \alpha = B/\sin \beta = C/\sin \gamma$$

where α, β and γ are respectively the angles between the lines of action of the forces B and C, A and C and A and B.

Landau damping. The damping of a space charge oscillation by a stream of particles moving at a speed slightly less than .the phase speed of the associated wave.

Landé factor. Symbol g. A factor introduced into the theory of electron energy changes in a magnetic field in order to obtain agreement with experimental results such as the anomalous ZEEMAN EFFECT. The value of g ranges from 1 for pure electron orbital momentum to 2 for pure spin momentum.

Landé interval rule. For sufficiently weak spin-orbit interaction, an atomic energy level splits into levels such that the interval between successive ones is proportional to the larger of the total angular momentum values.

Langmuir effect. The ionization occurring when atoms of low IONIZATION POTENTIAL come into contact with hot metal of high WORK FUNCTION. The effect can be used to produce intense ion beams of alkali metals.

Laplace equation. (1) The equation

$$\nabla^2 V \equiv \partial^2V/\partial x^2 + \partial^2V/\partial y^2 + \partial^2V/\partial z^2 = 0$$

where V is a POTENTIAL and DEL squared (∇^2) is the *Laplace operator* or *Laplacian*; there are equivalent equations in spherical and cylindrical co-ordinates.
(2) The equation

$$c = (\gamma p/\rho)^{\frac{1}{2}}$$

where c is the speed of sound in a gas of density ρ, pressure p and ratio of specific heats γ.

Laplace operator. *See* LAPLACE EQUATION (def. 1).

lapse rate. The rate of change of atmospheric temperature with altitude.

Larmor precession. The PRECESSION of the orbit of a charged particle when subjected to a magnetic field; the precession occurs about the direction of the field. For an electron revolving about a nucleus, the angular velocity of Larmor precession is

$$eH/(2mc)$$

where c is the speed of light, H the magnitude of the magnetic field strength and e and m respectively the electron charge and mass.

laser. A source of intense monochromatic coherent radiation in the ultraviolet, visible and infrared regions of the spectrum. The radiation results from STIMULATED EMISSION, the resulting photon repeating the process so that, provided enough excited atoms are available, a narrow beam of monochromatic radiation is produced. To ensure that there are sufficient excited atoms – a condition known as *population inversion* – power has to be supplied for example by OPTICAL PUMPING.
Gaseous, liquid and solid lasers have been devised. In the solid *ruby laser* chromium ions are optically pumped, i.e. excited by an intense light flash applied as shown (fig. L1), to achieve population inversion and stimulated emission is triggered by spontaneous emission of wavelength 694.3 nanometre. The full mirror reflects most of the light incident on it back along the tube, and the partial

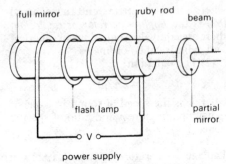

full mirror   ruby rod   beam

flash lamp   partial mirror

V

power supply

L1 Ruby laser

mirror reflects part of it. The part it transmits is the usable laser beam of peak power between 10 kilowatt and 100 kilowatt. The reflected beams produce further stimulated emission. In a *gas laser* population inversion is achieved by continuous electric discharge in a GAS DISCHARGE TUBE. In *semiconductor lasers* it is accomplished by recombination emission.

Since their introduction in 1960, lasers have been used in welding, surgery, holography, printing, optical communications and digital information reading; space defence applications are being studied. The name laser is an acronym for light amplification by stimulated emission of radiation.

latent heat. Symbol *L*. The total heat absorbed or produced during a change of phase at constant temperature. *See also* SPECIFIC LATENT HEAT.

latent image. *See* PHOTOGRAPHY.

lateral chromatic aberration. *See* CHROMATIC ABERRATION.

lateral magnification. Another name for MAGNIFICATION.

latitude. The angle between the plane of the equator and the normal to the Earth at the point whose latitude is required.

lattice. (1) A regular array of points in two and three dimensions. When atoms, molecules or ions are situated at such points, a CRYSTAL SYSTEM results.

(2) A regular pattern of fissile material and moderator in some NUCLEAR REACTOR TYPES.

lattice constants. A specification of the size and shape of the UNIT CELL of a crystal structure by the lengths of the cell edges and the sizes of their angles of intersection.

lattice energy. The energy per ion pair of an ionic crystal required to separate the crystal into individual ions at an infinite distance apart at absolute zero of temperature.

lattice vibrations. The periodic vibrations of atoms, ions or molecules in a crystal lattice about their mean positions. The amplitude of the vibrations increases with temperature and becomes so large at the melting point of the crystal that the lattice breaks down.

Laue diagram. The diffraction pattern which results when a beam of X rays or of some other kinds of radiation or particles is passed through a thin crystal on to a photographic plate behind the crystal. The type of crystal and its structure can be deduced from the pattern.

layer lattice. A type of crystal lattice in which the atoms in the layers are strongly bound but the bonding between layers is weak. Graphite for example has a layer lattice.

LCD. Abbrev. for LIQUID CRYSTAL DISPLAY.

LCM. Abbrev. for LEAST COMMON MULTIPLE.

LD 50. Short for median lethal dose. *See* DOSE.

lead acid battery. A type of ACCUMULATOR consisting of two lead plates dipping into dilute sulphuric acid. Discharge results in both plates being covered with lead sulphate; on charging the reaction is reversed. Each cell gives an electromotive force of about 2 volt. A 12 volt battery of six cells is commonly used in motor vehicles.

lead. (1) An electrical conductor.

(2) The angle or interval of time by which a particular phase of one periodically varying quantity is in advance of a similar phase in another such quantity. There may for example be a lead between an alternating current and the electromotive force producing it, and vice versa. *Compare* LAG.

lead equivalent. The thickness of metallic lead that would, under the same conditions, give the same protection against radiation as the material under consideration.

leading current. An alternating current that has a LEAD with respect to the electromotive force producing it.

leading voltage. An electromotive force that has a LEAD with respect to the current to which it gives rise.

leakage. The flow of a small electric current, known as a *leakage current,* through imperfect insulation.

leap year. *See* TIME.

least action principle. When a conservative dynamical system passes from one configuration to another, the ACTION of the system is a minimum.

least common multiple. The smallest number that every number of a given set of numbers will divide into exactly. For example the lowest common multiple of 2, 3 and 5 is 30.

least distance of distinct vision. The smallest distance of an object from the eye for clear vision of the object. For a young adult the average value is around 25 centimetre.

least energy principle. For stable equilibrium the total potential energy of a system must be a minimum.

least squares method. The determination of the most likely value from a set of observations by assuming that the sum of the squares of the DEVIATION of each observed value from the most likely value is a minimum.

least time principle. *See* FERMAT'S PRINCIPLE OF LEAST TIME.

Le Chatelier's principle. If a change is imposed on a system in equilibrium the system will alter in such a way as to counteract the change. The principle has wide application; an example is Lenz's law (*see* ELECTROMAGNETIC INDUCTION).

Leclanché cell. A primary CELL with a zinc rod cathode which is surrounded by a porous bag of manganese-dioxide/graphite mixture to prevent polarization. A carbon rod is used as anode and the electrolyte is between 10% and 20% ammonium chloride solution. The cell has an electromotive force of about 1.5 volt and in its dry form is extensively used. *See* DRY CELL.

LED. Abbrev. for LIGHT EMITTING DIODE.

LEED. *See* ELECTRON DIFFRACTION.

Lees disc method. A method of measuring the THERMAL CONDUCTIVITY of a poor conductor in the form of a thin disc of thickness d and face area A. The disc is held between two slabs of metal, the upper one being heated by contact with a steam chamber. The reading of a thermometer in each slab is recorded when it becomes constant; the rate of heat passage through the disc then equals the rate of heat loss from the bottom slab. To measure this the temperature of the bottom slab is raised above its previous value and a cooling curve plotted for it; its gradient $d\theta/dt$ at the steady temperature of the bottom slab can then be determined. The thermal conductivity is given by

$$(Msd\ d\theta/dt)/(A\ \delta\theta)$$

where M is the mass and s the specific heat capacity of the bottom slab and $\delta\theta$ is the difference of steady readings of the thermometers.

left-hand rule. *See* FLEMING'S RULES.

lemma. A result proved as a preliminary to the proof of a theorem.

length. *See* METRE.

lens. A piece of transparent substance, usually glass, plastic or quartz, with one or two curved surfaces. These surfaces are generally spherical (i.e. portions of spheres) although other shapes, for example cylindrical and parabolic, are used for special purposes. Various types of spherical lens are illustrated in fig. L2. The various types of spherical *converging lenses* are the *biconvex lens*, the *plano-convex lens* and the *concavo-convex lens*; these lenses bring the rays of a parallel beam of light to a real focus, i.e. they converge the beam. The various types of spherical *diverging lenses*

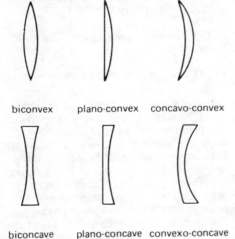

biconvex   plano-convex   concavo-convex

biconcave   plano-concave   convexo-concave

L2 Types of lens

are the *biconcave lens*, the *plano-concave lens* and the *convexo-concave lens*; these lenses cause the rays of a parallel beam of light to diverge, i.e. to appear to come from a virtual focus. *See also* IMAGE (formed by a thin spherical lens); LENS FORMULA; OPTICS SIGN CONVENTIONS.

lens formula. The formula

$$1/f = 1/u + 1/v$$

which on the sign convention real is positive and for rays close to and making small angles with the axis, applies to thin spherical lenses; f is the focal length of the lens and u and v are respectively the object and image distances from the lens. *See also* CENTRED OPTICAL SYSTEM; OPTICS SIGN CONVENTIONS.

lenticular. Shaped like or relating to a lens.

Lenz's law. *See* ELECTROMAGNETIC INDUCTION.

lepton. A collective name for ELEMENTARY PARTICLES which do not take part in STRONG INTERACTION. All leptons are FERMIONS. The known leptons are the electron, the muon, the neutrinos and the tau lepton together with their antiparticles, i.e. the positron, positive muon, antineutrinos and the positive tau. Three quantum numbers, known as *lepton numbers*, designated l_e, l_μ and l_τ are assigned as shown in the table. In weak interactions the sum of the l_e values for all the particles taking part is conserved, as is also the sum of the l_μ or l_τ values. Leptons do not have a QUARK substructure.

Lepton numbers

particle	symbol	particle	symbol	lepton numbers		
				l_e	l_μ	l_τ
electron	e	neutrino	ν_e	1	0	0
positron	e$^+$	anti neutrino	$\bar{\nu}_e$	−1	0	0
muon	μ^-	neutrino	ν_μ	0	1	0
positive muon	μ^+	anti neutrino	$\bar{\nu}_\mu$	0	−1	0
tau	τ	neutrino	ν_τ	0	0	1
positive tau	τ^+	anti neutrino	$\bar{\nu}_\tau$	0	0	−1

lepton number. *See* LEPTON.

Leslie's cube. A large cubic metal container with its four vertical sides having different colours and/or finishes on the outside. When filled with boiling water, it can be used to demonstrate the effect of surface nature on the emission of radiant heat by presenting each face in turn to a radiation detector at a constant distance.

LET. Abbrev. for LINEAR ENERGY TRANSFER.

lethal radiation dose. *See* DOSE (median lethal).

level. (1) The ratio of a quantity value to a reference value of the quantity.
(2) The logarithm of the ratio of a quantity value to a reference value of the quantity.

lever. A rigid bar which can turn about a pivot. The relative positions of load, effort and pivot determine the type.

levitation. The suspension of an object without visible support. *Electromagnetic levitation* is produced by generating eddy currents in an electric conductor by a time-varying magnetic field; it is the basis of a prototype cost-effective transport system.

Leyden jar. A glass jar with internal and external coatings of tin foil. The inner coating is charged by connection to a suitable voltage source and the outer coating is earthed. The glass acts as a DIELECTRIC. The whole jar can be regarded as a parallel plate CAPACITOR.

LF. Abbrev. for low frequency. *See* FREQUENCY BAND (table).

lift pump. A device in which a piston containing a valve V is moved up and down a cylinder by an external handle. At the base of the cylinder is another valve V′ leading to a pipe whose other end is in the liquid to be raised. On raising the piston, atmospheric pressure forces liquid into the cylinder past V′; on the downstroke water is forced through V and is delivered from the top of the cylinder on the next upstroke. The device is widely used for raising water from wells of depth less than the maximum height of water (approximately 10 metre) which atmospheric pressure can support.

light. ELECTROMAGNETIC RADIATION which produces a visual sensation when incident on the human retina. The wave length range of light is around 400 nanometre to 740 nanometre (*see* LUMINOSITY CURVES). The nature of light has been the subject of many theories, such as CORPUSCULAR THEORY, QUANTUM OPTICS and QUANTUM ELECTRODYNAMICS.

light chopper. *See* SPLIT BEAM SPECTROPHOTOMETER.

light emitting diode. A rectifying semiconductor device that differs from a normal semiconductor DIODE in that its electron-hole recombination process is associated with light emission. The electrical connections and physical construction of a light emitting diode are illustrated in fig. L3. The device is forward biased and a resistance is used to limit the current; light is emitted through the thin p-type region which therefore has minimum obstruction. The colour of the light varies with the semiconductor material used: gallium arsenide phosphide gives red or yellow light while gallium phosphide gives green. Since there is no time lag in the light production, rapid switching is possible. Light emitting diodes are widely used for displaying letters and numbers in self-luminous digital instruments.

light guide. A single fibre or array of fibres in a FIBRE OPTICS system.

lightness. The property of a colour determined by the amount of light it reflects. Colours of the same hue but different lightness are known as *shades*.

lightning. A high-energy luminous electric discharge between a charged cloud and a point on or connected to the Earth's surface: this is known as *forked lightning*.

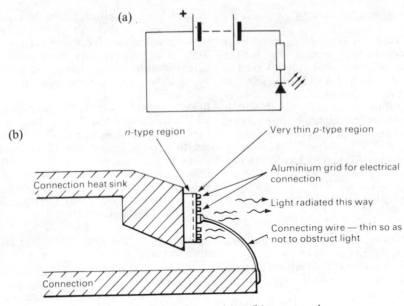

L3 LED (a) connections (b) construction

Alternatively the discharge may occur between two charged clouds or between oppositely charged layers of the same cloud: this is known as *sheet lightning*. The potential difference required to initiate a flash is about 10^8 volt. Generally there is a downward leader stroke, i.e. partial discharge, followed by an upward return stroke, the latter being much more luminous. The average current in a stroke is about 10 000 ampere but maximum values of around 20 000 ampere, associated with a temperature of about 30 000 K, have been obtained. A typical lightning flash consists of four or five strokes at about 40 millisecond apart.

Ball lightning comprises a small slowly moving luminous ball of plasma which is said to vanish with a loud bang. This type of lightning is the rarest and least understood.

lightning conductor. A sharply pointed metal rod attached to the top of a building and connected to the Earth's surface. A sharp point in the region of a charged cloud becomes strongly charged by INDUCTION and so ionizes the air above it. A

discharge between the cloud and the conductor therefore takes place more slowly and less violently than would a discharge between the cloud and Earth.

light pen. A device connected to the VISUAL DISPLAY UNIT of a computer. It is capable of sensing the information on the screen and can be used to impart information by 'drawing' lines on the screen.

light-sensitivity materials. Substances which absorb visible and ultraviolet photons with the production of excited electronic states or electron emission. *See* PHOTOELECTRIC EFFECT.

light year. A unit of length equal to the distance travelled by electromagnetic radiation in vacuo in one year, i.e. to

$$9.4607 \times 10^{15} \text{ metre}$$

It is used for expressing astronomical distances. *See also* PARSEC.

limit. The value approached by a mathematical function as its independent variable approaches some specified value.

limiting angle of prism. The largest angle of a prism of given transparent material, at its refracting edge, for which an emergent ray can be obtained. Its value is about twice the CRITICAL ANGLE for the prism material.

limit of resolution. Another name for RESOLVING POWER.

Linde process. A process for liquefying air by expanding it through a nozzle (*see* JOULE-KELVIN EFFECT). The cooled air is used to cool the incoming compressed air so that eventually liquefaction temperature is reached.

linear. (1) Characterized by one dimension only.
(2) Arranged in, involving or represented by a straight line.

linear absorption coefficient. *See* ABSORPTION COEFFICIENT.

linear accelerator. A type of particle ACCELERATOR in which electrons or protons are accelerated along a straight evacuated chamber. In older low-energy machines, the particles were accelerated in a *drift tube*, i.e. a series of cylindrical electrodes separated by gaps: a radio frequency electric field was applied to the cylinders whose lengths were such that the particles were accelerated at each gap, the velocity then remaining constant until the next gap. Modern high-energy machines do not use drift tubes but are usually travelling wave accelerators: particle acceleration is produced by the electric component of a travelling wave set up in a WAVEGUIDE; typical rates of energy gain are 7 mega-electronvolt per metre for electrons and 1.5 mega-electronvolt per metre for protons.

linear acoustics. *See* ACOUSTICS (def. 1).

linear amplifier. An amplifier whose output is directly proportional to its input.

linear attenuation coefficient. Symbol μ. A coefficient given by

$$(d\Phi/dl)/\Phi$$

where $d\Phi/dl$ is the rate of change of energy flux Φ with distance l traversed by the flux. The flux changes on its journey through a medium because of absorption and scattering by the medium.

linear energy transfer. The energy transferred per unit of path length by a charged particle of specified energy to atoms and molecules along its path. Linear energy transfer is proportional to the square of the charge on the particle and increases with decreasing particle velocity. The concept is mainly used in radiation protection work.

linear expansivity. Another name for coefficient of linear expansion. *See* COEFFICIENT OF EXPANSION.

linear momentum. Symbol *p*. A vector quantity given by the product of the mass of a particle and its velocity. The linear momentum of a body or particle system is the vector sum of the linear momenta of the individual members. *See also* CONSERVATION LAW.

linear motion with constant acceleration. Motion described by the four equations:

$$v = u + at$$
$$v^2 = u^2 + 2ax$$
$$x = ut + at^2/2$$
$$x = (u + v)t/2$$

where *u* and *v* are the initial and final linear speeds, *x* is the distance travelled in time *t* and *a* is the magnitude of the constant linear acceleration; speeds and distances in the direction opposite to that of the acceleration are taken as negative.

linear motor. A type of INDUCTION MOTOR in which the stator and rotor are linear and parallel rather than cylindrical and coaxial as in the conventional rotary machine. The purpose of the motor is to produce force or motion in a straight line.

linear relationship. A relationship between two variables that can be represented by a straight line plot.

linear scale. A scale with equally spaced intervals of equal value.

line defect. *See* DEFECT.

line frequency. The number of lines scanned per second by the electron beam in the cathode ray tube of a TELEVISION receiver.

line of force. An imaginary line whose direction at any point along its length is that of the force field at that point. Examples are electric lines of force, magnetic lines of force and gravitational lines of force.

line printer. A device which prints the output of a computer a whole line at a time rather than printing individual characters. Line printer speeds up to 50 lines a second are obtainable.

line spectrum. A SPECTRUM consisting of discrete lines. Such spectra result from the emission or absorption of photons due to electron transitions between different energy levels in atoms. *See also* SPECTRAL SERIES.

linkage. The amount of magnetic flux embraced by an electric circuit.

liquefaction of gases. The changing of substances from the gaseous to the liquid state. If the gas temperature is below the CRITICAL TEMPERATURE the gas can be liquefied by pressure alone; otherwise the LINDE PROCESS, or adiabatic expansion or adiabatic desorption may be employed.

liquid. A phase of matter intermediate between a gas and a solid and characterized by ease of flow and incompressibility. A liquid offers little resistance to shear stress and takes the shape of its container; unlike a gas it does not change its volume to fill the container. The intermolecular forces in liquids are larger than those in gases but smaller than those in solids. Diffraction studies show that liquids possess a short-range structural regularity which extends over several molecular diameters; these ordered bundles move around relative to each other. Theories of liquids are less well developed than those of solids and gases.

liquid air. A pale blue liquid resulting from the liquefaction of air. The blue colour is due to the liquid oxygen present.

liquid crystal. A substance which has the flow properties of a liquid but which has a more orderly arrangement of molecules than does a liquid. Substances forming liquid crystals have long molecules whose arrangement determines the type of liquid crystal. Liquid crystals have the property of plane polarizing light and their polarizing properties are affected by an electric field.

liquid crystal display. A digital display unit used for example in watches and calculators and consuming very little power. In the display unit a thin film of LIQUID CRYSTAL is sandwiched between transparent electrodes, each of which is glass thinly coated with metal as illustrated in fig. L4. The pattern in which the film is printed produces the required display. The planes of polarization of the polarizers (not shown) on either side of the liquid crystal display are crossed. As a result light falling on an unenergized liquid crystal is largely reflected back and it appears transparent. When an electric field is applied to particular components of a digit, part of the crystal is energized and that part appears dark due to the change in polarization produced in it; a digit is therefore seen. The circuitry is such that the required digits are rapidly connected in turn, and, through persistence of vision, six digits appear to be seen at the same time.

liquid drop model. A model of the nucleus in which the nucleons are represented by molecules of liquid, and the whole nucleus by a liquid drop. Just as the shape of a drop is maintained by SURFACE TENSION due to

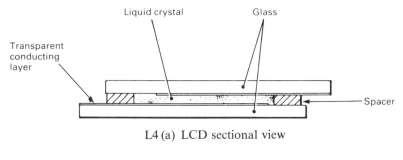

L4 (a) LCD sectional view

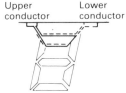

L4 (b) Transparent conductors for one segment

the interaction of molecules near the surface, so the shape of the nucleus is considered to be maintained by the forces of interaction between surface nucleons. The model is most applicable to heavy nuclei and is used in nuclear FISSION theory.

liquid helium. *See* HELIUM.

liquid in glass thermometer. *See* THERMOMETER.

liquid pressure. *See* PRESSURE.

Lissajous' figures. Traces made by a point which moves with simultaneous simple harmonic motions in two perpendicular directions. The figures may be observed on a cathode ray oscilloscope when the spot is subjected simultaneously to one sinusoidal signal horizontally and another one vertically. Fig. L5 shows the patterns obtained for vibrations of the same amplitude but different frequency ratios and phase differences; the frequency ratio is that of vertical to horizontal signal frequency.

litre. Symbol l. A metric unit of volume equal to 1 cubic decimetre.

Lloyd's mirror. A plane mirror used to produce INTERFERENCE fringes: monochromatic light from a slit suffers grazing reflection at the mirror and then interferes with light coming directly from the slit.

load. (1) The power delivered by an electric machine, circuit or device.

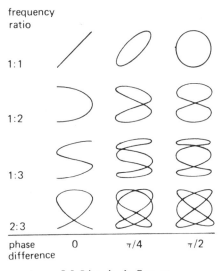

L5 Lissajou's figures

(2) The recipient or dissipator of electric power.

(3) *See* MACHINE.

loaded concrete. Concrete containing material of high atomic number such as barium, iron or lead. Its main use is as a radiation shield for nuclear reactors.

local fallout. *See* FALLOUT.

locus. The curve traced by a point moving so as to satisfy a given condition. For example a circle is the locus of a point which moves so that its distance from a fixed point, the centre, is constant.

lodestone. Another name for MAGNETITE.

logarithm. The power to which a number, called the *base*, has to be raised in order to equal a given number. For example any number x may be written as y^n; n is the logarithm to the base y of x, i.e.

$$n = \log_y x$$

When $y = 10$ the logarithms are called *common logarithm*. When $y = e = 2.718\,28...$ the logarithms are called *natural logarithms*, or *hyperbolic logarithms* or *Naperian logarithms* and are written as $\log_e x$ or $\ln x$. The logarithm of the product (ratio) of two numbers is the sum (difference) of the logarithms of the numbers. The logarithm of the power of a number is the product of the power and the logarithm of the number. Tables of logarithms to base 10 generally only contain the logarithms of numbers between 0 and 10. Logarithms of other numbers are obtained by adding the logarithm of the appropriate power of 10; thus

$$\log 200 = \log 100 + \log 2$$
$$= 2 + 0.301 = 2.301$$

and

$$\log 0.02 = \log 10^{-2} + \log 2 = \bar{2}.301$$

so written since only the 2 is negative, i.e.

$$\bar{2}.301 = -1.699$$

The number(s) to the left of the decimal point are known as the *characteristic* and the decimal part as the *mantissa*. Since the widespread adoption of electronic calculators the use of logarithm tables has declined, but the underlying theory is still important.

logarithmic. Denoting a relationship in which one variable is proportional to the LOGARITHM of another.

logarithmic series. The expansion

$$\log (1 + x) = x - x^2/2 + x^3/3 \,... \, (-1)^n x^{n-1}/n$$

$$\text{for } -1 < x \leqslant 1$$

logic circuit.. A circuit designed to perform a particular logical function such as 'and', 'either ... or', 'neither ... nor'. Normally the circuits are *binary* logic circuits, i.e. they operate between two distinct voltage levels; such circuits are extensively used in computers and are usually formed from an INTEGRATED CIRCUIT assembly. A collection of logic circuits is known as *logic network*.

logic network. *See* LOGIC CIRCUIT.

lone pair. A pair of electrons with opposite spins but occupying the same ATOMIC ORBITAL.

longitude. The angular distance between the Earth's meridian at the point under consideration and the standard meridian, which is a great circle passing through the poles and Greenwich and whose longitude is assigned the value 0°.

longitudinal chromatic aberration. *See* CHROMATIC ABERRATION.

longitudinal wave. A wave in which the vibrations of the transmitting medium lie along the direction of travel of the wave. An example is a sound wave.

long sight. Another name for HYPERMETROPIA.

Lorentz-Fitzgerald contraction. A contraction in length by a factor of

$$(1 - v^2/c^2)^{1/2}$$

in the direction of a body's motion, where v is the speed of the body relative to the ether and c is the speed of light. The contraction was postulated in order to account for the results of the MICHELSON-MORLEY EXPERIMENT. It is also predicted by the special theory of RELATIVITY.

Lorentz force. The force acting on a moving charge in a magnetic field.

Lorentz transformation. A set of equations relating space and time co-ordinates x', y', z', and t' in a frame of reference moving with constant speed v along the axis of another frame to the co-ordinates in this other frame. Each observer is considered to be at the origin of co-ordinates of the appropriate frame, and the observers are assumed to coincide at $t = 0 = t'$. The equations are:

$$x' = \beta(x - vt)$$
$$y' = y$$
$$z' = z$$
$$t' = \beta(t - vx/c^2)$$

where $\beta = (1 - v^2/c^2)^{1/2}$ and c is the speed of light. The equations are used in the special theory of RELATIVITY.

Lorenz method for resistance. An absolute method of determining resistance; the apparatus used is shown in fig. L6. Disc D is rotated at constant angular speed ω inside coil C and so an electromotive force E is produced as indicated. For zero deflection of galvanometer G, E must balance the potential drop across resistance R.

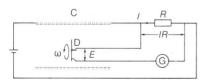

L6 Absolute method of resistance measurement

Thus

$$4\pi n I r^2 \omega \times 10^{-7}/2 = IR$$

where I is the current through the coil, r its radius and n its number of turns. Hence

$$R = 2\pi n r^2 \omega \times 10^{-7}$$

Loschmidt's number. Symbol L. The number of molecules in one cubic centimetre of an IDEAL GAS at standard temperature and pressure. Its value is

$$2.687\,19 \times 10^{19}$$

loudness. The magnitude of the sensation resulting from a sound reaching the ear. The relationship between loudness and intensity is complicated, depending on the sound frequency and duration; loudness is however very approximately proportional to the logarithm of intensity.

loudspeaker. A device for converting changing electric currents into sounds. In the most common form a small coil of wire is attached to a cardboard cone, the coil being situated in the field of a strong permanent magnet. The coil therefore vibrates when alternating current passes through it. These vibrations are transmitted to the cone and thus produce sound. For good speech reproduction the device should respond to frequencies in the range 150 hertz to 8000 hertz. For good music reproduction response in the 20 hertz to 20 000 hertz range is required.

low frequency. *See* FREQUENCY BAND (table).

low tension. Low voltage.

LS coupling. *See* COUPLING (def. 2).

LT. Abbrev. for LOW TENSION.

lubrication. The process of introducing a substance between solid surfaces in relative motion in order to reduce friction, wear, overheating and rusting. Oils and greases of various types are widely used but at high temperatures graphite is more suitable. Air bearings are being increasingly used but involve continuous pumping of the air to the bearings.

lumen. Symbol lm. The SI unit of LUMINOUS FLUX, equal to the luminous flux emitted by a uniform point source of intensity one CANDELA in a cone of unit solid angle.

luminance. Symbol L_v. The product of the luminous intensity per unit area of a surface viewed from a particular direction and the secant of the angle θ between the surface and that direction. Thus

$$L_v = \sec θ \; dI/dA$$

where I is the luminous intensity and A the area of the surface. The unit is candela per square metre.

luminescence. The emission of electromagnetic radiation from a substance as a result of any nonthermal process. Luminescence results from electron transitions in excited atoms and ions. Excitation may be produced by electromagnetic radiation, electronic bombardment, biological reactions, friction, radioactive decay and chemical reactions; the corresponding types of luminescence are respectively PHOTOLUMINESCENCE, ELECTROLUMINESCENCE, BIOLUMINESCENCE, TRIBOLUMINESCENCE, RADIOLUMINESCENCE and CHEMILUMINESCENCE. The radiation accompanying the return of the electrons to the ground state may be either FLUORESCENCE or PHOSPHORESCENCE. *See also* THERMOLUMINESCENCE. *Compare* INCANDESCENCE.

luminosity curves. Plots of the variation with wavelength of the response of the eye to an equal energy spectrum. The curve at low levels of illumination, i.e. the SCOTOPIC VISION curve, differs from that at high levels of illumination, i.e. from the PHOTOPIC VISION curve, as shown in fig. L7. The maxima of the curves are displaced from one another by about 0.5×10^{-7} metre. Although much great intensities are needed for photopic than for scotopic vision, the curves shown have been adjusted to the same maximum height.

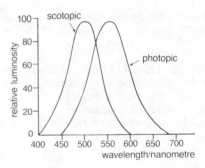

L7 Luminosity curves for equal energy spectrum

A FLICKER PHOTOMETER is used to obtain the photopic curve. A test patch is illuminated alternately with a fixed intensity of light at the wavelength of the maximum and a variable intensity of another wavelength. This intensity is adjusted until all sensation of flicker disappears, i.e. until the two lights appear equally bright. The reciprocal of this intensity is the relative luminosity for the wavelength used. The measurement is repeated for a series of wavelengths throughout the visible spectrum. For the scotopic curve, a threshold method is suitable: the energy at each wavelength for threshold visibility of the test patch is measured and the inverse of each of these readings plotted against wavelength.

luminous efficiency. The ratio of the LUMINOUS FLUX contained in a quantity of radiant flux to that quantity.

luminous emittance. Another name for LUMINOUS EXITANCE.

luminous exitance. Symbol M_v. The luminous flux leaving a surface per unit area. It is measured in lumen per square metre. *Compare* RADIANT EXITANCE.

luminous flux. Symbol Φ_v. The rate of flow of luminous energy. The SI unit is the LUMEN.

luminous intensity. Symbol I. The amount of light emitted from a point source per second in a given direction per unit solid angle. The SI unit is the CANDELA.

luminous paint. Paint containing a PHOSPHOR mixed with a small amount of radioactive material. Light is emitted as a result of bombardment of the phosphor by emissions from the radioactive material.

Lummer-Brodhun photometer. A PHOTO-METER, illustrated in fig. L8. The optical system used results in the illumination from the test source being seen as a spot surrounded by a region illuminated by the standard. The positions of the sources are adjusted until the field of view appears evenly illuminated. Measurement of these

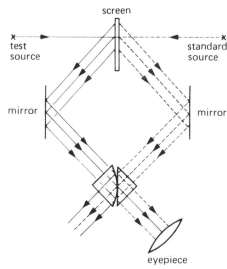

L8 Lummer-Brodhun photometer

distances of the sources from the screen enables their powers to be compared by applying the inverse square law in the usual way.

Lummer-Gehrcke plate. An interfero-meter consisting of a thick parallel-sided glass or quartz plate in which multiple reflections occur, giving rise to interference effects. The RESOLVING POWER is about 10^6 and so the instrument is suitable for studies of the HYPERFINE STRUCTURE of spectral lines.

lunar eclipse. *See* ECLIPSE.

lunar time. Time measured with respect to the Moon. The *lunar month* is the inter-val between successive new moons; 12 lunar months comprise a *lunar year*, which is equivalent to 354.3671 mean solar days (*see* SOLAR TIME).

lunation. Another name for lunar month. *See* LUNAR TIME.

lux. Symbol lx. The SI unit of ILLUMINANCE, equal to the illumination produced by a luminous flux of 1 lumen uniformly spread over an area of 1 square metre.

Lyman series. A spectral series in the ultraviolet region of the hydrogen spec-trum, with wavelengths λ given by

$$1/\lambda = R(1 - 1/m^2)$$

R is the RYDBERG CONSTANT and m is an integer greater than 1.

M

Mach angle. The semi-angle of the cone which is the envelope, i.e. the wave front, of the spherical pressure waves generated by a body moving at supersonic speed through a medium. It is given by $\sin^{-1}(1/M)$ where M is the MACH NUMBER.

machine. A device whereby a mass, i.e. *load*, may be moved by a much smaller force, i.e. *effort*, than the weight of the load. The ratio of load to effort is known as the *mechanical advantage* of the machine, and the ratio of distance moved by effort to that moved by load is known as the *velocity ratio*. The ratio of the work done on the load to that done by the effort is the *efficiency* of the machine. It equals the ratio of mechanical advantage to velocity ratio and is often expressed as a percentage: it cannot exceed 100%, which is the value for a frictionless, i.e. perfect machine.

Mach number. Symbol M. The ratio of the relative velocity of a body through the fluid in which it is situated to the local velocity of sound in the fluid. If M exceeds 1, the relative velocity is supersonic; if M exceeds 5, the relative velocity if hypersonic.

macroscopic. (1) Sufficiently large to be seen without using a microscope.

(2) Displaying properties associated with the statistical behaviour of large numbers of atoms or molecules.

magic numbers. The numbers

2, 8, 20, 28, 50, 82 and 126

They are so called because atomic nuclei containing these numbers of protons or of neutrons are more stable than other nuclei. The numbers correspond to filled shells in the nucleus. Magic number elements such as tin and calcium have relatively large numbers of isotopes.

magnet. A device for producing a magnetic field. A magnet is either temporary or permanent: thus the magnetic field of an ELECTROMAGNET exists only while current flows; on the other hand a magnetized piece of ferromagnetic material has a permanent magnetic field and is therefore a permanent magnet.

magnetic balance. A device for the direct determination of the force between magnetic poles. In a simple type a long bar magnet is balanced horizontally on a knife edge. Another pole is then placed above or below one of the magnet's poles so that the latter is deflected downwards. From the position of a rider which can be moved along the bar magnet to restore the horizontal balance, the force between the poles can be calculated.

magnetic bottle. An arrangement of magnetic fields used for containing the plasma in controlled FUSION, especially in a linear container.

magnetic circuit. A closed loop of lines of MAGNETIC FLUX.

magnetic constant. Sybmol μ_0. The absolute PERMEABILITY of free space. It has the value

$$4\pi \times 10^{-7} \text{ henry per metre}$$

magnetic cooling. A method for producing very low temperatures. Isothermal magnetization of a paramagnetic salt at low temperature followed by ADIABATIC DEMAGNETIZATION can produce a temperature as low as 10^{-3} K. If, at a temperature of

174

about 10^{-2} K, isothermal magnetic alignment of the nuclear spins of a substance with nuclear magnetic moments is followed by adiabatic demagnetization, the production of a temperature as low as 10^{-6} K is feasible.

magnetic crack detection. A technique for locating discontinuities on or near the surface of a ferromagnetic substance. The substance is painted with a dispersion of fine iron particles in oil and the distribution of magnetization investigated: irregularities indicate the sites of discontinuities since the iron particles congregate there.

magnetic cycle. A plot of the magnitude of magnetic induction against that of magnetic field strength. *See also* HYSTERESIS LOOP.

magnetic declination. *See* DECLINATION (def. 1).

magnetic dip. *See* DIP.

magnetic dipole. A north pole and a south pole of equal strength separated by a distance. Every magnetic pole is thought to exist as one member of a dipole, unlike a positive or negative electric CHARGE, which can have an independent existence.

magnetic dipole moment. The product of the pole strength of one of the two poles of a magnetic dipole and the separation of the poles. A small loop of wire of area A carrying a current I behaves as a magnetic dipole and has a magnetic moment IA, sometimes known as the *electromagnetic moment.*

magnetic domains. Tiny regions of strong magnetism existing inside ostensibly unmagnetized ferromagnetic material. *See* FERROMAGNETISM.

magnetic equator. A line on the Earth's surface joining all points of zero DIP. It is everywhere close to the equator. *See also* ISOCLINE.

magnetic field. A region in which magnetic

forces can be detected, for example the region surrounding a magnetized body.

magnetic field strength. Symbol H. A vector quantity equal to the MAGNETIC INDUCTION divided by the absolute PERMEABILITY of the medium. The magnitude is measured in ampere per metre.

magnetic flux. The product of the area under consideration and the component, normal to the area, of the average MAGNETIC INDUCTION over it. It is thus the surface integral of magnetic induction normal to the surface.

magnetic flux density. Another name for MAGNETIC INDUCTION.

magnetic hysteresis. *See* HYSTERESIS (def. 1).

magnetic inclination. Another name for DIP.

magnetic induction. Symbol B. A vector whose direction at a given point is that of the magnetic field at that point and whose magnitude is given by

$$F/(qv \sin \theta)$$

F is the magnitude of the force experienced by a moving charge q travelling with velocity magnitude v in a direction making an angle θ with the field direction. The magnitude of B is measured in tesla.

magnetic intensity. Former name for MAGNETIC FIELD STRENGTH.

magnetic meridian. A great circle passing through the point on the Earth's surface under consideration, and through the magnetic poles of the Earth.

magnetic mirror. A magnetic field of sufficient strength and suitably orientated to reverse the direction of movement of charged particles. Magnetic mirrors are used in a MAGNETIC BOTTLE.

magnetic moment. (1) Another name for MAGNETIC DIPOLE MOMENT. (2) *See also* SPIN.

magnetic monopole. A hypothetical iso-lated magnetic pole. *See* MAGNETIC DIPOLE.

magnetic pole. A region near either end of a permanent magnet. Lines of force of the magnet's field converge on one pole and diverge from the other. A freely sus-pended magnet comes to rest in the direc-tion of the Earth's magnetic field with its north-seeking pole pointing north and its south-seeking pole pointing south. Unlike magnetic poles attract each other and like ones repel each other. The force obeys the INVERSE SQUARE LAW of variation with dis-tance, just as does the force between elec-tric charges.

magnetic quantum number. *See* ATOMIC ORBITAL.

magnetic recording. The process of re-cording sound on MAGNETIC TAPE and then reproducing the sound from the tape. The principle of the *tape recorder* is illus-trated in fig. M1. The tape moves at cons-tant speed past a very narrow air gap in a ring of soft iron, which is magnetized by the current from the recording amplifier passing through the coil around the ring.

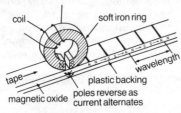

M1 Magnetic tape recording

Magnetism is thus induced on the tape as indicated, the polarity depending on the phase of the coil current and the strength on the current strength. In playback the magnetic tape is run past the same ring at the same speed as previously. The magnets on the tape cause induction of current in the coil surrounding the ring. This current is amplified and then fed to a loudspeaker thus reproducing the sound. *See also* SOUND TRACK.

magnetic screening. The protection of a space from magnetic effects by surround-ing it with a material of high relative PERMEABILITY.

magnetic shell. A thin sheet of ferro-magnetic material magnetized across its thickness. It can be regarded as an infinite number of small magnets.

magnetic storm. A disturbance of the Earth's magnetic field due to electrical dis-turbance resulting from sunspot activity.

magnetic susceptibility. The ratio of the magnetic dipole moment per unit volume to the magnetic field strength.

magnetic tape. A plastic tape which is coated with a layer of iron oxide and can be magnetized to record sound (*see* MAGNETIC RECORDING). Another application is the recording of binary information, in the form of magnetized dots, for use in com-puters.

magnetic variation. Another name for DECLINATION (def. 1).

magnetic well. An arrangement of mag-netic fields for containing a plasma in experimental FUSION REACTORS. *See also* MAGNETIC BOTTLE.

magnetism. The study of magnetic forces and fields.

magnetite. A black magnetic mineral consisting of ferro-ferric oxide. Pieces of the mineral were the earliest magnets.

magnetization. Symbol M. A vector de-fined as

$$B/\mu_0 - H$$

where B is the MAGNETIC INDUCTION, μ_0 the MAGNETIC CONSTANT and H the MAGNETIC FIELD STRENGTH.

magnetization curves. Plots of the magni-tude of MAGNETIZATION in a magnetic specimen against the magnitude of MAG-NETIC FIELD STRENGTH of the magnetizing field. If the specimen was demagnetized at the beginning of the investigation, the curves are called *normal curves*.

magneto. An electrical generator in which a permanent magnet provides the magnetic field.

magnetoacoustics. The study of the interaction between ultrasonic waves and magnetic fields.

magnetocaloric effect. The reversible heating and cooling of a specimen by changes of MAGNETIZATION.

magnetoelastic effects. The effects of stress upon the magnetic properties of a ferromagnetic material.

magnetohydrodynamics. The study of electromagnetic phenomena in electrically conducting fluids such as molten metal and plasma.

magnetometer. An instrument for investigating magnetic field strength.

magnetomotive force. The magnetic analogue of ELECTROMOTIVE FORCE. It is the line integral of $H \cos \theta ds$ round a closed path; H is the magnitude of the magnetic field strength and θ the angle it makes with the path element ds.

magneton. *See* BOHR MAGNETON; NUCLEAR MAGNETON.

magnetopause. The outer boundary of the Earth's magnetic field. It is approximately ten times the Earth's radius from the Earth's centre.

magnetoresistance. A change in the electrical resistivity of a substance due to the presence of an external magnetic field. It occurs in all metals but is most pronounced in ferromagnetic metals and their alloys.

magnetosphere. The region between the upper atmosphere and the MAGNETO-PAUSE.

magnetostatic potential. *See* POTENTIAL.

magnetostriction. A change in the physical dimensions of a ferromagnetic substance due to a change in its magnetization.

magnetron. An electron tube for producing microwaves. The tube contains a hot cathode surrounded by a coaxial cylindrical anode. When a sufficiently large magnetic field is applied parallel to the axis of the cylinder, the electrons emitted by the cathode are turned back towards it and rotate about it. This results in the induction of radio frequency fields in the resonant cavities of the anode; these fields act on the electrons also and so oscillations can occur.

magnification. The ratio of the size of an image formed by an optical system to the size of the object.

magnifying glass. A simple MICROSCOPE.

magnifying power. The ratio of the angle subtended at the eye by the final image formed by an optical system to the angle subtended by the object at the unaided eye.

magnitude. The brightness of a celestial object, the brightest objects having the smallest magnitudes. Magnitude is measured on a scale such that a brightness ratio of 100 to 1 corresponds to a magnitude difference of 5, so that a star of magnitude m is $100^{0.2}$, i.e. 2.512, times brighter than a star of magnitude $m + 1$.

magnon. A quantum of magnetic energy analogous to the photon. It is of use in the consideration of excitation or de-excitation by neutrons of waves of magnetic spin.

magnox. Any proprietary magnesium alloy used to encase fuel elements in some NUCLEAR REACTOR TYPES. For example magnox A contains 0.8% aluminium and 0.01% beryllium.

main sequence star. *See* HERTZSPRUNG-RUSSELL DIAGRAM.

mains frequency. The frequency of the alternating-current electricity supply obtained from the grid. In Britain it is 50 hertz and in the USA 60 hertz.

major axis. *See* ELLIPSE.

majority carrier. The type of carrier – either electrons or holes – responsible for transporting more than half the current in a SEMICONDUCTOR.

Malus' law. The intensity of light transmitted by a pair of POLARIZERS is proportional to $\cos^2\theta$ where θ is the angle between the axes of the polarizers.

manometer. A device for measuring fluid pressure. *See* PRESSURE GAUGE; MICROMANOMETER.

mantissa. *See* LOGARITHM.

many-body problem. The problem of solving the equations governing the interaction between a number of free particles whose initial positions and velocities are specified and which are subjected to known forces. For more than two bodies no rigorous solution can in general be found. Examples of many-body problems are planetary motion, many-electron systems and nucleon interactions.

mark space ratio. The ratio of pulse duration to the time between pulses in a pulse waveform.

Mars. The fourth planet in the solar system in order of increasing distance from the Sun. Its mass is about a tenth and its diameter about half of Earth's. Its axial period of rotation is about 24.5 hour and its orbital period is 687 day. The planet is reddish coloured and has two satellites. It has been visited by space probes.

mascon. A localized region of high gravity found on the Moon. Several have been located.

maser. An amplifier or oscillator which operates on the same principles and shows the same characteristics as the LASER, but in the microwave region of the spectrum. The name is an acronym for microwave amplification by stimulated emission of radiation.

mass. Symbol m. The quantity of matter in a body. It is a measure of a body's tendency to resist a change in motion. According to NEWTON'S LAWS OF MOTION, force applied is proportional to acceleration produced, mass being the constant of proportionality. Mass also determines the mutual attraction of two bodies by gravitational interaction (*see* GRAVITATION). By the theory of RELATIVITY, mass varies with velocity and is interconvertible with energy. *See also* RELATIVISTIC MASS; MASS-ENERGY RELATION.

mass defect. The difference between the sum of the masses of the individual nucleons in a nucleus and the mass of the nucleus. The difference in mass equals the energy holding the nucleons together.

mass-energy relation. A relation governed by the equation

$$E = mc^2$$

where E is the energy produced by a change m in mass and c is the speed of light in free space. Thus matter can be created or destroyed with corresponding decrease or increase of the energy. The CONSERVATION LAW for mass and for energy considered individually is not strictly true, but the total mass-energy is conserved. *See also* RELATIVITY.

mass number. Symbol A. The number of nucleons in a nucleus.

mass spectrometer. A device for measuring the relative abundance and the masses of isotopes present in a sample. Generally the sample is used to coat a filament. On heating the filament inside a GAS DISCHARGE TUBE, atoms of the specimen are ejected into the tube where they undergo electron bombardment and so become ionized. These ions emerge through a fine slit and enter a VELOCITY SELECTOR, as indicated in fig. M2. The selector is highly evacuated and is subjected to electric and magnetic fields at right angles so that only ions of constant velocity emerge through the end slit. These fixed velocity ions are now subjected to a magnetic field as indicated. They therefore follow a circular path of radius proportional to m/Q where m is the particle mass and Q its charge. Since the ions strike the photographic plate at a distance twice their path radius from the entrance slit, the separation of ions carrying the same charge is directly proportional to their mass difference. The concentration of ions of a particular mass and charge is proportional to the blackening which they produce on the photographic plate.

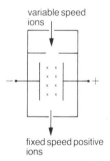

variable speed ions

fixed speed positive ions

x represents a magnetic field into the paper

M2 (a) Velocity selector in mass spectrometer

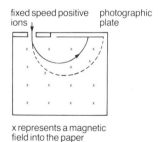

fixed speed positive ions   photographic plate

x represents a magnetic field into the paper

M2 (b) Magnetic deflection through 180°

matrix. A mathematical entity consisting of an array of numbers in rows and columns. Matrices follow certain defined mathematical rules and can be used in solving sets of simultaneous equations.

matrix mechanics. A form of QUANTUM MECHANICS designed for working with directly observable quantities.

maximum and minimum thermometer. A type of thermometer from which may be read the highest and lowest temperatures reached since setting the thermometer. As illustrated in fig. M3, each branch of a U tube contains a steel indicator I, shown separately in detail. When the temperature rises the alcohol in A expands, pushing the mercury in the U tube round so that the indicator on the maximum side is pushed up the tube. When the temperature falls the alcohol contracts, but the maximum indicator stays in place due to the action of the tiny spring attached to it; when the mercury thread contracts, the indicator on the minimum side is likewise pushed up. The indicators can be reset using a permanent magnet.

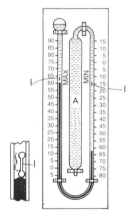

M3 Maximum and minimum thermometer

maximum density of water. Water reaches its maximum density of 1000 kilogramme per cubic metre at a temperature of 4° C. The explanation is that ice crystals have a very open three-dimensional tetrahedral structure, which is not completely replaced by the closer-packed water molecular structure until the temperature reaches 4° C. *See also* ANOMALOUS EXPANSION OF WATER.

maximum permissible dose.   *See* DOSE (def. 1).

maxwell. Symbol Mx. The electromagnetic unit of magnetic flux in CGS UNITS, equivalent to 10^{-8} weber.

Maxwell-Boltzmann distribution law. The probability of finding among a total of N ideal gas molecules, each of mass m, molecules having speeds between c and $c + \delta c$ is

$$(dN/N)/dc =$$

$$4\pi c^2 (m/(2\pi kT))^{3/2} \exp(-mc^2/2kT)$$

where k is the Boltzmann constant and T the thermodynamic temperature. *See also* MOLECULAR SPEEDS IN GAS.

Maxwell's bridge. An electrical bridge circuit, shown in fig. M4, used for measuring inductance. If keys K_1 and K_2 are closed in that order, the ballistic galvanometer will not deflect provided

$$R_1 R_3 = R_2 R_4$$

If K_2 is closed followed by K_2, the condition

$$L = R_2 R_4 C$$

is necessary for zero galvanometer deflection. The Rs represent resistance values and L and C inductance and capacitance values respectively. By trial and error suitable resistance values are found to give the

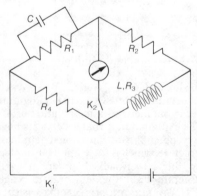

M4 Maxwell's bridge

double balance; L can then be calculated provided the value of C is known.

Maxwell's demon. An imaginary being, conjured up by Maxwell, able to work a trap door in a partition in a vessel containing gas initially at uniform temperature. The trap door is operated so that only fast molecules enter one side and only slow ones the other side of the partition, thereby separating the gas into regions of higher and lower temperature.

Maxwell's equations. Equations relating vector quantities applying at any point in a varying electric or magnetic field. They are:

$$\text{Curl } H = \partial D/\partial t + j$$
$$\text{div } B = 0$$
$$\text{curl } E = -\partial B/\partial t$$
$$\text{div } D = \rho$$

H is the MAGNETIC FIELD STRENGTH, D the ELECTRIC DISPLACEMENT, t the time, j the CURRENT DENSITY, B the MAGNETIC INDUCTION, E the ELECTRIC FIELD STRENGTH and ρ the VOLUME DENSITY OF CHARGE (*See also* CURL; DIVERGENCE).

Maxwell used the equations to show that where a varying electric field exists, it is accompanied by a varying magnetic field induced at right angles, and vice versa; the two form an ELECTROMAGNETIC FIELD. Each field vector obeys a wave equation and thus Maxwell provided the basis of the electromagnetic theory of light.

McLeod gauge. A mercury in glass pressure gauge in which a large known volume of gas at low pressure is compressed into a small volume and the new larger pressure measured. The gauge is an absolute instrument suitable for pressures as low as 1.3×10^{-3} pascal, but is unusable if easily condensed vapours are present.

mean. The average value of a set of numbers. *See also* GEOMETRIC MEAN.

mean anomaly.   *See* ANOMALY.

mean deviation.   *See* DEVIATION (def. 1).

mean free path. The average distance moved by a molecule in a gas between collisions. It equals

$$1/(2^{1/2}\pi nd^2)$$

where n is the number of molecules per unit volume and d is the molecular diameter.

mean life. Symbol τ. The average life time of an entity. For a radioactive nucleus it is the HALF LIFE divided by 0.693 15.

mean solar day. *See* SOLAR TIME.

mean speed. The value of $\sum_{r=1}^{n}(n_r v_r)/n$ where in a system of n particles, n_1 of them have speed v_1, n_2 of them speed v_2 and so on.

mean square speed. The value of

$$\sum_{r=1}^{n}(n_r v_r^2)/n$$

where in a system of n particles, n_1 of them have speed v_1, n_2 of them have speed v_2 and so on.

mechanical advantage. *See* MACHINE.

mechanical energy. The sum of the kinetic and potential energies of a body. It is constant for a conservative field of force.

mechanical equivalent of heat. Symbol J. A constant relating the CALORIE to the JOULE. The value is 4.1855 joule per calorie.

mechanical equivalent of light. A constant relating the WATT to the LUMEN. The value for radiation of wavelength 555 nanometre is 660 lumen per watt.

mechanical impedance. *See* IMPEDANCE (def. 3).

mechanical oscillation. *See* OSCILLATION.

mechanics. The branch of science concerned with the motion and equilibrium of bodies. It is subdivided into STATICS, DYNAMICS and KINEMATICS.

median lethal dose. *See* DOSE (def. 1).

medical physics. The application of physics to medicine, for example in RADIO-THERAPY, NUCLEAR MEDICINE and DIAGNOSTIC PHYSICS.

medium frequency. *See* FREQUENCY BAND (table).

mega-. (1) Symbol M. A prefix meaning 10^6, i.e. one million.
(2) A prefix used in computing to mean 2^{20}; thus 1 megabyte is equal to 1 048 576 BYTE.

megaphone. An instrument for amplifying and directing sound. It consists of a horn about a third of a metre long, into whose short end speech is directed.

megelectron volt. Symbol MeV. One million electron volt.

melting. The process of changing from solid to liquid.

melting point. Another name for FREEZING POINT.

membrane. A thin layer of tissue which covers a surface or divides a space or organ.

memory. The part of a computer that stores information in WORD or BYTE units for immediate use by the CENTRAL PROCESSING UNIT. The memory is now usually solid state; this has displaced magnetic *core store.* Solid-state memory is more compact and faster than core store. It may be *RAM* (*random access memory*) or *ROM* (*read-only memory*), both of which are fabricated in INTEGRATED CIRCUIT form. The contents of ROM are fixed whereas information can be written to and read from RAM. The information in RAM is lost when the supply voltage is removed.

meniscus. (1) A concave or convex upper surface of a liquid column.
(2) A convexo-concave or a concavo-convex LENS.

Mercury. The nearest planet to the Sun. Its mass is about 0.05 and its diameter 0.38 times Earth's. Its axial period of rotation is

58.6 day and its orbital period is 88 day. Mercury is difficult to observe and can only be seen as a morning or evening star.

mercury barometer.   *See* BAROMETER.

mercury cell.   A cell whose cathode consists of a mixture of compressed mercury II oxide and graphite. Either zinc or cadmium reacts with the mercury II oxide in concentrated potassium hydroxide solution to yield mercury. Since the electrolyte is not consumed, the cell energy density is high. This permits it to be made sufficiently small for use in calculators etc. Furthermore its stable voltage, high current delivery without performance loss and high shelf life render it very suitable for pacemaker use.

mercury in glass thermometer.   An instrument produced by introducing mercury into a fine glass capillary tube with a bulb attached at one end: a mercury reservoir is attached to the open end of the capillary and by alternately heating the bulb and allowing it to cool, mercury is drawn in until the bulb and about a third of the capillary are filled with mercury; the bulb is then heated so that any remaining air is expelled, leaving a small space at the top of the tube as a precaution against breakage should the thermometer be exposed to excessively high temperatures.

The thermometer is calibrated by marking the mercury meniscus position for immersion first in melting ice and then in steam in a HYPSOMETER, correcting to normal pressure. For a Celsius thermometer, the space between the marks is divided into 100 equal parts. Since the expansivity of mercury is temperature dependent and since the glass expansion is not negligible, thermometer readings can only be corrected to the gas scale by direct comparison with the readings of a CONSTANT PRESSURE or CONSTANT VOLUME GAS THERMOMETER.

mercury vapour lamp.   Essentially a GAS DISCHARGE TUBE in which the discharge occurs in mercury vapour. Both low- and high-pressure lamps are available; the 365 nanometre ultraviolet line is strong for both lamps, but the 253.7 nanometre line suffers reversal in the high-pressure lamp.

meridian.   A great circle passing through both the Earth's geographic poles and the point on Earth under consideration.

meson.   Any ELEMENTARY PARTICLE which has integral SPIN, participates in STRONG INTERACTION and has mass intermediate between that of a proton and that of an electron. A meson may be neutral of positively or negatively charged, with a charge magnitude equal to that of the electron; positive and negative mesons are antiparticles of each other. Mesons are responsible for the forces between nucleons in the nucleus. All are unstable, decaying in a variety of ways. It has been suggested that a tightly bound QUARK and antiquark constitute a meson.

mesopause.   The upper limit of the MESOSPHERE.

mesosphere.   The part of the atmosphere extending from about 55 kilometre to about 80 kilometre above the Earth's surface. *See* ATMOSPHERIC LAYERS.

metacentre.   The point of intersection of a vertical line through the CENTRE OF BUOYANCY of a tilted floating body with the line joining the centre of mass and the position of the centre of buoyancy for the body when upright. Provided the metacentre is higher than the centre of mass of the body, the body tends to return to the upright position.

metal-glass seal.   A type of seal required in many pieces of apparatus; frequently vacuum tightness is needed. It is necessary to make the seal when the glass is soft, i.e. at high temperature, and so, to prevent cracking on cooling, glass and metal of roughly the same expansion coefficient are ideally required. Soft glass and copper-coated nickel-iron alloy form a suitable combination for small jobs. When larger joints are to be made, such as of glass and metal tubes of several centimetres diameter in a vacuum plant, a tapered copper tube is used: the copper in contact with the glass is then thin enough to suffer the distortion resulting from the difference in expansion coefficients without cracking the glass.

metallic bond. The type of BOND occurring in metallic crystals: a regular lattice of positive ions is held together by a cloud of free electrons which move through the lattice.

metallizing. The process of covering insulating material with a film of metal in order to make it electrically conducting. The technique is of importance in solid-state electronics.

metallurgy. The study of metals and their alloys, including their structure, properties and industrial processes of smelting, refining and working.

metal rectifier. A plate of metal in contact with its oxide, or other suitable compound, which offers much greater resistance to current flow in one direction than in the opposite direction; examples are the COPPER OXIDE RECTIFIER and the SELENIUM RECTIFIER. Such devices are unsuitable for potentials above a few volt, so for high voltages a bank of rectifiers in series is used. The current passed depends on the area of contact.

metastable state. An excited state of long life (several millisecond) compared to most excited states, but of short life compared to the ground state. *See* ENERGY LEVEL.

meteor. A lump of matter which enters the Earth's atmosphere from space. The composition is variable but there is a preponderance of iron and nickel alloys. Due to friction from air particles, the surface temperature of a meteor reaches about 3000 K and so the meteor appears luminous. Generally a meteor leaves a trail of ionized gas in its wake.

meteorgraph. A device used in METEOROLOGY for recording temperature, relative humidity, pressure or wind speed or some combination of these. Often such devices are launched into the upper atmosphere by balloon or by kite.

meteorite. A very large METEOR, part of which strikes the Earth's surface, the remainder having been burnt off in its passage through the atmosphere. The mass reaching the Earth's surface may vary from a few gramme up to 65 000 kilogramme, as found at Grootfontein, South Africa.

meteorology. The science of the atmosphere, especially concerned with weather and climate.

method of mixtures. A calorimetric method in which a known mass of liquid at a known temperature is placed in a lagged calorimeter. A known mass of another substance at a different known temperature is transferred to the liquid, which is stirred and its temperature recorded. If the values of all the variables except one are known, that one can be calculated using the principle that the heat lost by the hot bodies equals the heat gained by the cold bodies (*see* HEAT). Some typical determinations are:

(a) Specific heat capacity of a solid – the heated solid is transferred to the liquid in the calorimeter;
(b) specific heat capacity of the calorimeter – a heated solid of known specific heat capacity is transferred to the liquid in the calorimeter;
(c) specific heat capacity of a liquid – the liquid is used in the calorimeter and a solid of known specific heat capacity transferred to it;
(d) specific latent heat of ice – dry ice at $0°$ C is transferred to water in the calorimeter;
(e) specific latent heat of steam – dry steam at atmospheric pressure is transferred to water in the calorimeter.

metre. Symbol m. The SI unit of length, defined (from October 1983) as the length of the path travelled by light in vacuum during a time interval of 1/299 792 458 of a second (*See* SPEED OF LIGHT). It was previously equal to 1 650 763.73 wavelengths in free space of the radiation corresponding to the transition between the $2p_{10}$ and $5d_5$ levels of the krypton-86 atom.

metre bridge. A form of WHEATSTONE BRIDGE in which two of the four resistors are replaced by a uniform resistance wire one metre in length, which can be tapped at any point along its length.

metre candle. An obsolete unit of illuminance, replaced by the LUX.

metric system. A system of weights and measures based on the metre, the litre and the kilogramme.

metric ton. Another name for TONNE.

metrology. The branch of physics concerned with the accurate measurement of mass, length, time and temperature and their direct derivatives.

meV. Symbol for milli-electronvolt.

MeV. Symbol for MEGELECTRON VOLT.

MF. Abbrev. for medium frequency. *See* FREQUENCY BAND (table).

MHD. Abbrev. for MAGNETOHYDRO-DYNAMICS.

mho. The reciprocal ohm, a former unit of conductance replaced by the SIEMENS.

mica. A mineral which is composed of complex silicates and readily splits into very thin plates. It has low thermal conductivity and high dielectric strength and so is widely used for electrical insulation.

mica capacitor. *See* CAPACITOR.

Michelson-Morley experiment. An experiment designed to measure the velocity of the Earth through the ETHER. As illustrated in fig. M5, light from the monochromatic source was split into two perpendicular beams by a half silvered mirror. Each

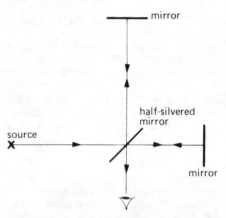

M5 Michelson-Morley interferometer

beam travelled to a mirror where it was reflected back along its previous path to the hald silvered mirror, where one beam was partially transmitted and the other partially reflected into the observer's eye, giving rise to interference FRINGES. If the Earth moved through the ether then rotation of the apparatus through 90° would result in a shift of the fringes. No such shift was detected in any position at any time. This result led to the downfall of the ether theory and paved the way for RELATIVITY theory.

micro-. Symbol μ. A prefix meaning 10^{-6}.

microbalance. A beam BALANCE capable of measuring masses as low as 10^{-8} gramme. A bulb is fixed to one end of the beam. Balance is secured by varying the air pressure in the balance case thus changing the force of buoyancy on the bulb. At constant temperature this force is proportional to the pressure, which is measured at balance by a manometer.

microcalorimeter. A differential calorimeter used for measuring very small quantities of heat, such as those associated with bacterial growth.

microdensitometer. A device for automatically measuring and recording small changes in TRANSMISSION DENSITY, such as those occurring on a photographic plate.

microelectronics. The branch of electronics concerned with components, circuits and devices of very small size.

micromanometer. A device for measuring very small pressure differences. An example is the *diaphragm gauge* in which the diaphragm displacement produced by the pressure difference is measured optically.

micrometer eyepiece. Generally a RAMSDEN EYEPIECE whose cross wires can be displaced by means of a MICROMETER SCREW, thus enabling small distances to be accurately measured.

micrometer screw. A device for the accurate measurment of distance. It employs a calibrated drum, which on rotation advances or retracts a screw of known pitch; from the difference of the drum readings, the distance moved can be accurately found.

micron. Symbol μ. A former name for the micrometre, i.e. 10^{-6} metre.

microphone. A device for converting sound energy to electric energy. It is used in telephones, sound recorders, broadcasting equipment etc. Various methods of energy transformation have been developed, resulting in variation in for example resistance or capacitance or electromagnetism or temperature or piezoelectric effect or magnetostriction or electrostriction. The aim always in mind is to ensure that the electric and acoustic vibrations correspond as closely as possible in order to ensure eventual faithful sound reproduction.

microscope. An optical instrument for producing a magnified image of a near object.

A *simple microscope* is a converging lens system, preferably corrected for CHROMATIC ABERRATION and for SPHERICAL ABERRATION. The object is usually situated at a distance from the lens less than its focal length so that a virtual erect magnified image is obtained at the least distance of distinct vision (*see* IMAGE (formed by a thin spherical lens)). The MAGNIFYING POWER equals the MAGNIFICATION.

A *compound microscope* is an instrument consisting essentially of two convex lens systems of short focal length, as illustrated in fig. M6. The object is situated at a distance from the objective just greater than its focal length so that a real enlarged inverted image is formed. This image is at a distance from the eyepiece just less than its focal length so that an errect magnified virtual image of the first image is formed at the LEAST DISTANCE OF DISTINCT VISION. It can be shown that the smaller the objective and eyepiece focal lengths, the greater the magnifying power of the instrument. However, difficulties of manufacture make the use of ultrashort focal length lenses impracticable and RESOLVING POWER considerations determine the amount of USEFUL ANGULAR MAGNIFICATION. For low-power work, transparent objects are illuminated by light from a strong source reflected on to the object by a mirror mounted below it. For higher-power work a substage condenser such as the ABBÉ CONDENSER is necessary. For a clear image the transparent object should be thin, since otherwise unfocused images from various

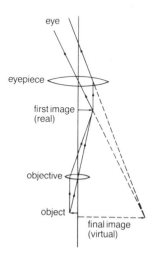

M6 Compound microscope

depths in the transparency are superimposed. For special purposes DARK FIELD ILLUMINATION has been introduced.

A *binocular microscope* is a compound microscope with two eyepieces.

A *stereoscopic microscope* consists of two compound microscopes, one for each eye, so that an impression of image depth is obtained.

An *interference microscope* is a compound microscope used for a specimen with no appreciable absorption. INTERFERENCE occurs between part of a beam which has passed through the object and another part which enters the microscope directly. Details in the object are much enhanced by the use of this instrument.

A *phase contrast microscope* is a type of interference microscope in which a path difference between beams is obtained by using a PHASE PLATE.

A *polarising microscope* is a microscope in which the object is illuminated by plane polarized light: the eyepiece is fitted with an ANALYSER which can be rotated. An object with the same properties throughout has a uniform appearance when viewed. A non uniform object may present an image of greater contrast than is obtainable using an ordinary microscope; study of this image can yield information about the molecular structure of the object. Insertion of a suitably mounted graduated quartz wedge in the beam from the microscope objective enables thickness measurements to be made on suitable specimens. A great advantage of the polarizing microscope is that, since staining is unnecessary, observations can be made on living cells and hence processes such as cell division can be observed.

An *ultraviolet microscope* is a compound microscope in which ultraviolet light is used to illuminate the object and so obtain an increase in RESOLVING POWER due to the shorter wavelength. It is necessary to record the image photographically.

An *X ray microscope* is a compound microscope in which the object is illuminated with X rays and so greater resolving power can be obtained than with the ultraviolet microscope. Physical recording of the image is required
See also ELECTRON MICROSCOPE.

microscopic. (1) Visible only with the aid of a microscope.

(2) Concerned with the behaviour of individual atoms or molecules rather than with that of matter in bulk. *Compare* MACROSCOPIC.

microscopy. The study of the construction and use of the various types of MICROSCOPE.

microwave. An electromagnetic wave of wavelength in the range 1 millimetre to 0.3 metre, i.e. between that of infrared radiation and that of radio waves.

microwave background. An isotropic distribution of MICROWAVE radiation throughout the universe. It corresponds to BLACK BODY radiation of temperature 2.7 K and is almost certainly a relic of the big bang postulated as the origin of the universe. *See* BIG BANG THEORY.

microwave spectrum. An emission or absorption spectrum in the MICROWAVE region. From such spectra information about molecular dimensions and moments of inertia can be deduced.

middle ear. A small air-filled cavity in the ear, housing three articulated bones called the *ossicles*; these are known individually as the *hammer*, the *anvil* and the *stirrup*. Their function is to transmit pressure fluctuations from the eardrum to the middle ear; the dimensions are such that, in the transmission process, the pressure at the eardrum is amplified between 40 and 90 times.

Mie scattering. The scattering of light by spherical particles of diameters comparable with the wavelength of the light. It is an extension of RAYLEIGH SCATTERING.

mil. (1) A thousandth of an inch.

(2) A thousandth of a litre.

(3) Short for CIRCULAR MIL.

mile. *See* Table 6A.

Milky Way system. Another name for our GALAXY.

Millikan's electronic charge determination. The first measurement of the charge on the ELECTRON. The apparatus used, illustrated in fig. M7, was mounted in a constant temperature enclosure. A and B represent horizontal circular plates each of about 20 centimetre diameter and separation 1.5 centimetre; A has a small hole at the centre. Occasionally one oil drop would pass through the hole in A. The illuminated drop was observed as a pin point of light through a low-power compound MICROSCOPE (not shown). The drop's terminal downward velocity was found for zero potential difference between A and B by timing the drop over a known distance indicated by a scale in the eyepiece. The observation was repeated in the presence of a potential difference between A and B which opposed the force due to gravity, and the upwards terminal velocity was measured. The observations were repeated on several drops, which could be given different charges by using a burst of X rays to ionize the air.

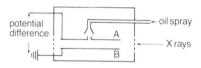

M7 Millikan's determination of electronic charge

For each drop, the first observation permits the drop radius to be calculated by equating the weight of the drop to the sum of the viscous force and the upthrust due to the air acting on it. Using this radius value, the second observation permits the charge to be calculated by equating the sum of the weight of the drop and the viscous force acting on it to the sum of the electric force and the upthrust due to the air. Millikan found that the charge was always an integral multiple of its lowest value. This lowest value was therefore taken as the charge on the electron.

Millikan's photoelectric experiment. An experiment in which vacuum-mounted clean lithium, sodium and potassium surfaces were illuminated in turn with monochromatic light from a spectrometer. The applied potential necessary to stop the PHOTOEMISSION from each surface was measured. The results were in accordance with EINSTEIN'S PHOTOELECTRIC EQUATION.

milli-electronvolt. Symbol meV. One thousandth of an electronvolt.

millimetre of mercury. The pressure that will support a column of mercury one millimetre high. It equals 133.3224 pascal.

millimetre of water. A unit of pressure sometimes employed when a water MANOMETER is used to measure small presures, for example venous system pressures. One millimetre of water is equivalent to 9.8 pascal.

minimum deviation. Symbol D. The deviation produced by a transparent prism when light passes symmetrically through it. If A is the prism angle then the refractive index of the material of the prism equals

$$\sin [(A + D)/2]/\sin (A/2)$$

Minkowski space-time. See RELATIVITY.

minor axis. See ELLIPSE.

minority carrier. The type of carrier – either electrons of holes – responsible for transporting less than half the current in a SEMICONDUCTOR.

minor planet. Another name for ASTEROID.

minute. (1) A period of time equal to 60 second. See also SOLAR TIME.
(2) Symbol '. A sixtieth of a degree of angle.

mirage. (1) An effect caused by the progressive decrease in air density as a hot ground surface is approached. Light from the sky is eventually totally internally reflected at an air layer of sufficiently low density, giving the illusion of a pool of water, as shown in fig. M8. In hot deserts a tree is often interpreted as its reflection in a pool of water.

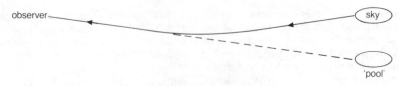

M8 Mirage

(2) The reverse effect to def. (1): it occurs when the ground is very cold, giving an air layer decreasing in density with height above the ground. An object on the ground then appears to be in the sky.

mirror. A highly polished boundary between two optical media at which light is reflected. Generally the brightness of the reflection is increased by silvering or aluminizing the boundary surface. *See also* CURVED MIRROR; IMAGE.

mirror formula. The formula

$$1/v + 1/u = 1/f = 2/r$$

where r is the radius of curvature of a spherical mirror, f its focal length and u and v respectively the object and image distances from the mirror. Distances to real objects and images are taken as positive, those to virtual objects and images as negative. *See also* IMAGE (in a spherical mirror).

mirror galvanometer. A galvanometer, the deflection of whose coil is determined from the deflection of a spot of light reflected from a mirror attached to the coil.

mirror image. The image of an object as it would appear in a plane mirror. If one object is the mirror image of another, they are identical except that, in general, they will not superimpose. For example a right hand is the mirror image of a left hand.

MISFET. *See* FIELD EFFECT TRANSISTOR.

MIST. *See* FIELD EFFECT TRANSISTOR.

MKS units. A system of units based on the metre as unit of length, the kilogramme a unit of mass and the second as unit of time, but differing from SI UNITS in the

value of the MAGNETIC CONSTANT: this is equal to 10^{-7} henry per metre in the MKS system. SI units have now replaced MKS units.

mmf. Abbrev. for MAGNETOMOTIVE FORCE.

mm Hg. Abbrev. for MILLIMETRE OF MERCURY.

Mobius strip. A surface produced by twisting a band through 180° and then joining the ends together. It is mathematically interesting since it is a single surface with a single bounding curve.

moderator. A substance, usually a light element such as graphite, used for slowing down free neutrons in a nuclear reactor so that they are more likely to cause FISSION in uranium-235 than to be absorbed by uranium-238. *See* THERMALIZE; THERMAL NEUTRONS.

modulation. The process of impressing one wave system, i.e. the *signal*, on another wave system, i.e. the *carrier*, so that the information contained in the signal is transmitted by the carrier wave. *See* AMPLITUDE MODULATION; FREQUENCY MODULATION; PHASE MODULATION; PULSE MODULATION.

modulus. The magnitude of a complex number. For example the modulus of $a + ib$ is

$$+(a^2 + b^2)^{1/2}$$

modulus of decay. The quantity α^{-1} in the equation

$$a = a_0 \, e^{-\alpha t}$$

describing oscillations undergoing DAMPING: a_0 is the initial amplitude of the oscillation, a the amplitude at time t and α the damping factor.

modulus of elasticity. Another name for ELASTIC MODULUS.

molar gas constant. The value of R in the perfect gas equation

$$PV = RT$$

where P, V and T are respectively the pressure, volume and thermodynamic temperature of one MOLE of gas. The value of R is 8.314 joule per kelvin.

molar heat capacity. The heat capacity of one MOLE of a substance.

molar polarization. The displacement of electrical centres which occurs when a molecule is subjected to an electric field and results in the production of a DIPOLE in the molecule. The ratio m/E is known as the *polarizability* of the molecule, where m is the dipole moment and E the magnitude of the electric field strength.

mole. Symbol mol. The basic SI unit of AMOUNT OF SUBSTANCE. It is that amount of any substance containing a number equal to the AVOGADRO CONSTANT of entities such as atoms, molecules, ions, electrons etc.

molecular beam. A collimated beam of atoms or molecules at low pressure. All the particles are moving in the same direction and there are few intermolecular collisions. To obtain such a beam a particle stream is passed through several apertures, any particle not passing through being removed by a vacuum pump. Molecular beams are used in spectroscopy.

molecular crystal. A crystal in which individual molecules are held together by VAN DER WAALS FORCES.

molecular energy levels. Energy levels corresponding to the various allowed states of rotation and vibration of the molecule.

molecular flow. The flow of a gas for which the MEAN FREE PATH of the gas molecules is large compared with the dimensions of the container and so very few intermolecular collisions occur.

molecular force. *See* INTERMOLECULAR FORCE.

molecular orbital. A WAVE FUNCTION of an electron in a molecule.

molecular speeds in gas. Such speeds conform to a MAXWELL-BOLTZMANN DISTRIBUTION LAW, as illustrated in fig. M9 for nitrogen at 0° C. As the temperature is raised, the maximum of the curve occurs at progressively higher speeds.

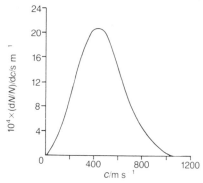

M9 Maxwell-Boltzmann distribution curve for molecular speeds

molecular spectrum. Another name for BAND SPECTRUM.

molecular weight. Another name for RELATIVE MOLECULAR MASS.

molecule. The smallest association of atoms that can be considered as a chemical unit of structure. The concept is of little use for ionic crystals or for many complex organic substances; for the latter the crystallographic unit is of greater significance.

moment. The turning effect of a force about an axis. The magnitude of the moment is the product of the magnitude of the force and the perpendicular distance from the axis to the line of action of the force. An object is in rotational equilibrium if the algebraic sum of the moments acting on it about any axis is zero. This is alternatively expressed by the statement that at equilibrium the sum of the clockwise moments about any axis equals the sum of the anticlockwise moments about the same axis. A clockwise moment is indicated ⤸ and an anticlockwise one is indicated ⤹ .

moment of couple. The product of the magnitude of either of the forces constituting a COUPLE and the perpendicular distance between the lines of action of the forces.

moment of inertia. Symbol I. The sum, for all the particles of a body, of the product of the mass of a particle and the square of its perpendicular distance from the axis under consideration; for a homogeneous body the summation is done by integration. *See also* ANGULAR MOMENTUM; ROTATIONAL ENERGY; ROTATIONAL MOTION.

moment of momentum. Another name for ANGULAR MOMENTUM.

momentum. *See* ANGULAR MOMENTUM; LINEAR MOMENTUM.

monatomic. Composed of independent single atoms; thus helium, argon and other inert gases are monatomic.

monitoring pill. A microminiature INTEGRATED CIRCUIT or circuits incorporated in a pill. When swallowed by a patient the pill transmits information continuously on conditions in the patient's tracts. The transmission is monitored by an external receiver and is very useful in the diagnosis of conditions which would otherwise present diagnostic problems.

monochord. A thin metal wire stretched horizontally over two bridges. The stretching force is applied by fixing the wire at one end, passing the other end over a pulley which is attached to the hollow box on which the sonometer is mounted, and then attaching weights to that end. The string may be set into vibration by striking, bowing or plucking. Its length can be varied by adjusting the bridge positions. The hollow box acts as a sounding board, thus increasing the volume of sound resulting from the vibrations. The apparatus is useful for verifying the relationship between the length, diameter, tension and vibration frequency of a wire. *See* VIBRATIONS IN STRINGS.

monochromatic radiation. Electromagnetic radiation of a single wavelength.

monochromator. A device used for producing single-wavelength electromagnetic radiation. Generally the wavelength is isolated from the spectrum of a polychromatic source.

monoclinic system. *See* CRYSTAL SYSTEM.

monolayer. A layer one atom or molecule thick.

Monte Carlo technique. A numerical method of solving scientific problems by constructing for each problem a random process whose parameters equal the required quantities and on which observations can be made by standard computational methods.

month. (1) *See* SOLAR TIME.
(2) *See* LUNAR TIME.

Moon. The Earth's only natural satellite, orbiting at a mean distance of 384 400 kilometre in approximately 29.5 day. The Moon's diameter is 3476 kilometre and its relative density 3.34. It has practically no atmosphere and is without surface water. The minimum night temperature is around 80 K and the maximum day temperature is about 400 K. The gravitational force due to the Moon markedly influences TIDES on Earth. The Moon was first reached by man in 1969.

Morse equation The empirical equation

$$V_r = D(1 - \exp(-a(r - r_0)^2))$$

where V_r is the potential energy of a

diatomic molecule for internuclear distance r, D is the dissociation energy of the molecule, r_0 is the equilibrium separation of the nuclei and a is a constant.

MOS. Abbrev. for metal-oxide SEMICONDUCTOR. *See* MOSFET.

mosaic electrode. The light-sensitive surface of a camera tube. *See* TELEVISON.

Moseley's law. For similar electron transitions in different elements, the frequency of a resulting line in an X RAY SPECTRUM is proportional to the square of the atomic number of the element giving rise to it.

MOSFET. A FIELD EFFECT TRANSISTOR employing MOS.

Mössbauer effect. A phenomenon in which GAMMA RAY emission of much smaller line width than usual occurs. It is limited to certain solids for which the recoil energy associated with the gamma emission is taken up by the solid as a whole. The effect is most easily observed in the 14.4 kilo-electronvolt radiation from iron-57. It has been used for example to investigate the strength of atomic binding and to check theories of ether drift.

MOST. Short for MOSFET.

motion in circle. *See* CIRCULAR MOTION.

motion in straight line. *See* LINEAR MOTION WITH OR WITHOUT CONSTANT ACCELERATION.

motion of projectile. *See* PROJECTILE TRAJECTORY.

motion under gravity. *See* PROJECTILE TRAJECTORY.

motor. *See* ELECTRIC MOTOR.

motor rule. *See* FLEMING'S RULES.

moving coil ammeter. *See* MOVING COIL INSTRUMENT.

moving coil galvanometer. *See* MOVING COIL INSTRUMENT.

moving coil instrument. An electric measuring instrument, illustrated in fig. M10. Direct current is passed through the flat rectangular coil of wire suspended vertically between the curved pole pieces of an electromagnet. The vertical sides of the coil thus experience opposite forces, causing the coil to rotate until restrained by the torsion in the suspension. The angle turned through depends on the magnitude of the current and is measured by the deflection of a beam of light reflected at the small mirror attached to the suspension; the instrument is then known as a *moving coil galvanometer*. As an alternative to the mirror arrangement, a pointer can be attached to the coil and its displacement over a graduated scale noted; the pointer instrument is known as a *moving coil ammeter*. The sensitivity of the latter can be controlled by using a SHUNT. Connecting a high resistance in series with the instrument in order to reduce the current drawn from the circuit enables it to be used as a *moving coil voltmeter*.

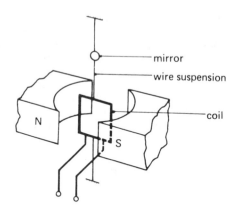

M10 Moving coil instrument

moving coil loudspeaker. An instrument used to obtain sound energy from the electric energy produced by a microphone. As illustrated (fig. M11), a speech coil S is wound on a cylindrical former F positioned symmetrically in the radial magnetic field.

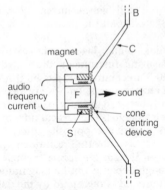

M11 Moving coil loudspeaker

Cardboard cone C is rigidly attached to F and loosely connected to a baffle board B. Current passed through S produces vibrations in it, thereby disturbing the large mass of air in C and thus producing a loud sound. The purpose of B is to reduce low-frequency interference effects.

moving coil voltmeter. *See* MOVING COIL INSTRUMENT.

moving iron instrument. An electric measuring instrument, illustrated in fig. M12. The piece of soft iron is pivoted so that it can move into a fixed coil of wire. When current is passed through the coil the iron is attracted to the coil and moves towards it, until restrained by the torsion in a spring. The angle turned through depends on the magnitude of the current and is indicated by the displacement of the pointer over the graduated scale. Since the

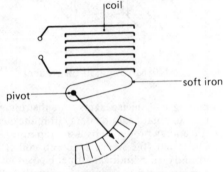

M12 Moving iron instrument

attraction is independent of the current direction, the instrument can be used for both alternating and direct current. It can also be used to measure either alternating or direct voltage if a high resistance is connected in series with it.

multielectrode valve. A valve containing two or more sets of electrodes within a single envelope. Each set of electrodes has its own independent stream of electrons, but the sets may have one or more common electrodes, such as a common cathode.

multiple reflection. A phenomenon occurring when two or more mirrors reflect the same beam of light several times in succession, resulting in the formation of a number of images.

multiplet. (1) A group of ELEMENTARY PARTICLES of identical spin but different electric charge and/or STRANGENESS. Thus a neutron and a proton belong to a multiplet.
(2) A spectral line which is split as a result of energy level splitting.

multiplication factor. The ratio of the total number of neutrons produced by FISSION in a nuclear reactor in a given time to the total number absorbed or leading out in the same time.

multiplicity. The number of ways of vectorially coupling the orbital and spin angular momentum vectors of an atom. It is represented by $2S + 1$, where S is the total spin angular momentum quantum number. An even (odd) number of electrons leads to an odd (even) value of $2S + 1$.

multipolar. Denoting or having a FIELD MAGNET with more than two poles.

mu-meson. Former name for MUON.

muon. Symbol μ. A particle of mass 207 times that of the electron and of charge equal to the electron's. The muon is a LEPTON. It is the analogue of the electron in the second GENERATION. It has a lifetime of 2×10^{-6} second and decays into an electron and neutrinos.

musical scale.   An ordered stepwise series, arranged in order of frequency, of musical sounds. The frequency ratios between the various notes within an octave is repeated every octave.

mutarotation.   A time change in the optical activity of a solution. The phenomenon is exhibited by freshly prepared solutions of some sugars due to a change of the sugar to another optically active form.

mutual capacitance.   The extent to which two capacitors can affect each other. It is expressed as the ratio of the amount of charge transferred to one to the corresponding potential difference of the other.

mutual inductance.   *See* ELECTROMAGNETIC INDUCTION.

myoneural junction.   The SYNAPSE between a nerve fibre and a muscle fibre. Transmission across such a junction produces an ACTION POTENTIAL in the muscle fibre similar to but longer lasting and less confined than that in a NEURON.

myopia.   An eye defect in which parallel rays of light are focused to a point in front of the retina when the eye is at rest, i.e. the eyeball is too long. Although near objects are clearly seen, distant ones require the aid of a concave lens for distinct vision.

N

nabla. Another name for DEL.

nadir. The point opposite an observer's ZENITH on the CELESTIAL SPHERE.

NAND circuit. A combination of an 'and' and a 'neither ... nor' LOGIC CIRCUIT.

nano-. Symbol n. A prefix meaning 10^{-9}.

Napierian logarithms. See LOGARITHM.

natural abundance. See ABUNDANCE.

natural convection. Another name for free convection. See CONVECTION.

natural dosage of radiation. The radiation dosage due to cosmic rays, natural radioactive materials in soil and rocks and small amounts of radioisotopes, principally potassium-40, occurring naturally in the body. The total natural dosage per annum from all sources is 1.25×10^{-3} sievert: 0.5×10^{-3} sievert from cosmic rays and the same from the Earth and 0.25×10^{-3} sievert from potassium-40.

natural frequency. See FREE OSCILLATIONS.

natural logarithms. See LOGARITHM.

natural radioactivity. See RADIOACTIVITY.

near infrared and ultraviolet. The parts of the spectrum of electromagnetic radiation closest to the visible region. Near infrared emission and absorption are usually associated with transitions between vibrational energy levels of a molecule. See also INFRARED RADIATION; ULTRAVIOLET RADIATION.

near point. The closest point to the eye at which clear vision is possible. See LEAST DISTANCE OF DISTINCT VISION.

near sight. Another name for MYOPIA.

near ultraviolet. See NEAR INFRARED AND ULTRAVIOLET.

nebula. A cloud of interstellar gas and dust. Many have been observed. Some are mainly visible by their emitted radiation and others by the radiation they reflect; some absorb radiation.

nebular hypothesis. Any of a group of theories based on the assumption that the solar system originated from the condensation of a nebula.

Néel temperature. See ANTIFERRO-MAGNETISM.

negative feedback. See FEEDBACK.

negative glow. See GAS DISCHARGE TUBE.

negative principal points. Another name for ANTIPRINCIPAL POINTS.

negative resistance. The description of the negative slope of some voltage against current plots. Among devices showing this phenomenon are magnetrons, silicon controlled rectifiers and tunnel diodes.

negative specific heat capacity. The SPECIFIC HEAT CAPACITY of a substance from which it is necessary to extract heat in order to raise its temperature. A saturated vapour is an example.

negatron. Another name for ELECTRON.

neon. Symbol Ne. A gas obtained from the atmosphere as a byproduct in the liquefaction of air. It is used in NEON TUBES and FLUORESCENT LAMPS.

neon tube. A gas discharge tube containing neon at low pressure; it glows red. The voltage across a tube remains constant over a range of currents and so, for currents up to 100 milliampere, a tube can be used as a voltage stabilizer. Electrodeless neon tubes readily glow in the presence of high-voltage high-frequency currents and so may be used for the detection of such currents.

Neptune. The outermost GIANT PLANET, distance 4.5×10^6 kilometre from the Sun. Its diameter is 48 400 kilometre, its mass 17.2 times that of Earth, its orbital and axial rotation times 164.79 year and about 16 hour respectively. The temperature is around $-205°$ C and the atmosphere contains mainly methane and hydrogen.

neptunium. Symbol Np. The first synthetic transuranic element to be produced.

Nernst calorimeter. An apparatus, illustrated in fig. N1, used to measure the SPECIFIC HEAT CAPACITY of a metal speciment S, mass m. C is a coil which can be electrically heated; F is a layer of foil to minimize radiation heat losses. Conduction and convection heat losses are eliminated by evacuating the enclosure. The amount of heat supplied is IVt where I is the current flowing through C, t the time for which it flows and V the voltage across C. The rise in temperature T of the

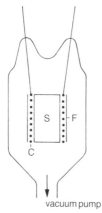

N1 Nernst calorimeter

specimen, corrected for radiation losses, may be found from resistance measurements on the coil. The specific heat capacity is given by

$$IVt/(mT)$$

Nernst effect. The production of a potential difference mutually perpendicular to a conductor or semiconductor along which there is a temperature gradient, and to an applied magnetic field. The effect is analagous to the HALL EFFECT.

Nernst glower. A rod composed of rare earth oxides which, when heated, is a useful source of infrared radiation.

net radiometer. An instrument for measuring the difference in intensity between radiation striking and leaving the Earth's surface.

network. A number of electrical components connected together in order to form a number of interrelated circuits. The components include resistors, inductors and capacitors; the *network parameters* are the actual values of these components. If these values are independent of current, they are said to be *linear*, otherwise *non-linear*; if current will pass in both directions they are said to be *bilateral*, otherwise *unilateral*. A network without a source or sink of energy is labelled *passive*, otherwise *active*. If more than two conductors meet at a point, it is known as a *branch point* and a conducting path between two branch points is known as a *branch*. Any closed conducting loop in the network is known as a *mesh*.

Neumann's law. Another name for Faraday-Neumann law. *See* ELECTROMAGNETIC INDUCTION.

neuron. A single nerve cell.

neutral current. WEAK INTERACTION where no change occurs in the charges of the participants.

neutral equilibrium. *See* EQUILIBRIUM STABILITY.

neutral filter. A filter which absorbs radiation of all wavelengths equally, i.e. it reduces the radiation intensity without changing the relative spectral distribution.

neutralization. The provision of negative FEEDBACK in an amplifier sufficient to neutralize any positive feedback and so prevent oscillations.

neutral temperature. The temperature of the hot junction of a THERMOCOUPLE, the cold junction being at 0° C, for which the thermoelectric electromotive force is a maximum.

neutrino. A stable ELEMENTARY PARTICLE with zero rest mass, a spin of ½ and a speed equal to that of light. Three types of neutrino exist, one in each GENERATION; for each type there is a corresponding antineutrino. The existence of the neutrino was originally postulated to explain energy conservation in beta decay.

neutron. An ELEMENTARY PARTICLE of zero charge and a rest mass of $1.674\,92 \times 10^{-27}$ kilogramme. Neutrons are present in the nuclei of all atoms except hydrogen. A free neutron is unstable, decaying to a proton and an electron. The nature of the neutron was first established by Chadwick (*see* CHADWICK'S EXPERIMENT).

neutron diffraction. A technique for determining the crystal structure of a solid by using it to diffract a beam of neutrons. It is an alternative technique to X RAY DIFFRACTION in some investigations.

neutron excess. Another name for ISOTOPIC NUMBER.

neutron flux density. The product of the number of free neutrons per unit volume and their mean speed. In a NUCLEAR REACTOR designed for power production, the neutron flux density is in the range 10^{16} to 10^{18} per metre squared per second.

neutron number. Symbol N. The number of neutrons present in the nucleus of an atom. It is the difference of the MASS NUMBER and the ATOMIC NUMBER.

neutron star. A star of less than 20 kilometre diameter formed after its nuclear energy sources have been exhausted and gravitational collapse of the matter in the star has occurred. When the density exceeds 10^7 kilogramme per cubic metre, protons and electrons may fuse together to form neutrons. For stars not exceeding a mass twice that of the Sun, strong repulsive forces between the neutrons halt the contraction and a stable neutron is formed. *See also* PULSAR.

new candle. Another name for CANDELA.

Newland's law of octaves. If the elements are arranged in order of their relative atomic mass, every eighth element is chemically similar; this is comparable with the intervals of the MUSICAL SCALE. *See also* PERIODIC TABLE.

newton. Symbol N. The SI unit of force, equal to the force that gives a mass of 1 kilogramme an acceleration of 1 metre per second per second.

Newtonian fluid. A fluid which obeys Newton's law of viscosity. *See* VISCOSITY.

Newtonian force. A force whose magnitude is inversely proportional to the square of the separation of the points between which the force acts.

Newtonian mechanics. *See* NEWTON'S LAWS OF MOTION.

Newtonian telescope. A type of reflecting astronomical telescope, illustrated in fig. N2. Light from the object under observation is incident on the concave mirror and

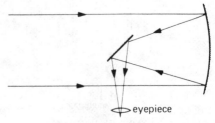

N2 Newtonian telescope

is then reflected into the eyepiece by a small plane mirror.

Newton's formula. For CENTRED OPTICAL SYSTEMS,

$$pq = ff'$$

where p and q are respectively the object and image distances from the appropriate focal point and f and f' are the focal lengths of the system.

Newton's law of cooling. The rate at which a body loses heat is proportional to the temperature difference between the body and its surroundings. The law applies strictly only if there is forced CONVECTION, but is quite well obeyed for small temperature differences under conditions of free CONVECTION.

Newton's law of gravitation. The magnitude F of the force of attraction between two bodies of mass m_1 and m_2 and separated by a distance x is

$$Gm_1m_2/x^2$$

G is the gravitational constant, equal to

$$(6.6732 \pm 0.0031) \times 10^{-11} \text{ newton}$$

metre squared per kilogramme squared. The direction of the force is along the line joining the bodies.

Newton's laws of motion. Three laws as follows.
1. Every body will continue in a state of rest or of uniform motion in a straight line unless compelled to change that state by an external impressed force.
2. The rate of change of linear momentum is proportional to the force applied. Thus

$$F = \mathrm{d}(mv)/\mathrm{d}t$$

where m represents mass and t time; F and v are respectively the force and velocity magnitudes. Hence

$$F = m \, \mathrm{d}v/\mathrm{d}t + v\mathrm{d}m/\mathrm{d}t$$

using the rule for differentiation of a product, i.e.

$$F = ma + v \, \mathrm{d}m/\mathrm{d}t$$

where a is the magnitude of the acceleration. For most nonrelativistic cases the mass remains constant, i.e.

$$\mathrm{d}m/\mathrm{d}t = 0$$

hence $F = ma$

In these cases the magnitude of the force is measured by the magnitude of the acceleration it produces. The direction of acceleration is that of the force.
3. Every action on a body produces an equal and opposite REACTION.

These laws are the basis of *Newtonian mechanics*. They are moreover predicted by RELATIVITY theory for all velocities relative to the observer which are small compared with that of light.

Newton's rings. Circular interference FRINGES seen when a plano-convex lens is placed on a flat glass plate and is illuminated and observed as shown in fig. N3. The centre of the fringe system is the point of contact of plate and lens. It can be

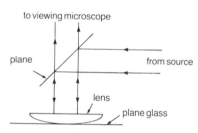

N3 Apparatus for producing Newton's rings

shown that the radius of the nth ring from the centre is $(n\lambda R)^{1/2}$ where λ is the illuminating wavelength and R the radius of curvature of the lens. Although there is no optical path difference at the centre, this point is dark due to the 180° phase change which occurs at an optically denser medium. If the source is not monochromatic, coloured rings are seen since the ring radius is wavelength dependent.

Nicholson's hydrometer. *See* HYDROMETER.

nichrome. A heat-resistant alloy used in electric heating elements. It is of variable composition but is generally about 62% nickel, 23% iron and 15% chromium.

nickel iron accumulator. Another name for EDISON ACCUMULATOR.

Nicol prism. A prism made by cutting two pieces of calcite in particular directions and cementing them together with Canada balsam. As illustrated in fig. N4, the incident unpolarized light undergoes DOUBLE REFRACTION in the first component. The extraordinary ray passes through the Canada balsam and emerges from the second component. The ordinary ray is totally internally reflected at the Canada balsam and so the rays are separated. The direction of polarization of the extra-ordinary ray is with vibrations parallel to the short diagonal of the rhomb-shaped section viewed end on. Such prisms are thus used for producing and analysing plane polarized light.

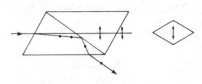

N4 Nicol prism

nife cell. A type of accumulator consisting of a positive nickel and a negative iron plate dipping into sodium hydroxide solution.

night blindness. A condition associated with a shortage of RHODOPSIN in the retinal rods of the eye. It is thought that vitamin A deficiency inhibits the ability to resynthesize rhodopsin which has been broken down by the action of light.

NMR. Abbrev. for NUCLEAR MAGNETIC RESONANCE.

nodal points. *See* CENTRED OPTICAL SYSTEM.

node. (1) A point of zero displacement of a STANDING WAVE. The separation between adjacent nodes is half a wavelength. *See*

also VIBRATIONS IN PIPES; VIBRATIONS IN PLATES; VIBRATIONS IN RODS; VIBRATIONS IN STRINGS.

(2) A point at which the wave function value is zero in an ATOMIC ORBITAL.

(3) A terminal of any branch of an electric NETWORK or a common terminal of two or more branches.

(4) The meet of the orbit of a celestial body with the ECLIPTIC.

noise. (1) Any undesired sound.

(2) Any unwanted disturbance in an electric or communication system. If the disturbance has a wide frequency spectrum it is known as *white noise* and is mainly composed of random noise arising from transient disturbances, and thermal noise. White noise produces loudspeaker hiss and television screen snow. If the noise is due to a single momentary disturbance, or to several such at intervals, it is known as *impulse noise*. Impulse noise is responsible for loudspeaker clicks.

noise factor. The ratio of the actual NOISE at a circuit output to that at the input.

nonconservative force. A force such that the work done by it in returning an object to its starting point by a closed path is not zero. The force applied in pushing an object round a closed path on a rough table is an example of a nonconservative force.

nondegeneracy. The normal state of matter, i.e. its state at moderate temperature and stress.

non-Euclidean geometry. Any logical self-consistent mathematical system concerned with shape but not based on the parallel postulate of Euclid (*see* EUCLIDEAN GEOMETRY). Non-Euclidean geometry is used in RELATIVITY theory.

noninductive winding. A type of WINDING, illustrated in fig. N5. The winding turns back on itself so that any current induced in one half of the wiring will be equal but opposite to that induced in the other half.

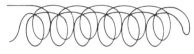

N5 Noninductive winding

This is because the current follows the same path but in reverse directions in the two parts of the wire.

nonlinear acoustics. *See* ACOUSTICS (def. 1).

nonlinear distortion. A combination of harmonic and amplitude DISTORTION in an electric signal.

nonlinear optics. The study of the effects caused by very high intensity light beams. An example is the increase in the refractive index of glass in the path of an intense LASER beam; as a result the beam is narrowed and so its intensity and that of its associated electric field increase, sometimes to such a value that the glass is shattered.

non-Newtonian fluid. A fluid that does not obey Newton's law of viscosity. *See* VISCOSITY.

nonohmic resistance. A resistance which does not obey OHM'S LAW. Semiconductor rectifiers are examples.

nonpolar. Denoting a substance lacking any permanent DIPOLE MOMENT. Polar compounds will not dissolve in nonpolar solvents, but such solvents readily dissolve other nonpolar substances.

nonradioactive tracers. Nonradioactive substances, frequently unusual isotopes, whose atoms can be identified often with the aid of a mass spectrograph. For example the average lifetime of red blood cells has been determined as about 4 month using the nonradioactive nitrogen isotope ^{15}N: over three days subjects were fed glycine which had been synthesized using ^{15}N which was then taken up by the haemoglobin. Blood samples were then analysed for ^{15}N at regular intervals until the amount of it returned to the normal value of 0.36%.

nonreactive. Of negligible REACTANCE.

nonreactive load. A load for which the alternating current is in phase with the terminal voltage.

nonsaturated mode. *See* SATURATED MODE.

nonvortical field. A vector field whose CURL is everywhere zero.

normal. A line perpendicular to a given line or surface. The normal to a curved line or surface at any point is the perpendicular to the tangent to the line or surface at that point.

normal boiling point. The boiling point at a pressure of 1.013 25 × 10^5 pascal.

normal distribution. Another name for GAUSSIAN DISTRIBUTION.

normal force. The force between two bodies in contact in a direction perpendicular to the surface of contact.

normalization. The introduction of a numerical factor into the equation $y = f(x)$ so that the area, if finite, under the corresponding graph becomes unity. The process is of importance in quantum mechanics and statistics.

normal melting point. The melting point at a pressure of 1.013 25 × 10^5 pascal.

normal temperature and pressure. Another name for STANDARD TEMPERATURE AND PRESSURE.

normal vibration. An internal oscillation of a molecule in which all the atoms execute simple harmonic motion, move in phase and have the same vibration frequency.

note. (1) A musical sound of specified frequency.
(2) A symbol in a musical score representing frequency and time value of a musical sound.

NOT circuit. A basic LOGIC CIRCUIT which inverts the input signal, i.e. a binary 1 output (input) is associated with a binary 0 input (output).

nova. A close BINARY STAR in which an explosion occurs (on a white dwarf companion), producing a luminosity increase of up to 100 000 times. The luminosity then slowly decreases to its original value.

n-p-n transistor. *See* TRANSISTOR.

NTP. Abbrev. for NORMAL TEMPERATURE AND PRESSURE.

n-type conductivity. Semiconductor conductivity resulting from a flow of electrons.

n-type semiconductor. An extrinsic SEMICONDUCTOR for which the density of conduction electrons exceeds that of mobile holes.

nuclear barrier. A region of high potential energy through which a charged particle must pass in order to enter or leave an atomic nucleus.

nuclear energy. The energy released during a nuclear FUSION or FISSION process. *See also* BINDING ENERGY.

nuclear energy change. Another name for Q VALUE.

nuclear fission. Another name for FISSION.

nuclear force. Another name for STRONG INTERACTION.

nuclear fusion. Another name for FUSION (def. 2).

nuclear heat of reaction. Another name for Q VALUE.

nuclear isomers. Nuclei of the same MASS NUMBER and ATOMIC NUMBER but with different radioactive properties. Isomers represent different energy states of the nucleus.

nuclear magnetic resonance. A method of investigating the spins of atomic nuclei. A sample, usually liquid or solid, is subjected to a strong variable magnetic field of magnitude about 2 tesla and also to a radio frequency field of 1–100 megahertz at right angles to the magnetic field. Since a nucleus with spin has an associated magnetic moment, the external magnetic field produces a splitting of energy levels in such nuclei; the difference in the energy levels depends on the magnitude of the magnetic field strength. For certain values of the field, the energy level difference equals the radio frequency energy and so the latter is strongly absorbed by the sample. A signal is therefore produced in the small detector coil surrounding the sample. The nuclear magnetic resonance spectrum is the plot of detector signal against magnitude of magnetic field strength.

The technique is used for the measurement of nuclear magnetic moments, for precise magnetic field measurement, for the determination of the location of nuclei in molecules and for chemical analysis.

nuclear magneton. The unit of nuclear magnetic moment, equal to

$$eh/(4\pi mc)$$

where e is the electronic charge, h the Planck constant, m the rest mass of the proton and c the speed of light. The value is

$$5.05 \times 10^{-27} \text{ joule per tesla}$$

nuclear medicine. The study of the use of radioisotopes in the diagnosis of disease, notably cancer. A radioisotope behaves in the body in the same way as its non-radioactive counterpart and will, in particular, accumulate in the same regions. An abnormally high level of radioactivity in any of these regions indicates the presence of cancer cells in them. The radioactivity can be measured in a variety of ways, for example using a GAMMA CAMERA, GEIGER COUNTER or SCINTILLATION COUNTER. The thyroid, kidneys and liver are the main subjects of investigation by the method.

nuclear physics. The branch of physics concerned with the study of atomic nuclei.

nuclear polarization. The tendency of nuclear spins to be aligned in a particular direction.

nuclear potential. The potential energy of some specified particle as a function of its distance from the nucleus.

nuclear power station. A power station in which nuclear energy is converted into electric energy in a nuclear reactor. *Compare* HYDROELECTRIC POWER STATION; THERMAL POWER STATION.

nuclear radius. The distance from the centre of a nucleus at which the density of nuclear matter drops sharply. It is of the order

$$10^{-15} \times A^{\frac{1}{3}} \text{ metre}$$

where A is the RELATIVE ATOMIC MASS of the nucleus.

nuclear reaction. A reaction between a nucleus and a particle bombarding it whereby a new nucleus is formed and one or more particles ejected. Thus alpha bombardment of nitrogen results in the formation of oxygen and a proton:

$$^{14}_{7}N + ^{4}_{2}He \longrightarrow ^{17}_{8}O + ^{1}_{1}H$$

nuclear reactor. An assembly in which a self-sustaining neutron CHAIN REACTION due to FISSION can be maintained and controlled. Several types of fuel, moderator, coolant and cladding have been developed. Present-day main choices are shown in the table. The best choice for a particular purpose is a subject of debate.

nuclear reactor types. Five main classifications exist: thermal, fast, breeder, power and research. A *power reactor* is designed for maximum power output whereas the main purposes of a *research reactor* are experimental. In the *thermal reactor* the maximum neutron energy is about 0.1 electronvolt. In a *fast reactor* the neutron energy range is 10 kilo-electronvolt to 2 mega-electronvolt. A *fast breeder reactor* is designed to produce more FISSILE material than it consumes: the CORE is surrounded by a *blanket* of natural uranium in which some neutrons from the fast reactor cause the production of plutonium; the plutonium can be used to enrich the fuel. Most existing reactors are thermal, the type considered to be safest and easiest to control; they do not however function as breeder reactors. *See also* HETEROGENEOUS REACTOR; HOMOGENEOUS REACTOR.

nuclear reactor vessel. The principal vessel surrounding the reactor core.

nuclear recoil. The mechanical recoil suffered by a residual nucleus following the disintegration of a larger nucleus.

nuclear spin quantum number. *See* SPIN.

nuclear waste. Another name for RADIOACTIVE WASTE.

nuclear weapon. A weapon in which an explosion results from uncontrolled nuclear FISSION or FUSION. The first weapon was the *fission bomb*, also known as the *A bomb* or *atom bomb*. In this device two masses of uranium-235, each less than the CRITICAL

Main choices for nuclear reactor

cladding	coolant	fuel	moderator
aluminium	carbon dioxide	natural uranium metal	graphite
magnesium alloy	helium	natural uranium dioxide	water
stainless steel	water	enriched uranium dioxide	heavy water
zirconium alloy	heavy water	enriched uranium carbide	
	liquid sodium	plutonium oxide	

MASS, were brought together to form a mass greater than the critical mass. The resulting chain reaction had an explosive effect of 20 kilotons of TNT, although the uranium mass was only a few kilograms. Subsequently uranium was replaced by plutonium with even greater explosive effect. An enormous further increase in explosive power was obtained with the development of the *fusion bomb*, also known as the *H bomb* or *hydrogen bomb*. In this weapon a fission bomb explosion is used to increase the temperature of a hydrogen-containing material, such as lithium deuteride, to the point at which nuclear fusion occurs.

nucleonics. The technology of nuclear physics and its applications.

nucleon number. Another name for MASS NUMBER.

nucleons. Constituents of an atomic nucleus. A nucleon is either a proton or a neutron.

nucleophilic. Possessing, or having an affinity for, positive electric charge.

nucleor. The hypothetical core of a nucleon, thought to be surrounded by a PION cloud.

nucleus. The most massive part of an atom, carrying a positive charge of Ze where Z is the ATOMIC NUMBER of the element and e the magnitude of the charge on an electron. For nuclei of the same element Z is fixed, but the MASS NUMBER, A, may vary within limits and hence so may the number of neutrons in nuclei of the same element (*see* ISOTOPES). A given nucleus is represented by its chemical symbol with Z and A as subscript and superscript respectively; thus $^4_2$He and $^3_2$He represent isotopes of helium.

The nucleus is held together by STRONG INTERACTION and theories such as the LIQUID DROP MODEL and the SHELL MODEL have been proposed to explain nuclear

structure. The diameter of a nucleus is of the order of 10^{-15} metre, the diameter of an atom being of the order of 10^{-10} metre. The mass of a nucleus is always less than the sum of the rest masses of its constituent nucleons; the greater the MASS DEFECT, the more stable the nucleus (*see* BINDING ENERGY). The nuclei of most naturally occurring substances, other than radioactive ones, are stable. Artificial nuclei may result from the bombardment of stable nuclei with high-energy charged particles.

nuclide.   (1) A nucleus characterized by its ATOMIC NUMBER and its NEUTRON NUMBER, i.e. nuclide refers only to a particular nuclear species.
(2) The atom to which a specified nucleus belongs.

null method.   A method in which a zero reading is obtained by balancing one quantity against another. The method is used for example in the WHEATSTONE BRIDGE.

numerator.   The number placed above the line in a common FRACTION; thus for 7/8, the numerator is 7.

number of poles.   The number of different electrical conducting paths that a switching device closes or opens simultaneously.

numerical aperture.   Symbol NA. The quantity $n \sin i$ where n is the refractive index of the medium in which light strikes a microscope objective and i is half the angle subtended by the objective at the object. The greater the numerical aperture, the better the RESOLVING POWER of the microscope.

Nusselt number.   A dimensionless quantity defined as $hl/(\lambda\theta)$ where h is the rate of loss of heat per unit area of a hot body immersed in material medium, the temperature difference between body and medium being θ; l is a typical dimension of the body and λ the thermal conductivity of the medium.

nutation. The oscillation of the axis of a rotating body about its mean position. The phenomenon is shown by a spinning top and by the Earth.

Nyquist noise theorem. A theorem stating

$$dP/df = kT$$

where dP/df is the rate of change of power P, due to thermal noise, with frequency f at non extreme temperature T; k is the Boltzmann constant.

O

OASM system. A system of units based on the ohm, ampere, second and metre.

object. What is viewed using an optical system. An object is said to be *real* if light actually comes from it and *virtual* if light only appears to come from it. Thus if rays from one optical system are intercepted by a second optical system before they have come to a focus, then the real IMAGE from the first system becomes the virtual object for the second system.

objective. The optical component of an optical system nearest to the object.

object lens. Another name for OBJECTIVE.

object space. The space between OBJECT and OBJECTIVE.

oblate. Denoting the ELLIPSOID obtained by rotating an ellipse about its minor axis.

observable. Signifying something measurable.

obtuse angle. An angle lying between 90° and 180°.

occlusion. The trapping of small quantities of one substance in another. Small amounts of solution are often occluded in crystals formed from the solution. Gas atoms or molecules may be occluded in interstitial positions in a solid.

occultation. The disappearance of one celestial body behind another while the first body is being observed from Earth.

octagon. An eight-sided POLYGON.

octahedron. A POLYHEDRON with eight faces. If the faces are congruent equilateral triangles, the octahedron is said to be *regular*.

octal notation. A system of representing numbers using the base 8, i.e. integers 0–7 are employed. Thus the number represented by 175 in octal notation is

$$1 \times 8^2 + 7 \times 8^1 + 5 \times 8^0 = 125$$

in decimal notation. Each digit in an octal number can be represented by a three digit number in BINARY ARITHMETIC. For example binary 101 is

$$1 \times 2^2 + 0 \times 2^1 + 1 \times 2^0 = 5$$

Octal is therefore used in computers, each digit corresponding to one byte.

octant. A segment of a circle subtending an angle of 45° at the centre of the circle, i.e. it is an eighth of the circle.

octave. A frequency interval between two notes such that the frequency ratio of the notes is 2 : 1. An octave covers 8 notes of a scale in music.

octet. A group of eight electrons in the shell of an atom. This configuration is very stable and is found in inert gas atoms except helium.

ocular. Another name for EYEPIECE.

odd-even nucleus. An atomic nucleus containing an odd number of protons and an even number of neutrons.

odd-odd nucleus. An atomic nucleus containing odd numbers of both protons and neutrons. Most such nuclei are unstable.

oersted. Symbol Oe. An electromagnetic unit of magnetic field strength in CGS UNITS. It is the field strength which produces a force of magnitude one dyne or unit magnetic pole.

ohm. Symbol Ω. The SI unit of electric resistance equal to the resistance between two points of a conductor (which is not a source of electromotive force) whose potential difference is one volt and between which flows a current of one ampere. This ohm replaced the *international ohm*, which was defined as the resistance at 0° C of a column of mercury of uniform cross section and length 106.3 centimetre and mass 14.4521 gramme.

ohmic. Denoting a resistance or other electric device which obeys OHM'S LAW.

ohmic loss. The power dissipation in an electric circuit due to resistance.

ohmmeter. An instrument, illustrated in fig. O1, for measuring electric resistance. The value of resistance R is adjusted so that with TT shorted the reading of voltmeter V is full scale. When the short is replaced by unknown resistance X, the reading falls and the scale is precalibrated so that X can be read directly on it.

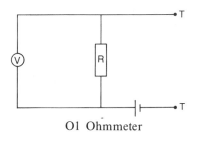

O1 Ohmmeter

Ohm's law. At constant temperature, the current flowing through a conductor is directly proportional to the potential difference applied across it. The constant of proportionality is the RESISTANCE of the conductor.

oil immersion objective. An objective used in a MICROSCOPE in order to increase the

NUMERICAL APERTURE. A few drops of cedar wood oil are placed on the upper slide covering the object and the objective lowered so that its lower surface is immersed in the oil. By suitable choice of object distance, SPHERICAL ABERRATION due to the objective can be reduced.

Olber's paradox. A paradox concerned with the darkness of the night sky, which should, for an infinite unchanging and static universe, be uniformly bright. However since these assumptions on the nature of the universe are erroneous, the paradox is resolved.

old sight. Another name for PRESBYOPIA.

omega minus particle. Symbol Ω^-. A negatively charged elementary particle with a mass 3276 times that of the electron. It is classified as a HYPERON.

opacity. The reciprocal of the TRANSMITTANCE of a medium.

open circuit. *See* CIRCUIT.

open cluster. *See* CLUSTER.

opera glasses. *See* GALILEAN TELESCOPE.

open pipe. *See* VIBRATIONS IN PIPES.

operating point. The point on the CHARACTERISTIC CURVE for a TRANSISTOR or VALVE representing the voltage and current for the operating conditions being used.

operational amplifier. A high gain, high input impedance direct coupled amplifier usually supplied as a complete packaged unit. Such amplifiers have wide instrumentation and control applications.

operator. A mathematical symbol representing a particular operation. For example the differential operator d/dt represents differentiation with respect to the variable t.

ophthalmoscope. A portable instrument for viewing the retina of the EYE; the usual form is shown in fig. O2. A battery operated light source B together with a lens L whose position can be adjusted are contained in a handle supporting the plane mirror M, through a hole at whose centre observations are made. As shown, the subject's retina R is illuminated by an unfocused diverging light beam. Light scattered from the retina of a normal relaxed eye emerges from the eye as a parallel beam, which is focused by E, the observer's relaxed eye. To correct for any deviations from normal refraction by the subject's eye, provision is made for introducing a suitable correcting lens behind the hole in the mirror; a clear view of the subject's retina can thus always be obtained.

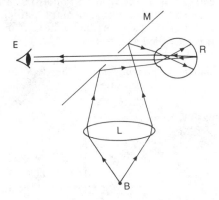

O2 Ophthalmoscope

Oppenheimer-Phillips process. A reaction in which a low-energy deuteron gives its neutron to a nucleus without entering it.

opposition. A situation arising when a celestial body lies on the line, not produced, joining two other celestial bodies. If the body lies on the produced line, *conjunction* is said to occur. This is illustrated in fig. O3 for an outer planet such as Jupiter or Mars. At full moon the Sun and Moon are in opposition with respect to Earth; at new moon they are in conjunction.

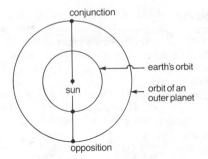

O3 Opposition and conjunction

optical activity. The property of some crystals, liquids, solutions and vapours of rotating the plane of polarization of plane polarized light (*see* POLARIZATION (def. 1)). The amount of rotation is in proportion to the length of substance traversed and, where appropriate, to the concentration; it also depends on the wavelength of the polarized light and on the temperature of the substance. The rotation may be either DEXTROROTATORY or LAEVOROTATORY. Optical activity is measured in a POLARIMETER. It is expressed by the *specific optical rotary power*, written $[\alpha]_\lambda^t$ and equal to $\alpha V/(lm)$ where α is the observed rotation, V the volume, m the mass of dissolved or pure substance and l the path length; λ and t denote wavelength and temperature respectively. Optical activity is thought to result from the asymmetric arrangement of molecules in a crystal or of atoms in a molecule.

optical axis.   (1) The line joining the centres of the entrance and exit pupils of a CENTRED OPTICAL SYSTEM. The cardinal points lie on it.

(2) The direction, in a crystal exhibiting DOUBLE REFRACTION, in which the ordinary and extraordinary rays are not separated, i.e. in which they travel with the same speed.

optical bench. A track on which optical components can be mounted and moved. It is used extensively in optics experiments.

optical centre. The axial point through which a ray, directed before entering an optical system towards a nodal point, actually passes. For thin lenses the optical centre coincides with the geometric centre of the lens.

optical character recognition. A technique of feeding conventional printed characters into a computer. A special type face is usually necessary.

optical crown. Any of various types of low DISPERSION optical glass.

optical defects of eye. *See* MYOPIA; HYPER-METROPIA; PRESBYOPIA; ASTIGMATISM.

optical density. Another name for TRANS-MISSION DENSITY.

optical distance. Another name for OPTI-CAL PATH.

optical distortion. *See* DISTORTION (def. 2).

optical flat. A surface which nowhere deviates from flatness by an amount greater than the wavelength of light. Such surfaces are essential for many optical devices and experiments and are tested by interference methods.

optical flint. Any of various types of high DISPERSION optical glass.

optical glass. Glass which is manufactured to a tight specification with respect to absence of defects such as strain, inhomogeneity, colour, incorrect density, refractive index and dispersion.

optical instrument. Any device, incorporating optical components, which improves seeing.

optical isomerism. The phenomenon exhibited by substances whose physical properties, apart from OPTICAL ACTIVITY, are identical. The phenomenon is very common in substances which contain asymmetric carbon atoms, for example the two forms of lactic acid, for which the molecular structure of one isomer is the mirror image of the other.

optical maser. Another name for LASER.

optical path. The product of the actual distance traversed by a light ray and the refractive index of the medium in which it is travelling. If the ray travels through several media, the total optical path is the sum of the optical paths in the various media.

optical pumping. The process of using optical radiation to excite phase-coherent radiation by MASER action. *See also* LASER.

optical pyrometer. *See* PYROMETER.

optical rotatory dispersion. A plot of OPTI-CAL ACTIVITY, as measured by the specific optical rotary power, against wavelength. It yields information on molecular structure.

optical window. *See* ATMOSPHERIC WINDOWS.

optic axis. *see* OPTICAL AXIS.

optic nerve. The nerve bundle responsible for transmitting electric signals generated in the rods and cones of the retina to the brain.

optics. A branch of physics concerned with the study of the production, propagation, measurement and properties of light. The ray treatment of light is called *geometric optics* and the wave treatment is called *physical optics*.

optics sign conventions. The two main conventions are '*real is positive*' and the *Cartesian sign convention*. For the former, any distances measured from an optical component are considered positive if they end on a real entity and negative if they end on a virtual one. In the Cartesian convention, light is drawn moving from left to right and positive distances are those measured from an optical component to an entity in the direction of the incident light; negative distances are those measured from an optical component against the direction of the incident light; transverse distances above the optic axis are positive and those below it negative.

optoelectronics. (1) The study of an electric field on optical phenomena. *See* KERR EFFECT; STARK EFFECT.

(2) The technology concerned with the generation of optical signals from electrical signals and the detection of such signals, used for example in transmission of signals along optical fibres.

orbit. (1) The path, often an ellipse, of one celestial body around another.

(2) The path, often visualized as a circle or ellipse, of an electron around the nucleus of an atom.

orbital. A wave function in an atom or molecule. *See* ATOMIC ORBITAL; MOLECULAR ORBITAL.

order. *See* DIFFERENTIAL EQUATION.

order-disorder transformation. (1) The transformation in a solid solution from a state in which atoms of the various components take up preferred sites in the structure (i.e. an *ordered* phase) to a state in which the atoms are distributed at random over these sites (i.e. a phase of *disorder*).

(2) The reduction in the static dielectric constant of a solid dielectric with increasing temperature, corresponding for dipolar materials to a change from a state of orientational order to one of disorder.

order of diffraction. A whole number which characterizes a diffraction spectrum by its displacement from the undeviated beam. In general for each order there are two spectra, one on either side of the undeviated beam.

order of interference. An integer which characterizes an interference fringe by the number of wavelengths in the path difference between the beams giving rise to it.

order of magnitude. An approximate magnitude to within a factor of 10; thus 0.01 and 0.04 are of the same order of magnitude but 10 and 1000 differ by 2 orders of magnitude.

ordinary ray. *See* DOUBLE REFRACTION.

ordinate. The y co-ordinate of a point on a two-dimensional Cartesian graph, i.e. it is the distance of the point from the x axis measured from the origin along the y axis.

origin. The point for any system of co-ordinates where all the co-ordinates are zero.

orthochromatic film. A photographic film which is sensitive to both green and blue light.

orthogonal (1) Mutually perpendicular, for example orthogonal lines intersect at 90°.

(2) Having or involving a set of mutually perpendicular axes, as in orthogonal crystals.

orthorhombic system. *See* CRYSTAL SYSTEM.

oscillating universe. *See* BIG BANG THEORY.

oscillation. A periodic energy variation in an electrical, mechanical or atomic system.

An *electric oscillation* may be demonstrated by charging a CAPACITOR connected in parallel with an inductor. The charged capacitor discharges through the inductor which starts to store magnetic energy at the expense of the electric energy of the capacitor and continues to do so until the capacitor is fully discharged. The capacitor then starts to recharge but in the opposite direction until all the energy is in the electric field once more. The current then reverses and the whole process occurs in reverse order and then starts again so that continuous electric oscillation results, as indicated in fig. O4.

A *mechanical oscillation* involves the

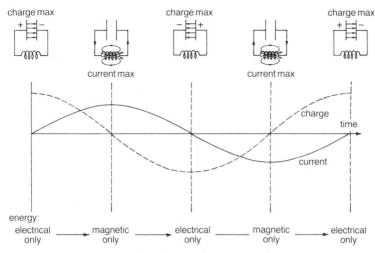

O4 Electric oscillations and energy exchanges

exchange of KINETIC and POTENTIAL ENERGY. For example a pendulum loses kinetic energy after passing through the mid-point of its swing, which is the position of maximum speed, but gains potential energy until at the top of its swing all its energy is potential. As it swings back the reverse process occurs, as illustrated in fig. O5.

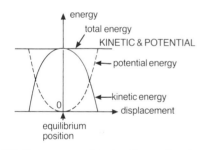

O5 Energy associated with mechanical oscillations

The oscillation of a spring may be demonstrated by suspending a mass m from the end of an elastic thread or helical spring so that an extension e is produced, as shown in fig. O6. On increasing e by pulling the mass down slightly, the mass will perform simple harmonic oscillations about its equilibrium position O with a period of

$$2\pi[(m + \lambda s)e/(mg)]^{\frac{1}{2}}$$

λ is roughly $1/3$, s is the mass of the spring, often negligible, and g is the acceleration due to gravity.

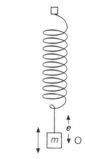

O6 Oscillation of spring

The oscillation of a liquid may be demonstrated by introducing the liquid into a U tube and depressing the liquid level in one arm by blowing gently down it. The liquid levels in the two arms will then oscillate for a short time about their rest level O, as shown in fig. O7. It can be demonstrated that the oscillation is simple harmonic of period

$$2\pi(h/g)^{\frac{1}{2}}$$

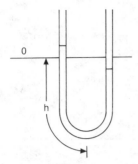

O7 Oscillation of liquid

$2h$ is the total length of the liquid and g is the acceleration due to gravity.

It can be shown that for a rigid body performing small oscillations about a fixed axis the oscillation is simple harmonic of period

$$2\pi(k^2/(gh))^{1/2}$$

h is the perpendicular distance of the centre of mass from the axis, k is the RADIUS OF GYRATION about the axis and g is the acceleration due to gravity.

See also FREE OSCILLATIONS; FORCED OSCILLATION.

oscillator. An electric circuit in which electric OSCILLATIONS occur freely.

oscillogram. The record produced by an OSCILLOGRAPH, or the reading from an OSCILLOSCOPE.

oscillograph. An OSCILLOSCOPE equipped to make a permanent record of the observations.

oscilloscope. An instrument which produces a visual image of one or more rapidly varying quantities. The CATHODE RAY OSCILLOSCOPE is the most widely used type.

osmometer. Any apparatus for measuring OSMOTIC PRESSURE.

osmosis. The preferential transmission by a semipermeable membrane of certain substances in solution. For example an aqueous solution of sugar in a thistle fun-

nel sealed by a piece of parchment and placed in water, gains water and retains the sugar; parchment therefore allows the passage of water molecules but not sugar molecules.

osmotic pressure. Symbol Π. The minimum pressure which, applied to a solution, prevents OSMOSIS. For dilute solutions of nonelectrolytes,

$$\Pi V = RT$$

where V is the volume of solution containing unit mass of solute, R the UNIVERSAL GAS CONSTANT and T the absolute temperature (*compare* EQUATION OF STATE for an ideal gas). An electrolyte solution exhibits a higher osmotic pressure than the equation predicts since electrolytes ionize in solution.

ossicles. *See* MIDDLE EAR.

Ostwald viscometer. An instrument, shown in fig. O8, for measuring the coefficient of VISCOSITY η_1 of a liquid by comparing its rate of flow through a capillary tube with that of a liquid of known coefficient η_2. Liquid is introduced at A, sucked up above X and the time of fall between fixed marks X and Y measured. The procedure is then repeated for an equal volume of the liquid of known viscosity coefficient. Then

$$\eta_1/\eta_2 = t_1\rho_1/(t_2\rho_2)$$

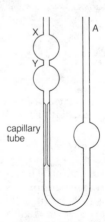

O8 Ostwald viscometer

where t and ρ represent time and density respectively and subscripts 1 and 2 refer to the two liquids.

Otto cycle. The thermodynamic cycle, shown in fig. O9, upon which the operation of spark ignition engines is based. PQ corresponds to adiabatic compression of an explosive air-petrol mixture by a piston. Q is the firing point and along QR the temperature rises rapidly at constant volume. RS corresponds to adiabatic expansion, the piston being pushed back. Along SP the temperature falls at constant volume; as the piston returns, the exhaust gas is swept out and a fresh petrol-air mixture taken in to restart the cycle.

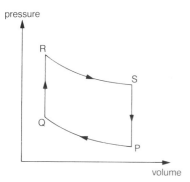

O9 Otto cycle

ounce.   *See* Table 6D.

output.   (1) The signal, current, voltage etc. delivered by an electric circuit or device.
   (2) The terminals or other place where the signal is delivered.
   (3) The processed data delivered by a computer.
   (4) The part of the computer system that converts the data into usable form. Examples are the printer, the punched card producer, magnetic tape and visual display unit.

output impedance.   The impedance presented to the load by a circuit or device.

output transformer.   A transformer which

couples an output circuit, usually an amplifier, to a load.

overcurrent release.   A cut-out device which can be set to operate when the current exceeds a selected value.

overdamping.   *See* DAMPING.

overshoot.   *See* PULSE, fig. P13.

overtone.   Another name for PARTIAL.

overvoltage release.   A cut-out device which can be set to operate when the voltage exceeds a selected value.

Owen's bridge.   A type of WHEATSTONE BRIDGE alternating current network, shown in fig. O10. When there is no signal in the detector, the conditions

$$L_2 = R_1 R_4 C_3 \text{ and } R_1 C_3 = R_2 C_4$$

are satisfied. To use the bridge, values for R_1, C_3 and C_4 are chosen and R_2 is varied until the detector signal is as small as possible. R_4 is then varied until the signal is a minimum. By adjusting R_2 and R_4 in turn, zero signal may be obtained and L_2 can then be calculated.

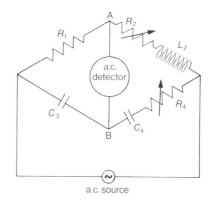

O10 Owen's bridge

oxygen.   Symbol O. An element forming 28% of the atmosphere by volume, manufactured by liquefying air. Oxygen is essential for combustion and for the respiration of plants and animals. It has wide industrial and medical uses.

oxygen point. The temperature of equilibrium between liquid and gaseous oxygen at a pressure of $1.013\,25 \times 10^5$ pascal. The temperature, 90.188 K, is taken as a fixed point in the INTERNATIONAL TEMPERATURE SCALE.

ozone. Symbol O_3. An allotrope of OXYGEN.

ozone layer. Another name for OZONOSPHERE.

ozonizer. An apparatus for producing ozone by maintaining an electric discharge in a stream of OXYGEN.

ozonosphere. A layer in the Earth's atmosphere containing OZONE. It extends from 15 kilometre to 30 kilometre above the Earth's surface and absorbs the Sun's higher-energy ultraviolet radiation.

P

pachimeter. An instrument for measuring the elastic shear limit of a solid material.

packing fraction. A quantity defined as

$$(A_r - A)/A$$

where A_r is the RELATIVE ATOMIC MASS of an isotope and A its mass number. The value for ^{16}O is zero, becoming negative for greater values of A up to $A = 180$ and reaching a minium for $A = 50$. Positive values indicate instability and so substances with A values in the range 0-16 are used in nuclear FUSION processes and those with A values greater than 180 in nuclear FISSION processes.

pair production. The production of an elementary particle and its antiparticle from a photon entering the field of an atomic nucleus. Thus a gamma ray photon of energy greater than 1.02 megaelectron volt, passing close to an atomic nucleus, may be converted into an electron and a positron. *See also* ANNIHILATION.

palaeomagnetism. The study of the residual magnetization of rocks in order to elucidate the past behaviour of the Earth's magnetic field. Iron compound bearing rocks, formed at high temperatures, were magnetized by the Earth's field while they cooled and so the intensity and direction of the residual magnetism yields information about the intensity and direction of the Earth's field at the time of the rocks' formation; this time can be found by RADIO-ACTIVE DATING. Such studies show that the Earth's polarity has reversed, i.e. north and south poles have interchanged, many times in the past.

panchromatic film. A photographic film sensitive to all wavelengths of the visible spectrum.

paper capacitor. *See* CAPACITOR.

parabola. A CONIC of eccentricity equal to 1. The equation of the parabola illustrated (fig. P1) is

$$y^2 = 4ax$$

having its focus at the point $(a, 0)$; its directrix is the line $x = -a$. The vertex is at the origin and the axis coincides with the x axis. A line drawn from the focus to any point on the parabola makes the same angle with the normal to the parabola at that point as does a line drawn from the point parallel to the x axis: a property of significance for the PARABOLIC MIRROR.

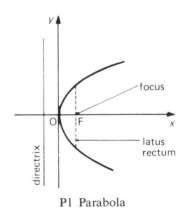

P1 Parabola

parabolic. Having the shape of a PARABOLA or of a PARABOLOID OF REVOLUTION.

213

parabolic mirror. *See* CURVED MIRROR; PARABOLOID OF REVOLUTION.

paraboloid of revolution. The shape obtained by rotating a PARABOLA about its axis. Thus cross sections perpendicular to the axis are circles and cross sections in planes parallel to the axis are parabolas. Rays of light incident parallel to the axis on a reflecting surface of such a shape will all pass through the focus, and a source placed at the focus will give rise to an accurately parallel beam of light. Reflecting paraboloids of revolution therefore find optical and radio TELESCOPE application and are also used in search lights, electric fires etc.

parallax. (1) The apparent change in the separation between two objects when they are viewed from different positions. Objects in line with the eye appear to move apart as the observer moves to left or right, unless the objects coincide. This is the basis of the *no parallax test* for finding an image position in elementary optics. *Parallax error* can occur in measuring instruments when the pointer is a small distance in front of the scale. Hence in many instruments a mirror is placed behind the pointer, whose position is read when it is in line with its image, thus eliminating the error.
(2) *See* ANNUAL PARALLAX.

parallel. (1) Denoting lines and planes that are the same distance from each other at all points.
(2) Denoting a mode of connection of electric apparatus such that the current divides between them. *See* CAPACITOR; RESISTOR; CELL. *Compare* SERIES.

parallel axes theorem. The MOMENT OF INERTIA of a body about any axis is equal to the sum of its moment of inertia about a parallel axis through the centre of mass and the product of the mass of the body and the square of the separation of the axes.

parallelepiped. A solid whose faces are all parallelograms.

parallelogram. A quadrilateral which has both pairs of opposite sides parallel. Its area is the product of the length of a side and the perpendicular distance of that side from its opposite side.

parallelogram rule. If a VECTOR is completely represented in magnitude and direction by a side of a parallelogram and another vector is similarly represented by an adjacent side, then the resultant of the vectors is represented in magnitude and direction by the parallelogram diagonal drawn from the meet of the adjacent sides. Common examples of the application of the rule are to forces and velocities.

parallel plate capacitor. *See* CAPACITOR.

paramagnetism. The phenomenon exhibited by those substances which, under the influence of an applied magnetic field become weakly magnetized in the direction of the field but lose this directional magnetization when the field is removed. A paramagnetic substance is regarded as an assembly of MAGNETIC DIPOLES which are normally directed at random due to thermal agitation. For small fields and high temperatures, the magnetization produced is proportional to the field strength; at low temperature or high field strength, saturation is approached. The dipoles are thought to arise from electron magnetic moment due to spin and orbital motion in atoms or molecules containing unpaired electrons. *See* CURIE'S LAW; CURIE-WEISS LAW. *See also* FERROMAGNETISM; DIAMAGNETISM.

parameter. A term in a mathematical expression; each value of the term is associated with many sets of values of the other variables. For example the expression

$$x^2 + y^2 = r^2$$

describes a circle radius r centre the origin. Each different value of the parameter r corresponds to a particular circle so that the expression describes a family of concentric circles.

parasitic capture. The capture by an atomic nucleus of a neutron without resulting in the fission of the nucleus.

parasitic oscillations. Unwanted oscillations which may occur in a circuit. These oscillations are usually of much higher frequency than those for which the circuit was designed.

paraxial rays. Light rays close to the OPTICAL AXIS of an optical system.

parent nuclide. *See* DAUGHTER PRODUCT.

parity. Symbol P. A symmetry property of a wave function. If the signs of all the co-ordinates in a wave function are changed and the value of the wave function stays the same, the parity is said to be $+1$; however if the value of the wave function changes sign after the substitution, then the parity is said to be -1. The parity of the total wave function describing a system of elementary particles is conserved in STRONG INTERACTION and in ELECTROMAGNETIC INTERACTION but not in WEAK INTERACTION: for example in beta decay the emitted electrons always spin in the opposite sense to the direction of motion and so parity is not conserved.

parking orbit. Another name for STATIONARY ORBIT.

parsec. Symbol pc. A unit of length used in astronomy, equal to the distance corresponding to an ANNUAL PARALLAX of 1 second of arc, i.e. to 3.086×10^{16} metre of 3.26 light year.

partial. Any pure tone component of a complex tone, the frequency being an integral multiple of that of the FUNDAMENTAL, which is not a partial (*compare* HARMONIC). Some musicians however use partial and harmonic as synonyms.

partial derivative. A derivative with respect to one variable only, and other variables present being treated as constants. The symbol ∂ is used. Thus if

$$y = 3x^2z^3$$

then

$$\partial y/\partial x = 6xz^3 \text{ and } \partial y/\partial z = 9x^2z^2$$

partial pressure. The pressure that a gas, present in a mixture of gases occupying a fixed volume, would exert if it alone occupied the volume. *See also* DALTON'S LAW OF PARTIAL PRESSURES.

particle physics. The branch of physics concerned with the properties, interactions and structure of ELEMENTARY PARTICLES.

particle velocity. Symbol u. The alternating component of the velocity of a medium transmitting sound, i.e. the total velocity of the medium less the velocity not due to sound propagation.

particle-wave duality. A phenomenon applying to all physical entities. The choice between wave description and particle description is entirely a matter of convenience. The wave and particle aspects are linked through the relations

$$E = h\nu \text{ and } p = h/\lambda$$

where the energy E and momentum p refer to a particle while the frequency ν and wavelength λ refer to a wave. The PLANCK CONSTANT h appears in both equations in accordance with RELATIVITY theory.

parton. An ultimate fundamental particle postulated as a basic unit of other fundamental particles. In the simplest theory the parton is a QUARK.

pascal. Symbol Pa. The SI unit of pressure, equal to a pressure of 1 newton per square metre.

Pascal's law. Whenever an external pressure is applied to any confined fluid at rest, the pressure is increased at every point in the fluid by the amount of the external pressure.

Paschen-Back effect. An effect similar to the ZEEMAN EFFECT but concerned with magnetic fields sufficiently strong for both the electron orbital and electron spin angular momentum vectors to each separately take up their possible orientations relative to the field direction. The resulting hyperfine structure produced in the spectral lines differs from that of the Zeeman effect.

Paschen series. A series in the spectrum of the hydrogen atom. It is defined by the equation

$$1/\lambda = R(1/3^2 - 1/n^2)$$

where λ is the wavelength of a line, n is any integer greater than or equal to 4 and R is the RYDBERG CONSTANT. The lines are in the infrared part of the spectrum.

Paschen's law. The breakdown voltage for a discharge between electrodes in gases is a function of the product of gas pressure and electrode separation.

passive component. An electronic component incapable of an amplifying function. Examples are resistors, capacitors and inductors.

patching. A technique used in an analog COMPUTER to allow temporary connection of circuits.

Pauli exclusion principle. The principle that no two electrons in an atom can have the same set of quantum numbers. Thus if two electrons in an atom have the same n, l and m values, i.e. share an ATOMIC ORBITAL, then they must have opposite spins since only the values $\pm\frac{1}{2}$ are available for the spin quantum numbers. Two such electrons are said to be *spin paired*.

Pauli spin matrices. A set of matrices introduced by Pauli in connection with electron spin in nonrelativistic WAVE MECHANICS.

p.d. Abbrev. for POTENTIAL DIFFERENCE.

peak factor. The ratio of the peak value (i.e. AMPLITUDE) of an alternating quantity to its ROOT MEAN SQUARE value. For a sinusoidal quantity the peak factor is $2^{\frac{1}{2}}$.

peak value. Another name for AMPLITUDE (def. 1).

pelleting. Another name for SEDIMENTATION.

Peltier effect. A THERMOELECTRIC EFFECT in which passage of an electric current through a junction between two different solids causes heat to be produced or absorbed at the junction, according to the direction of the current. Thus semiconductor junctions can be used as cooling elements.

p-e model of nucleus. A model based on the assumption that nuclei are composed of protons and electrons. It failed to account for several experimental observations and was therefore abandoned.

pencil of rays. A slender cone or cylinder of rays traversing an optical system and limited by a STOP.

pendulum. A body suspended so as to be free to oscillate. Pendulums have been used in many determinations of the acceleration due to gravity.

A *simple pendulum* is a small mass which is suspended from a point by a thin thread and swings in a vertical plane. For small oscillations its period of oscillation is given by

$$2\pi(l/g)^{\frac{1}{2}}$$

where l is the thread length and g the acceleration due to gravity.

A *conical pendulum* is a simple pendulum whose bob swings in a horizontal circle. For circles of small radius the period is the same as that of the simple pendulum of the same length.

A *compound pendulum* is a rigid body of any convenient shape, such as a bar, which oscillates about an axis passing through any point of the body other than the centre of mass. For small oscillations the period of swing is given by

$$(k^2 + h^2)^{\frac{1}{2}}/h$$

where k is the RADIUS OF GYRATION about a parallel axis through the centre of mass and h is the distance of the centre of mass from the axis of oscillation. A compound pendulum with a nearly vertical axis of rotation is known as a *horizontal pendulum* and is used for finding the variation in the direction of gravity with time. the SEIS-MOGRAPH employs a massive horizontal pendulum.

See also BALLISTIC PENDULUM; FOU-CAULT PENDULUM; KATER'S PENDULUM; SIMPLE EQUIVALENT PENDULUM; TORSION PENDULUM.

pendulum clock. *See also* CLOCKS.

pennyweight. *See* Table 6D.

pentode. A THERMIONIC VALVE with five electrodes. It is a TETRODE modified by the inclusion between anode and screen grid of a third grid which is given a negative potential with respect to both anode and screen grid. Low-velocity secondary electrons from the anode are thus prevented from reaching the screen grid.

penumbra. (1) A region of partial shadow surrounding the UMBRA.
(2) The outer lighter portion of a SUN-SPOT.

percolation limit. The concentration of magnetic element, in a disordered crystalline alloy having one constituent with a magnetic moment, above which the ferromagnetic state replaces the state of random orientation of atomic spins.

perfect fluid. A fluid whose coefficient of VISCOSITY is zero.

perfect gas. Another name for IDEAL GAS.

perigee. The point at which a body orbiting Earth is closest to Earth. *Compare* APOGEE.

perihelion. The point at which a body orbiting the Sun is closest to the Sun. *Compare* APHELION.

perimeter. The distance round a plane figure. Thus the perimeter of a square of side a is $4a$ and the perimeter of a circle of radius r is $2\pi r$.

period. The constant interval between identical states of a system whose properties vary periodically. For a system in which a quantity varies with time, the period is the time taken to complete a CYCLE; thus for a body moving round a circle with angular velocity ω the period is $2\pi/\omega$. The period of a crystal lattice in a particular direction is the separation of similar lattice points along that direction.

periodic table. A classification of chemical elements in order of ATOMIC NUMBER. The sequence of elements breaks up into seven horizontal periods and eight vertical groups. The elements in each group have similar electron configurations and therefore show marked similarities in behaviour.

periodic time. *See* PERIOD.

periscope. An apparatus for viewing objects when there is no direct line of sight to the eye. In its simplest form, illustrated in fig. P2, the instrument comprises two parallel mirrors at 45° to the direction of view. Light from the object is turned through 90° by the top mirror, strikes the lower mirror and is again turned through 90° to enter the observer's eye. Total internal reflection by prisms, rather than mirror reflection, is used in better quality instruments such as submarine periscopes.

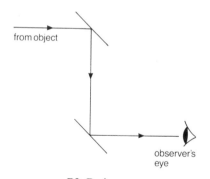

from object

observer's eye

P2 Periscope

permanent gas. A gas which cannot be liquefied by pressure alone at normal temperatures, i.e. a gas requiring cooling to reach its CRITICAL TEMPERATURE.

permanent magnet. A magnetized mass of ferromagnetic substance, stable against reasonable handling and of high REMANENCE. A definite demagnetizing field is required to remove the magnetism.

permanent set. The strain remaining in a material after the removal of all stress.

permeability. (1) *absolute permeability*. Symbol μ. The ratio of MAGNETIC INDUCTION to the external MAGNETIC FIELD STRENGTH inducing it. The unit is henry per metre.
 (2) *relative permeability*. Symbol μ_r. The ratio of the absolute permeability for a given medium to the absolute permeability of a vacuum.

permitted dosages of radiation. Dosages laid down by the International Commission on radiological protection. For a person exposed to occupational radiation hazards, the yearly dose must not exceed 5×10^{-2} sievert, of which not more than 3×10^{-2} sievert must be received in any period of 13 consecutive weeks. Anyone working in the vicinity of a radioactive area must not receive more than 1.5×10^{-2} sievert per annum. The maximum allowed dosage for the general population is 5×10^{-3} sievert per person per year.

permittivity. (1) *absolute permittivity*. Symbol ε. The ratio of ELECTRIC DISPLACEMENT to the ELECTRIC FIELD STRENGTH producing it. The unit is farad per metre.
 (2) *relative permittivity*. Symbol ε_r. The ratio of the absolute permittivity for a given medium to the absolute permittivity of a vacuum. The relative permittivity of the dielectric medium of a capacitor is also called the *dielectric constant*.

permutation. An ordered selection of a number r of entities from a set of n entities, denoted by ^np_r and equal to

$$n!/(n - r)!$$

(*see* FACTORIAL). Thus 6 permutations are possible from the letters A B C and are

AB, BA, AC, CA, BC, CB

Compare COMBINATION.

Perrin tube. A spherical evacuated glass bulb with a side arm, illustrated in fig. P3. A magnet placed as shown will deflect the electron beam at R from hot cathode C into the side arm P, causing the negatively charged leaf of the gold leaf electroscope E to rise even further and thus demonstrating directly the negative charge on the electron.

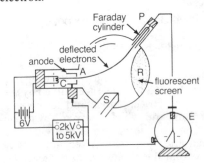

P3 Perrin tube

perpendicular. Denoting lines or planes at right angles to each other.

perpetual motion. Continuous motion without any supply of energy. It would only be possible in the absence of frictional forces. The production of useful work by a perpetual motion machine would violate the first law of THERMODYNAMICS.

persistence. (1) The interval of time after excitation ceases during which light is emitted from the screen of a CATHODE RAY TUBE.
 (2) The faint luminosity shown by certain gases for an appreciable time after the cessation of an electric discharge in them.

personal equation. A systematic error of measurement made by an experienced observer.

perturbation theory. A method of solving difficult equations that are only slightly different from ones already solved. For example the effect on the known orbit of a single planet of the presence of other planets is susceptible to this treatment. The technique is also much used in quantum mechanics.

peta-. Symbol P. A prefix meaning 10^{15}.

Petzval surface. *See* CURVATURE OF FIELD.

phase. (1) The state of vibration of a periodically varying system at a particular time. Systems vibrating with the same frequency are said to be *in phase* if their maximum and minimum values occur at the same time; otherwise they are said to be *out of phase*. In the expressions

$$y = A\sin nt, y = A\sin(nt + B)$$

for two sinusoidal vibrations, B represents the *phase difference* between them. *See also* SIMPLE HARMONIC MOTION.

(2) Any of the apparent changes in the shape of the Moon in the course of its orbit round the Earth.

(3) A homogeneous part of a system divided from other parts of the system by definite boundaries. Thus a mixture of ice crystals in water contains two phases but a solution of salt in water is single phase.

phase angle. The angle between two vectors representing two sinusoidal alternating quantities of the same frequency.

phase contrast microscope. *See* MICROSCOPE.

phase difference. *See* PHASE (def. 1).

phase discriminator. A detector circuit in which phase variations in the input wave cause amplitude variations in the output wave.

phase modulation. Modulation in which the carrier wave phase is varied about its unmodulated value by an amount proportional to the signal amplitude and at the frequency of the signal, the carrier wave amplitude remaining constant.

phase plate. A transparent plate carrying an annular groove so that light passing through the groove differs by a quarter of a wavelength in optical path from light

passing through the whole plate. *See also* MICROSCOPE (phase contrast).

phase rule. For a system in equilibrium,

$$P + F = C + 2$$

where P is the number of PHASES in the system, C the number of chemically distinguishable substances present and F the least number of independent variables defining the system.

phase shift. Any change occurring in the phase of one periodic quantity or in the phase difference between two or more such quantities.

phase splitter. A circuit in which a single input waveform produces two output waves of specified phase difference.

phase velocity. The velocity of propagation of any one phase state in a steady train of sinusoidal waves. *Compare* GROUP VELOCITY.

phon. A unit of loudness of sound, defined with reference to a standard tone of frequency 1000 hertz. The intensity of the standard is varied until it is judged to have the same loudness as the sound under investigation, and this intensity is recorded; if the measured value is n decibel above the threshold of hearing, the loudness is said to be n phon.

phonon. A quantum of thermal energy associated with lattice vibration of frequency f in a crystal. It equals hf, where h is the PLANCK CONSTANT.

phosphor. A substance exhibiting LUMINESCENCE.

phosphorescence. (1) LUMINESCENCE persisting after removal of the exciting source.

(2) Luminescence whose decay, on removal of the exciting source, is temperature dependent.

phot. A unit of intensity of illumination equal to one lumen per square centimetre.

photocathode. A cathode which emits electrons as a result of the PHOTO-ELECTRIC EFFECT.

photocell. Any light-electricity TRANSDUCER. The earliest kind was a vacuum tube containing a PHOTOCATHODE and anode; when light fell on the cathode, a current flowed when the anode was maintained at a positive potential with respect to the cathode. Modern photocells consist of a bar or thin polycrystalline film of SEMICONDUCTOR with suitable contacts. Illumination causes an increase in conductivity due to the production of charge carriers; the current is recorded.

photochromic substance. A substance which changes colour when light falls on it. Sometimes, when the light is cut off, the original colour returns.

photoconductivity. The increase of the electric conductivity of certain solids, usually semiconductors such as selenium, when exposed to electromagnetic radiation. It occurs when the photons have sufficient energy to raise electrons from a filled band to the conduction band. *See also* BAND THEORY.

photodetachment. The removal by a photon of an electron from a negative ion to give a neutral atom or molecule.

photodiode. A semiconductor DIODE which is sensitive to light. When operated the diode has REVERSE BIAS. Minority carriers flow in the circuit and constitute a dark current. When the junction is illuminated more hole-electron pairs are produced, which are then swept across the junction, constituting the light current.

photodisintegration. The disintegration of an atomic nucleus by a gamma ray photon or an X ray photon.

photoelasticity. The study of the effects of stress in transparent materials on light traversing them. Under stress normally isotropic transparent materials may exhibit DOUBLE REFRACTION; marked effects are therefore produced on illuminating them with polarized light. Complex stress in structures can be investigated by stressing a model made in material showing photoelasticity, passing polarized light through the model and observing the birefringent stress patterns.

photoelectric cell. Another name for PHOTOCELL.

photoelectric constant. The ratio of the PLANCK CONSTANT to the charge on the electron, i.e. h/e.

photoelectric effect. The ejection of electrons from a solid as a result of irradiation with electromagnetic radiation of suitable frequency. The number of emitted electrons depends on the incident radiation intensity, not its frequency. Electrons are only emitted when the frequency of the radiation exceeds a certain value, which is characteristic of the substance and is known as the *photoelectric threshold*. For most solids the threshold value lies in the VACUUM ULTRAVIOLET region of the spectrum, although for some metals, for example sodium, potassium, caesium and rubidium, it lies either in the near ultraviolet or visible region. According to *Einstein's photoelectric equation*,

$$E = h\nu - \phi$$

where E is the maximum electron kinetic energy, ν the frequency of the incident radiation, h the PLANCK CONSTANT and ϕ the WORK FUNCTION. The electrons of maximum kinetic energy are those most loosely bound in the solid. More strongly bound ones may also be ejected but with energies less than E. *See also* PHOTOIONIZATION.

photoelectric threshold. *See* PHOTOELECTRIC EFFECT.

photoelectron. An electron emitted as a result of the PHOTOELECTRIC EFFECT or PHOTOIONIZATION.

photoelectron spectroscopy. ELECTRON SPECTROSCOPY applied to photoelectrons.

photoemission. The emission of electrons by a substance as a result of bombardment by photons, as in the PHOTOELECTRIC EFFECT and in PHOTOIONIZATION.

photofission. Another name for PHOTODISINTEGRATION.

photography. The production of more or less permanent images by exposing a suitable material to radiation, followed by treatment, usually chemical, of the material. The suitable material is usually a silver halide embedded in emulsion on film. By a PHOTOELECTRIC EFFECT, exposure of the film causes conversion of silver ions to silver atoms, forming a *latent image* in the emulsion. By treatment with a mild reducing agent, known as a *developer*, opaque specks of metallic silver are formed round each silver atom of the latent image. Subsequent fixing produces a permanent image on the film; this is known as a photographic *negative* since it is darkest where the original subject was lightest. To produce a *positive* from the negative, another emulsion, coated on photographic paper, is exposed to light which has passed through the negative. The resulting positive also requires chemical treatment to produce a permanent record.

For *colour photography* a film consisting of three layers of emulsion, one red sensitive, one green sensitive and one blue sensitive, is used. The final image in the print or transparency is formed from the COMPLEMENTARY COLOURS of red, green and blue in the correct proportions. Colour photography is thus based on the SUBTRACTIVE PROCESS of colour mixing.

photoionization. The ionization of atoms or molecules by electromagnetic radiation. Only a photon of energy greater than the first IONIZATION POTENTIAL I_1 of an atom, or molecule, can remove an electron from it. The difference in photon energy and first ionization potential is shared between the ion and electron kinetic energies. The ion's share is negligible since it is very much more massive than the electron. Hence, if E is the kinetic energy of the electron,

$$E = h\nu - I_1$$

where ν is the frequency of the incident radiation and h is the PLANCK CONSTANT. The equation is Einstein's photoelectric equation applied to a single atom or molecule. If an incident photon has energy greater than the second ionization potential, I_2, it may remove a more strongly bound electron from the neutral atom or molecule. The kinetic energy of such an electron will be $h\nu - I_2$, which is less than E. See also PHOTOELECTRIC EFFECT.

photoluminescence. LUMINESCENCE resulting from irradiation by electromagnetic radiation. Absorption of this radiation raises atoms or molecules of the luminescent substance to excited states. As the atoms or molecules return to the ground state, either directly or via an intermediate excited state, radiation of longer wavelength than the exciting radiation is emitted. Practical use is made of the phenomenon in fluorescent paints and materials. It is also used in detergent whiteners, which absorb ultraviolet radiation and then emit blue light over a long time period, thus giving white fabric a blue cast and counteracting any yellowing.

photolysis. The chemical decomposition or dissociation of molecules due to the absorption of electromagnetic radiation.

photometer. An instrument employed in PHOTOMETRY. It is mainly used for comparing light sources, typically an unknown with a standard. Both visual and physical photometers are available. For the former the eye is used to make the comparison, while for the latter a PHOTOCELL, THERMOPILE and BOLOMETER may be employed. See also LUMMER-BRODHUN PHOTOMETER; FLICKER PHOTOMETER; GREASE SPOT PHOTOMETER; SPECTROPHOTOMETER.

photometry. The measurement of light intensity and amounts of illumination. The radiation may either be evaluated according to its visual effects, i.e. using the human eye, or its energy may be determined. When the eye is used the adjective *luminous* precedes the physical quantity, whereas for energy measurement the adjective *radiant* precedes the quantity.

photomicrography. The recording of microscope images on photographic media; the end product is a *photomicrograph*.

photomultiplier. A device for detecting photons, illustrated in fig. P4. Light incident on the photocathode of the tube causes electron emission by the PHOTO-ELECTRIC EFFECT. These electrons are accelerated by a potential difference to the first electrode where, by SECONDARY EMISSION, they cause the emission of 2 to 5 electrons per incident electron. The process is repeated at the second electrode, which is at a positive potential with respect to the first electrode, and so on. In a ten-stage tube, the number of electrons reaching the final collector plate for every single electron emitted from the photocathode may be of the order of 10^6. The multiplier is thus a very sensitive light measuring device. It is frequently used in SCINTILLATION COUNTER instruments.

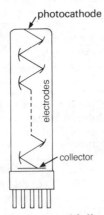

P4 Photomultiplier

photon. The QUANTUM of electromagnetic radiation. It has energy $h\nu$ where h is the PLANCK CONSTANT and ν the frequency of the radiation. The rest mass of the photon is zero.

photonuclear reaction. Another name for PHOTODISINTEGRATION.

photopic vision. Vision by the eye when the retinal cones are the receptors mainly used (*see* COLOUR VISION). This type of vision occurs at normal luminance levels and colours are perceived. *See also* SCOTO-PIC VISION; LUMINOSITY CURVES.

photosensitive pigments in eye. *See* COLOUR VISION; IODOPSIN; RHODOPSIN.

photosphere. The surface region of the Sun. It is the boundary of several hundred kilometre thickness between an opaque zone and the transparent solar atmosphere.

photosynthesis. Synthesis occurring due to the action of light. A very important example is the conversion by plant leaves of atmospheric carbon dioxide to carbohydrates in the presence of light.

phototransistor. A bipolar junction TRANSISTOR with floating base electrode. The base signal is supplied by excess carriers produced by illumination of the base. The emitter current depends on the illumination until equilibrium is established between base recombination and carrier generation, when the emitter current saturates.

phototrophic substance. Another name for PHOTOCHROMIC SUBSTANCE.

photovoltaic effect. The production of an electromotive force between two layers of different materials when electromagnetic radiation is incident on the surface layer. An example is the cuprous oxide/copper cell illustrated in fig. P5. The wire mesh enables electrical contact to be easily made to the oxide surface. It is assumed that an extremely thin barrier exists between the oxide and the copper. As a result electrons

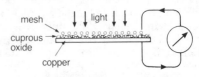

P5 Photovoltaic cell

liberated in the oxide by the radiation can readily pass to the copper but are blocked from returning. The electromotive force tending to return the electrons produces the current in the external circuit. Other combinations, such as the SELENIUM CELL, also show the effect.

physical change. Any change not involving the production of different chemical compounds.

physical colour. The colour sensation produced by a stimulus in relation to the spectral distribution of the stimulus.

physical optics. *See* OPTICS.

physics. The study of matter and energy without reference to chemical changes occurring. Traditional physics covers heat, light, sound, electricity and magnetism. Modern physics extends the study to atomic, nuclear and particle physics, relativity and quantum mechanics.

physiological acoustics. *See* ACOUSTICS (def. 1).

physiological colour. The colour sensation produced by a stimulus in relation to the response of the eye to the stimulus.

physiological effects of acceleration. The effects of acceleration on the human body. The subject has greatly increased in importance since the start of the space age. On rocket takeoff an astronaut can experience an acceleration six times that due to gravity, making lying on a specially designed couch essential during the acceleration period. Even so, arms and legs feel leaden and are difficult to raise, internal organs are compressed, breathing requires much effort and loose facial skin is drawn tight against the skull. When in orbit the acceleration vanishes: long-term effects of this are under study.

physiological optics. The analytical assessment of the reception of light by the eye and the processing of the resulting signals by the nervous system.

physisorption. *See* ADSORPTION.

pi. Symbol π. The ratio of the circumference of a circle to its diameter, equal to 3.141 592 653...

pick-up. A TRANSDUCER which converts recorded information into electric signals. Thus the mechanical vibration produced in a pick-up by its contact with the grooves in a record will stress a PIEZOELECTRIC CRYSTAL or ceramic device and so cause an electromotive force in it. In a magnetic pick-up the vibration causes movement in a small induction coil, thus changing the magnetic flux through it and hence the current in it. Each kind of pick-up thus provides an electrical signal for the audio system.

pico-. Symbol p. A prefix meaning 10^{-12}.

piezoelectric crystal. A crystal showing the PIEZOLECTRIC EFFECT.

piezoelectric effect. An effect observed in some asymmetrical crystals such that when subjected to stress they develop a potential difference across a pair of opposite faces. The sign of the potential for compressive stress is opposite to that for stretching. Moreover such crystals, when placed in an electric field, expand along one axis and contract along another. Rochelle salt and quartz show the piezoelectric effect strongly.

piezoelectric oscillator. An oscillator formed from a suitably cut piezoelectric crystal which is mounted between two electrodes. The oscillator is most conveniently set into vibration by connecting it to a source of undamped electric oscillations. This can be done in various ways; for example the oscillator circuit tuned to very near the crystal frequency can be coupled to the crystal, which will then maintain the oscillator frequency without drift at the crystal frequency.

pi meson. Another name for PION.

pinch effect. The magnetic attraction between parallel conductors carrying currents in the same direction. The phenomenon is used in the confinement of the hot plasma in a FUSION REACTOR.

pincushion distortion. *See* DISTORTION (def. 2).

p-i-n diode. A semiconductor DIODE with a region of almost INTRINSIC SEMICONDUCTOR between the n-type and p-type regions.

pion. A type of MESON principally concerned in nucleon-nucleon forces. It is the lightest meson. It occurs in three varieties, with electric charge relative to the proton of $+1, 0$ and -1 respectively and written π^+, π^0, π^-.

Pirani gauge. A gauge for measuring low gas pressures in the range 1 pascal to 10^{-3} pascal. It consists of an electrically heated wire mounted in the gas. The conduction of heat from the wire by the gas, and hence the temperature and resistance of the wire, depend on the gas pressure. Therefore by measuring the potential difference necessary to keep the wire resistance constant at different gas pressures, the pressures can be found.

pitch. (1) The frequency of the pure tone judged by a normal ear to occupy the same place in the musical scale as the sound under investigation. Pitch is influenced by the loudness of the sound: increase in sound intensity decreases the pitch for low frequencies and increases it for high ones.
(2) The distance apart of successive threads on a screw or of successive teeth on a gear wheel.

Pitot tube. A measuring device for flow velocity. It is inserted into a horizontally flowing fluid as shown in fig. P6. The fluid entering the narrow inner tube T will come to rest at some point B. Then by the BERNOULLI EQUATION,

$$p_B = p_A + 0.5\ \rho v^2$$

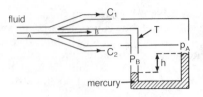

P6 Pitot tube

where p_A and p_B are the pressures at A and B respectively; ρ and v are respectively the fluid density and flow velocity. The pressure at C_1 and C_2 is p_A, hence the pressures on the mercury are as shown. Thus if h is the difference in mercury levels,

$$p_B - p_A = g\rho'h$$

where ρ' is the mercury density and g the acceleration due to gravity. Equating the two expressions for $p_B - p_A$ gives

$$g\rho'h = 0.5\ \rho v^2$$

i.e.

$$v = (2g\rho'h/\rho)^{\frac{1}{2}}$$

The flow velocity may therefore be calculated. The tube should be carefully constructed to avoid eddies, which would produce serious errors. Such tubes have been used to study the velocity of arterial blood flow. They have many industrial applications. *Compare* VENTURI METER.

planck. The unit of ACTION, equal to 1 joule second.

Planck constant. Symbol h. A universal constant of value

6.626 196 $\times\ 10^{-34}$ joule second

Planck function. Symbol Y. The function $-G/T$ where G is the GIBBS FREE ENERGY and T the thermodynamic temperature.

Planck radiation formula. The expression

$$8\pi ch\ \delta\lambda/[\lambda^5 \exp(ch/(k\lambda T)) - 1]$$

for the energy per unit volume emitted between wavelengths λ and $\lambda + \delta\lambda$ by a BLACK BODY at thermodynamic temperature T; h is the PLANCK CONSTANT, k the BOLTZMANN CONSTANT and c the speed of light.

Planck's law. The energy of electromagnetic radiation occurs in small individual packets of photons, each of energy $h\nu$ where h is the PLANCK CONSTANT and ν the frequency of the radiation. This law is the basis of QUANTUM THEORY.

plane geometry. *See* GEOMETRY.

plane mirror. A polished flat metal surface or similar device. *See also* IMAGE (in a plane mirror).

plane of flotation. The plane in which the liquid surface intersects a stationary body floating in the liquid.

plane of symmetry. A plane such that the two parts into which it may be considered to divide a system are mirror images of each other in that plane.

plane polarization. *See* POLARIZATION (electromagnetic).

plane progressive wave. A wave described by the equation

$$y = a \sin 2\pi(t/T - x/\lambda)$$

where at time t the displacement is y at a distance x from the origin; a, T and λ are respectively the amplitude, period and wavelength of the wave. For a given value of x, the equation gives the time variation at that x value; for a given value of t, the equation gives the wave form at that instant.

planet. A massive body revolving around a star and visible only by the light it reflects from the star: it is not self-luminous. From Earth, eight other planets are visible.

planetary electron. An electron orbiting around the nucleus of an atom.

planetary motion. *See* KEPLER'S LAWS.

planetoid. Another name for ASTEROID.

plano-concave lens. *See* LENS.

plano-convex lens. *See* LENS.

plasma. (1) A region in a GAS DISCHARGE TUBE containing ionized gas, consisting of electrons and positive ions in approximately equal numbers.
(2) A very high temperature highly ionized substance in which nuclear FUSION can occur. It mainly consists of electrons and atomic nuclei.

plastic deformation. A phenomenon which occurs when a material is stretched beyond its ELASTIC LIMIT. It is caused by movement of crystal planes, which occurs at the site of a lattice DEFECT.

plastic flow. A phenomenon which occurs in materials stretched beyond the YIELD POINT. The end point of plastic flow is when breaking occurs.

plate tectonics. A theory which postulates that the Earth's crust contains rigid regions known as *plates* which have moved throughout time to give the current continent positions. Six major plates and a number of smaller ones have been proposed. Seismic and volcanic activity is thought to occur at the plates' margins, where also material disappears or is produced. TERRESTRIAL MAGNETISM observations lend support to the theory.

platinum resistance thermometer. A type of RESISTANCE THERMOMETER.

plethysmography. *See* IMPEDANCE PLETHYSMOGRAPHY.

plum pudding atomic model. A model in which the nucleus was considered to occupy most of the atom and electrons to be scattered throughout it like currants in a plum pudding. The concept was shattered by GEIGER AND MARSDEN'S EXPERIMENT.

Pluto. The outermost planet in the solar system. Its diameter is about a fifth of Earth's, its period of orbital rotation and of axial rotation respectively about 248 year and 6 day. The temperature is around 63 K.

plutonium. Symbol Pu. An element which does not occur naturally but can be produced by neutron bombardment of uranium. The fissile isotope plutonium-239 is at present being produced in quite appreciable amounts in all fission NUCLEAR REACTOR TYPES, along with smaller quantities of other plutonium isotopes. A detailed inventory is kept of production in order to prevent military misuse of it. The substance is highly toxic and since the half life of plutonium-239 is 24 400 year, it is a very dangerous substance.

pneumatics. The branch of physics dealing with the dynamic properties of gases.

p-n junction. The region where a P-TYPE and an N-TYPE SEMICONDUCTOR meet. The behaviour of such a junction depends on the geometry, bias conditions and doping level in each semiconductor region. Usually opposite types of the same material are used to produce a simple p-n junction, but dissimilar materials are sometimes used, the junction then being known as a *heterojunction.* If REVERSE BIAS is applied to the p-type component, a depletion layer is produced at the junction and very little current flows until breakdown occurs. Under bias in the opposite direction, i.e. forward bias conditions, a current flows in the external circuit since carriers are attracted across the junction into the region of opposite type.

p-n model of the nucleus. The model based on the assumption that nuclei are composed of neutrons and protons.

p-n-p transistor. *See* TRANSISTOR.

PO box. *See* POST OFFICE BOX.

point contact transistor. An early form of transistor, now obsolete.

point defect. *See* DEFECT.

point group. One of 32 groups of symmetry operations, corresponding to the 32 crystal classes. *See* CRYSTAL SYSTEM.

points, action at. A phenomenon which arises because the surface density of charge at a point on a conductor is much greater than for a smooth part of the conductor. The associated high electric field strength at the point is conducive to starting an electric discharge from the point, i.e. to promoting action at it. *See also* LIGHTNING CONDUCTOR.

poise. Symbol P. The unit of viscosity coefficient in CGS UNITS. It equals 0.1 pascal second.

Poiseuille's equation. The equation

$$Q = \pi(p_1 - p_2)r^4/(8\eta l)$$

where Q is the rate of flow of a liquid flowing steadily through a circular pipe radius r and length l, η is the coefficient of VISCOSITY and p_1 and p_2 are respectively the pressure at the two ends of the tube. The equation is used in the determination of viscosity coefficient by measurement of steady flow rates.

Poisson's ratio. Symbol μ. The ratio of lateral to longitudinal strain in a rod stretched by in-line forces applied to its ends, the sides being free to contract. If there is little change in volume then the ratio is 0.5; in practice it is often less.

polar co-ordinates. Co-ordinates used to locate the position of a point by its distance from a fixed point, the *pole*, and its angular displacement from a line. As illustrated in fig. P7(a), in two dimensions the co-ordinates

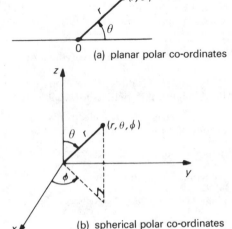

(a) planar polar co-ordinates

(b) spherical polar co-ordinates

P7 Planar and spherical polar co-ordinates

are (r, θ), r being the *radius vector* and θ the *azimuth* of the point. For three dimensions (b), *spherical polar co-ordinates* are illustrated: the point is located by a radius vector r, a *colatitude* θ measured from a vertical line and a *longitude* θ measured from a horizontal line. *See also* CYLINDRICAL CO-ORDINATES.

polarimeter. An instrument for determining the specific optical rotary power of substances showing OPTICAL ACTIVITY. Light is polarized by passing it through a NICOL PRISM or other polarizer. It then falls on the sample contained in a transparent cell. Light emerging from the cell falls on another Nicol prism, the *analyser*, which can be rotated over a calibrated scale. The emerging light is viewed through an eyepiece. The analyser is rotated until the light has maximum intensity and the amount of rotation of the plane of polarization is read off from the scale.

polariscope. Any instrument for viewing objects in polarized light.

polarizability. *See* MOLAR POLARIZATION.

polarization. (1) *electromagnetic polarization*. The restriction of the direction of vibration in ELECTROMAGNETIC RADIATION. Normally such radiation consists of transverse vibration of electric and magnetic fields, whose vibration directions are in all directions perpendicular to that of propagation. Under certain conditions the vibration directions are restricted. If the electric vector is restricted to one direction only, the radiation is said to exhibit *plane polarization* (*see also* BREWSTER'S LAW; DOUBLE REFRACTION). If the electric vector tip describes a circle around the direction of propagation, the radiation is said to exhibit *circular polarization*; if an ellipse is described the radiation is said to exhibit *elliptical polarization*.

(2) *electric polarization*. The phenomenon causing the current from a simple cell to fall appreciably from its initial value. For example if connection is made externally between a zinc plate and a copper plate immersed in dilute sulphuric acid, a layer of hydrogen bubbles soon collects on the copper plate; this results in an increase in the cell's internal resistance and an opposing electromotive force. To prevent polarization, gas deposition must be avoided. This is accomplished in different ways for different cell types: thus in the LECLANCHÉ CELL a chemical depolarizer is used to react with the hydrogen produced.

(3) *electrostatic polarization*. The formation of dipoles in a dielectric material by applying an electric field across it. Under the action of the field, electric charges on molecules align themselves so that each molecule may become a dipole. *See also* MOLAR POLARIZATION.

polarizer. A crystal or collection of crystals used to produce plane polarization of light. *See* POLARIZATION (electromagnetic); NICOL PRISM; POLAROID.

polarizing angle. *See* BREWSTER'S LAW.

polarizing microscope. *See* MICROSCOPE.

Polaroid. A type of transparent film containing many very small doubly refracting crystals aligned with their axes parallel. Incident light is doubly refracted, one plane polarized component being absorbed and the other transmitted. Stray polarized light arising, for example, by reflection is reduced by the film, which is therefore used in sun glasses to reduce glare. *See also* DOUBLE REFRACTION.

Polaroid camera. A camera which yields finished positive prints or transparencies about 10 seconds after exposure for monochrome, and about 60 second after exposure for colour.

pole. (1) *See* MAGNETIC POLE.
(2) The mid-point of a SPHERICAL MIRROR.

pole face. An end surface of the core of a magnet, through which surface passes the useful magnetic flux.

pole piece. Either of the pieces of ferromagnetic material attached to the ends of a permanent magnet or electromagnet in an electric device.

polyatomic.   Denoting a substance whose molecules are composed of several atoms.

polychromatic radiation.   Radiation comprising a mixture of wavelengths. *Compare* MONOCHROMATIC RADIATION.

polygon.   Any plane figure with three or more sides.

polyhedron.   A geometric solid whose plane faces are POLYGONS. For a simple polyhedron, *Euler's theorem* states that the number of faces plus the number of vertices equals the number of edges plus two. If all the faces of a polyhedron are congruent, it is said to be *regular*. There are five types of regular polyhedron: tetrahedron, hexahedron, octahedron, dodecahedron and icosahedron.

polynomial.   a mathematical expression of the type

$$a_0 + a_1x + a_2x^2 + a_3x^3 + ...$$

where x is a variable and the as are constants.

polythene.   A thermoplastic material with good insulating properties. A polythene rod becomes negatively charged when rubbed with a cloth.

population.   A term used in statistics to refer to the situation under consideration, whether or not a collection of people is involved.

population inversion.   *See* LASER.

positive charge.   The type of charge acquired by a cellulose acetate rod when it is rubbed with a cloth.

positive column.   Another name for positive glow. *See* GAS DISCHARGE TUBE.

positive electron.   Another name for for POSITRON.

positive feedback.   *See* FEEDBACK.

positive glow.   *See* GAS DISCHARGE TUBE.

positive hole.   *See* HOLE.

positive ion.   *See* CATION.

positive rays.   Streams of positive ions as produced in a GAS DISCHARGE TUBE. *See also* CANAL RAYS.

positron.   A particle having charge and mass equal in magnitude to that of the electron but whose charge is of opposite sign, i.e. it is the ANTIPARTICLE of the electron. Annihilation with an electron results in the production of energy and new HADRON and QUARK varieties.

positronium.   An electron-positron pair. If the particle spins are parallel it is known as an *orthopositronium* and as a *parapositronium* if the spins are antiparallel. The mean life of the former is about 10^{-7} second and it decays to three photons; for the latter the mean life is less than 10^{-7} second and it decays to two photons.

post office box.   A form of WHEATSTONE BRIDGE in which resistance coils are arranged in a special box. Each end of a coil is connected to an individual brass block so that the coil can be shorted by inserting a brass plug into a hole drilled between the two blocks.

potassium-argon dating.   A RADIOACTIVE DATING technique for geological samples. Natural potassium contains 0.001 18% of radioactive potassium-40 which decays, with a half life of 1.28×10^9 year, to the stable argon-40 isotope. By measuring the potassium-40 to argon-40 ratio, an estimate of age up to 10^7 year can be made.

potential.   A term used in a variety of contexts to denote the work necessary to move a unit of a particular quantity from infinity to the point in question. In the case of *electrostatic potential* the quantity is electrostatic, i.e. unit charge. With *magnetostatic potential* the quantity is magnetostatic, i.e. unit pole, and with *gravitational potential* it is gravitational, i.e. unit mass. Gravitational potential is always negative but electrostatic and magnetostatic potential may be positive or negative. Potential is a SCALAR quantity and is a function only of the position of the point considered, since only conservative force fields are involved.

potential barrier. A region in a field of force such that an entity on which the field acts encounters opposition to its movement.

potential difference. The difference in POTENTIAL between two points. It equals the work done in moving unit quantity between the points by any path.

potential divider. A circuit, illustrated in fig. P8, used to obtain a known fraction of an applied voltage V. This voltage is connected across a chain of resistors connected in series; only two are shown but the same principle applies. Tapping as shown gives a voltage V_2 across R_1 of

$$R_1 V/(R_1 + R_2)$$

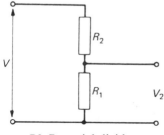

P8 Potential divider

potential energy. Symbol U. The energy possessed by a body or system due to its position. Thus a body of mass m situated at height h above a reference plane in a gravitational field has potential energy of mgh, where g is the acceleration due to gravity.

potential function. (1) A function satisfying the LAPLACE EQUATION.
(2) A scalar function describing the forces acting on any particle of a conservative

system. It obeys the equation

$$F = -\nabla V$$

where F is the resultant force vector, ∇ is DEL and V is the potential function.

potential gradient. The rate of change of potential in the direction giving a maximum value.

potential transformer. Another name for VOLTAGE TRANSFORMER.

potential well. A region in a field of force surrounded by a higher potential region, the transition between the two being abrupt.

potentiometer. (1) A type of POTENTIAL DIVIDER, illustrated in fig. P9, which is used to measure potential difference. It consists of a straight length of uniform resistance wire XY with an accumulator connected across it. A standard cell is connected as shown and the sliding contact S moved along XY until there is no galvanometer deflection. The potential difference between X and S then equals the electromotive force of the standard cell since no current flows through it; the voltage drop per centimetre of wire is thus accurately found. The cell is then replaced by the unknown voltage, always taking care to connect like polarities to X. The new balance length is found, and is multiplied by the previously determined voltage drop per centimetre to give the unknown voltage.
(2) A circuit component consisting of a resistance wound on a former and provided with a sliding contact. As a result a

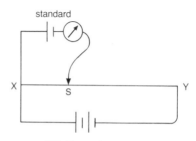

P9 Potentiometer

controlled variable voltage may be obtained between the slider and an end of the resistance when a potential difference is applied between the ends of the resistance.

potentiometric. Denoting an experimental technique which depends on measurements of potential. Thus a potentiometric titration is one in which the end point is determined by following the electrode potential of an electrode in the mixture.

pound. The unit of mass in IMPERIAL UNITS. *See* Table 6D.

poundal. The unit of force in IMPERIAL UNITS. *See* Table 6E.

pound force. Symbol lbf. A unit of force. *See* Table 6E.

pound weight. A unit equal to g POUNDAL where g is the local acceleration due to gravity measured in feet per second. It differs from the POUND FORCE in that the standard rather than the local value of the acceleration due to gravity is used for the pound force.

powder photography. An X RAY ANALYSIS method in which the specimen is a randomly orientated crystalline powder.

power. (1) Symbol P. The rate of doing WORK, i.e. of expending energy.
(2) The reciprocal of the focal length in metre of an optical device.
(3) The number of times a number or expression is multiplied by itself; for example

$$y \times y \times y \times y = y^4$$

giving y to the power of 4.

power amplification. The ratio of the power at the output terminals of an amplifier to that at the input terminals.

power amplifier. An amplifier having a power gain. Usually its output is applied to a LOUDSPEAKER or to an AERIAL.

power factor. The ratio of the true power

to the *apparent power* in an alternating current circuit, the apparent power being the product of the ROOT MEAN SQUARE voltage and the root mean square current in the circuit.

power of accommodation. The change in power of the EYE due to alteration in the focal length of the crystalline lens, which is produced by changes in ciliary muscle tension. The maximum power of accommodation of the human eye is about 4 dioptre.

power pack. A circuit supplying power to another circuit. For maximum efficiency the electric IMPEDANCE of each circuit should match.

power reactor. *See* NUCLEAR REACTOR TYPES.

power transistor. A transistor dissipating more than about 1 watt. Such transistors are used for switching and amplification. For the higher powers they require some form of temperature control.

Poynting's theorem. The rate of energy transfer from ELECTROMAGNETIC RADIATION is proportional to the product of the electric and magnetic field strengths associated with the radiation.

Poynting vector. The VECTOR PRODUCT of the electric and magnetic field strengths at any point. Its surface integral thus gives the rate of energy transfer from electromagnetic radiation associated with the fields.

preamplifier. An amplifier used as an earlier stage to the main amplifier.

precession. The motion of a body spinning on an axis OA say, while A is moving round another axis through O. Precessional motion can occur for a top or a GYROSCOPE and is shown by the Earth. *See also* LARMOR PRECESSION.

predator prey relations. An example of an automatic type of FEEDBACK system: if predators increase, prey are killed off faster and so their numbers diminish. Predators then cannot find sufficient food and so they die or move. The prey population then increases and so the cycle is maintained.

presbyopia. A defect of the EYE in which near objects cannot be focused clearly. It is caused by hardening of the eye's crystalline lens so that the ciliary muscle cannot increase its power sufficiently. The defect can be corrected by wearing convex spectacle lenses for close work.

pressure. Symbol p. The force per unit area. In a fluid at rest, the fluid pressure at any point at a given depth is the same in all directions. In a liquid, the *liquid pressure p* increases uniformly with depth h as given by the formula

$$p = \rho gh$$

where ρ is the liquid density and g the acceleration due to gravity. In a gas the pressure decreases exponentially with height. The SI unit of pressure is the pascal. *See* Table 2. *See also* ATMOSPHERIC PRESSURE; ABSOLUTE PRESSURE.

pressure coefficient. The coefficient of pressure increase with absolute temperature of a gas at constant volume. It is given by

$$(p - p_0)/[p_0(T - 273)]$$

where p and p_0 are the pressures at temperatures T K and 273 K respectively. It has practically the same value of $1/273$ K^{-1} for all gases. *See also* COEFFICIENT OF EXPANSION.

pressure gauge. An instrument for measuring pressure. In *primary gauges* the pressure is balanced against a known force. For example in the liquid column MANOMETER, the pressure difference between two gases above the liquid, density ρ, in the arms of a U tube is given by $g\rho h$ where h is the difference of height in the two arms and g is the acceleration due to gravity. The McLEOD GAUGE is another primary gauge example. *Secondary gauges* depend on the measurement of some

physical property that varies with pressure. Examples are the PIRANI GAUGE and the IONIZATION GAUGE.

pressure head. The head, i.e. height of liquid, capable of exerting a given pressure.

pressure transducer. A device using pressure to give some sort of signal which is then converted into another type of signal, often electric. For example in an electromanometer the deformation of the transducer element, under the action of pressure, is transformed into an electric signal which is amplified and recorded.

pressurized water reactor. A NUCLEAR REACTOR in which water is subjected to pressure to prevent boiling and is used as both coolant and moderator.

Prevost's theory of exchanges. A body at a given temperature radiates the same amount of energy no matter what the temperature of its surroundings. The theory is consistent with the observations that a body loses energy to colder surroundings and gains energy from hotter surroundings, and tends to a state of thermal equilibrium for which equal amounts of energy are emitted and received.

primary. (1) The body around with another body orbits. Thus in the solar system the Sun is the Earth's primary. (2) Short for PRIMARY WINDING.

primary cell. *See* CELL (def. 1).

primary colours. For additive COLORIMETRY, a set of three coloured lights which when mixed together in equal proportions produce white light. For subtractive colorimetry, a set of three coloured pigments which when mixed together in equal proportions produce black pigment. *See also* COLOUR VISION.

primary cosmic rays. *See* COSMIC RAYS.

primary electrons. Electrons incident on a substance from which they may release secondary electrons. *See* SECONDARY EMISSION.

primary standard. A standard used nationally or internationally as the basis of a unit. *Compare* SECONDARY STANDARD.

primary winding. The winding on the input side of a TRANSFORMER.

prime number. A number which has no factors other than itself and one.

principal axis. Another name for OPTICAL AXIS.

principal focus. The point to which light parallel to the optical axis is converged or from which it appears to be diverged by an optical system.

principal points. *See* CENTRED OPTICAL SYSTEM.

principal quantum number. *See* ATOMIC ORBITAL.

printed circuit. An electric circuit in which the conducting interconnections and some components are made by thin channels of metal coated on an insulating board. Initially the whole board is covered with a conducting film; the required parts are then covered with protecting film and the unwanted ones etched away. Double-sided printed circuit boards are also produced, with contacts to connect the circuits if desired.

prism. (1) A solid geometric figure having two faces consisting of congruent parallel POLYGONS. The other faces are formed by joining corresponding vertices of the polygons.
(2) A piece of transparent material such as quartz, glass or rock salt, cut in the shape of a prism and used for deviating and/or dispersing suitable incident radiation. The commonest type, shown in fig. P10, is the triangular glass prism used for light. Prisms are also used for producing total internal reflection (*see* TOTAL REFLECTING PRISM) and for inverting images. *See*

also LIMITING ANGLE OF PRISM; MINIMUM DEVIATION.

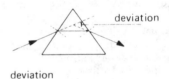

P10 Deviation by prism

prismatic. Denoting an optical instrument using one or more prisms.

prismatic colours. Colours produced when daylight is dispersed by a prism. *See* VISIBLE SPECTRUM.

prism binoculars. *See* BINOCULARS.

prism dioptre. Symbol *P*. A quantity defined by the equation

$$P = 100 \tan \theta$$

where θ is the angle of deviation produced by a THIN PRISM.

probability. A numerical expression for the likelihood that an event will occur. If an experiment can result in *n* equally likely but mutually exclusive outcomes, *r* of which correspond to the occurrence of some event E, then the probability of E occurring is *r/n*; for example the probability of drawing a face card in a single random draw from a well-shuffled pack of playing cards is 12/52, i.e. 1/4, since there are 12 face cards in the pack and 52 mutually exclusive equally likely outcomes. The probability of E occurring *q* times in succession is $(r/n)^q$; thus the probability of a tossed coin coming down heads in each of three consecutive throws is $(1/2)^3$, i.e. 1/8, since the probability of it coming down heads in a single throw is 1/2. A very important application of probability theory is to genetics, for example in assessing the risk to children when one or both of the parents has a defective gene of some kind.

probe. A resonant conductor inserted into a WAVELENGTH or CAVITY RESONATOR in order to either inject or extract energy.

product. The result of multiplying two or more numbers or quantities together.

program. A set of instructions in suitable form for feeding into a COMPUTER.

progression. Another name for SEQUENCE.

progressive wave. A disturbance, either continuous or transient, which travels through a medium or space. The resulting displacements of the medium are small and the medium returns to its initial state after the disturbance has passed. *See also* PLANE PROGRESSIVE WAVE. *Compare* STANDING WAVE.

projectile trajectory. The path of an object for a given initial velocity. If an object starts from Earth with velocity V at an angle θ to the horizontal, its initial vertical and horizontal speeds are respectively $V \sin \theta$ and $V \cos \theta$. If time t elapses before the body returns to Earth, it travels a horizontal distance (i.e. has a *range of projectile*) of $tV \cos \theta$ since no acceleration acts horizontally. By applying the equations governing LINEAR MOTION WITH CONSTANT ACCELERATION to the vertical movement, it can be shown that the path of the object is a PARABOLA and that for a given velocity magnitude the range is a maximum when θ is 45°.

The flight of a golfball is a demonstration of the effects of air resistance and back spin on the trajectory of a projectile, as illustrated in fig. P11. The sold curve gives

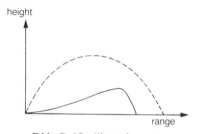

P11 Golfball's trajectory

the golfball's trajectory, and the dashed one the trajectory for a particle with the same initial velocity but moving under gravity only. *See also* ESCAPE VELOCITY.

projector. An optical instrument used for showing slides on a screen; its essential features are shown in fig. P12. The slide S is placed in front of condensing lens C. This lens forms an image of bright point source P at the objective lens L so that the maximum amount of light passes through L. L is of short focal length, in the range 10-20 centimetre, so that a highly magnified image of the slide is obtained on screen A.

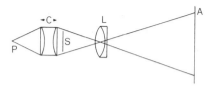

P12 Projector

prominence. An eruption of gas in the Sun's upper CHROMOSPHERE, causing matter to be ejected into space in vast streamers. Many prominences appear to stand motionless over long periods; these have SUNSPOT associations.

prompt neutron. A neutron produced in a NUCLEUS REACTOR by primary fission rather than by decay of a fission product.

propagation loss. The energy loss due to absorption, scattering and beam-spreading from a beam of electromagnetic radiation.

propagation vector. A vector whose direction is that of propagation of a sinusoidal wave and whose magnitude is $2\pi/\lambda$ where λ is the wavelength.

propellant. (1) The fuel, including the oxidant, used in a space rocket.
(2) The explosive charge used to fire a bullet or shell.
(3) An inert gaseous substance, liquefied under pressure, used to expel the contents of an aerosol can.

proper fraction. *See* FRACTION.

proper function. Anglicized form of eigenfunction. *See* CHARACTERISTIC FUNCTION.

proper value. Anglicized form of eigenvalue. *See* CHARACTERISTIC VALUE.

proportional counter. A detector of ionizing radiation for which the size of the output pulse is proportional to the number of ions formed in the initial ionizing event. It operates in a voltage region lying between that of the GEIGER COUNTER and of the IONIZATION CHAMBER.

proprioceptor. An organ found in all skeletal muscles. Its function is to sense what is happening in the muscle and report back to the central nervous system, using electric impulses. If appropriate, electric signals to change the action are then transmitted back to the muscle. The system thus effectively employs negative FEEDBACK.

protective relay. A RELAY that causes the opening of a CIRCUIT-BREAKER in order to disconnect faulty apparatus from the supply and thus protect the apparatus from the damaging effects of overloads and internal faults.

proton. A stable positively charged ELEMENTARY PARTICLE. Protons are responsible for the charge on a nucleus: the nucleus of the hydrogen atom is a proton and all other nuclei contain protons. The proton charge has the same magnitude as the electronic charge. The rest mass of the proton is

$$1.672\ 62 \times 10^{-27}\ \text{kilogramme}$$

A proton is thought to be built from three QUARK particles.

proton microscope. A device similar to the ELECTRON MICROSCOPE but using a beam of protons instead of electrons. This results in better resolving power and contrast.

proton number. Another name for ATOMIC NUMBER.

proton synchrotron. A particle accelerator of large radius, capable of accelerating protons to very high energies: 500 giga-electronvolt has been reached by an American machine. The proton synchrotron uses a varying frequency electric field in contrast to the fixed frequency one used in the SYNCHROTRON. In other respects the machines are basically similar.

pseudo vector. (1) A quantity which under space rotation transforms as a vector, but under space inversion transforms as a vector together with a change of sign.
(2) A quantity which transforms as a FOUR VECTOR under Lorentz transformation but has an additional sign change under one or both of space reflection and time reflection.

psi particle. *Early name for* J/PSI PARTICLE.

psychrometer. A type of WET AND DRY BULB HYGROMETER.

psychrometry. The measurement of atmospheric humidity.

Ptolemaic system. A system based on the concept that the components of the universe, such as the Sun, Moon etc., revolved around the Earth. It was superseded by the COPERNICAN SYSTEM.

p-type conductivity. Conductivity produced in a SEMICONDUCTOR by a flow of HOLES.

p-type semiconductor. An extrinsic SEMICONDUCTOR in which the mobile HOLE density exceeds that of conduction-band electrons.

pulsar. A star which is a source of regular fluctuating electromagnetic radiation, usually of radio frequency but also of optical, gamma ray and X ray frequencies. A pulsar is probably a NEUTRON STAR whose directions of radiation are governed by its magnetic field. Since the star is rotating, the radiation would only be observable from Earth for certain orientations of the star.

pulsating current. A unidirectional current of regularly varying magnitude.

pulsating star. A star radiating with variable intensity due to regular volume variations in its surface atmosphere.

pulse. A sudden increase in magnitude of a physical quantity, shortly followed by a rapid decrease; examples are a single wave and a wave train. In practice the graph of disturbance magnitude against time is never simple. A typical so-called *rectangular pulse* is illustrated in fig. P13. The various associated quantities are as marked.

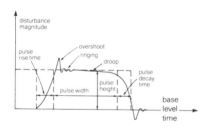

P13 Rectangular pulse

pulse height analyser. An electronic circuit for sorting voltage impulses according to their amplitude. *See also* SCINTILLATION COUNTER.

pulse modulation. MODULATION in which information is transmitted by controlling the amplitude, duration, position or presence of a series of electric pulses.

pump. Any of several kinds of device.
(1) A machine for raising fluid from a lower to a higher level.
(2) A machine for imparting energy to a fluid by for example increasing the pressure exerted by the fluid.
(3) A machine for transporting a fluid from one place to another.

See also FILTER PUMP; FORCE PUMP; LIFT PUMP; SODIUM PUMP; VACUUM PUMP.

punched card. A card on which data is coded in the form of holes, different hole patterns corresponding to different symbols. The system is sometimes used in a COMPUTER.

pupil. A central opening in the iris through which light enters the EYE. The pupil adjusts its size according to the incident light intensity, so preventing overloading of the retinal system. An automatic camera shutter performs a similar function for a photographic film.

pure spectrum. A spectrum in which the various images of the source slit (*see* SPECTROMETER) do not overlap, i.e. they are monochromatic.

P wave. The part of an electrocardiograph recording which precedes auricular contraction. *See* ELECTROCARDIOGRAPHY.

pyknometer. A glass device, illustrated in fig. P14, used for measuring liquid relative density. The liquid is sucked into the tube until it reaches the graduation mark, thus giving a precise volume of liquid whose mass is found in the usual way. The technique described for the RELATIVE DENSITY BOTTLE is then followed.

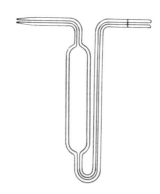

P14 Pyknometer

pyramid. A solid geometric figure whose base is a POLYGON and whose other faces are triangles with a common apex. A pyramid is said to be regular if its base is a regular polygon.

pyroelectricity. The development, on heating, of electric charges of opposite sign on opposite faces of certain crystals such as tourmaline.

pyrometer. An instrument for measuring high temperatures, often at a distance.

The *optical pyrometer* determines the temperature of a radiant source from the colour of the radiation. The colour may be judged by eye, but is generally matched with that of a standard radiator, such as a heated filament, whose temperature can be varied.

The *colour pyrometer* is a type of optical pyrometer in which a calibrated wedge-shaped filter transmits only red and green light. The temperature is obtained from the wedge setting where the source appears white.

The *radiation pyrometer* is a pyrometer in which thermal radiation from the hot body is focused on a sensitive THERMO-COUPLE whose electromotive force is measured.

The *thermocouple pyrometer* permits direct measurement of a temperature by immersing a THERMOCOUPLE in the hot substance.

pyrometry. The science of high temperature measurement, usually by investigation of the emitted radiation.

Pythagoras' theorem. In a right-angled triangle, the square of the length of the hypotenuse is equal to the sum of the squares of the lengths of the other two sides, i.e. in the triangle ABC having a right angle at A,

$$BC^2 = AB^2 + AC^2$$

Q

QCD. Abbrev. for QUANTUM CHROMO-DYNAMICS.

QED. Abrev. for QUANTUM ELECTRO-DYNAMICS.

Q factor. Short for QUALITY FACTOR.

QRS complex. The variation of electric potential of the heart with time during ventricular contraction.

quadrant. A segment of a circle subtending an angle of 90° at the centre of the circle, i.e. it is a quarter of the circle.

quadrant electrometer. An electrometer in which a light metal vane is suspended inside four metal quadrants by a torsion wire. Opposite pairs of quadrants are connected and the potential difference to be measured is applied between the pairs. When the vane is given a potential large compared with that of the quadrants, its deflection is proportional to the quadrant potential difference. For many but not all applications the instrument is obsolescent.

quadratic equation. An equation of the form

$$ay^2 + by + c = 0$$

where y is the variable and a, b and c are constants. The solutions are

$$y = (-b \pm (b^2 - 4ac)^{1/2})/(2a)$$

quadrature. The relation between two periodic quantities of similar wave form when there is a phase difference of 90° between them. The quantities are said to be *in quadrature*.

quadrilateral. Any plane figure bounded by four straight lines.

quadripole. An electric network with only four terminals, i.e. two input and two output terminals. If on interchange of input and output terminals the electric properties remain unchanged, the quadripole is said to be *symmetrical*.

quadrupole. A set of four electric charges of equal magnitude, or of two magnetic dipoles of equal moment.

quality. (1) The *timbre* of a sound, resulting from the presence of HARMONICS.
(2) The fidelity of reproduction of a sound.

quality factor. A number assigned to a system undergoing FORCED OSCILLATIONS. It is defined as 2π times the maximum energy contained in the system divided by the average energy dissipated per oscillation. The higher the quality factor, the sharper the RESONANCE. It is thus desirable for radio receivers to have a high quality factor since a highly selective response is required.

quantity of charge. The integral of electric current passing with respect to time.

quantity of heat change. The product of the mass of a body and its specific heat capacity and its temperature change.

quantized. Denoting a physical quantity which can only take certain discrete values in a particular system. The quantity therefore changes in steps and cannot vary continuously.

quantum. The smallest amount of energy by which a system can change.

quantum chromodynamics. A theory in particle physics describing interaction of COLOUR. The carriers are GLUONS.

quantum discontinuity. The discontinuous absorption or emission of energy which accompanies a quantum jump.

quantum electrodynamics. The quantum mechanical theory of ELECTROMAGNETIC INTERACTION between particles and between electromagnetic radiation and particles. The carriers are PHOTONS.

quantum electronics. The application of quantum mechanics to the investigation of microwave power generation in solid crystals.

quantum field theory. A theory which utilizes appropriate operators, obeying certain commutation relations, to represent all the physical observables in a system. The total energy, momentum, charge etc. of the field is built up additively from the individual contributions to each of these variables of the particles present.

quantum gravity. A quantum mechanical version of the general theory of RELATIVITY. At present it is incomplete

quantum mechanics. A mathematical theory developed from QUANTUM THEORY and concerned with the mechanics of atomic and related systems in terms of measurable quantities. It embraces WAVE MECHANICS and MATRIX MECHANICS. In relativistic quantum mechanics the simple theory was extended so that the principle of relativity is also satisfied; the properties of electron spin then followed naturally from the relativistic form of SCHRÖDINGER'S EQUATION.

quantum number. Any of several numbers which together characterize the state of an atomic or molecular system. The numbers are directly related to the eigen functions (see CHARACTERISTIC FUNCTION) associated with SCHRÖDINGER'S EQUATION. Elementary particles also have quantum numbers

to describe their properties. *See also* ATOMIC ORBITAL.

quantum optics. Optics based on the concept that light is emitted from a source in the form of discrete packets of waves called quanta or photons, which individually act like particles but which in large numbers can behave like waves. The subject is thus essentially mathematical. The analysis leads to conclusions in agreement with experimental results.

quantum response of eye. A subject concerned with the minimum number n of photons necessary to stimulate a retinal receptor. The probability P that n photons will be absorbed during one flash for different values of the incident intensity I at the retina may be calculated for different values of n. The fraction of times a subject sees a light of a certain intensity I', which is proportional to I, in a number of trials is a measure of the probability of the receptors responding. As a result experimental plots of P against $\log I$ can be obtained. By comparing their shapes with those of the theoretical curves, n can be found. Such experiments on different subjects, using SCOTOPIC VISION, yielded values of n in the range 1–8. It is to be noted that $n = 1$ does not imply that, for this subject, a single quantum absorbed by a single receptor produced a sensation of sight, since for the optic nerve to transmit an electric impulse from the retina, several receptors need to be activated. *Compare* COINCIDENCE CIRCUIT.

quantum statistics. Statistics concerned with the distribution of particles of a given type amongst the various quantized energy levels available. There are two types of such statistics: BOSE-EINSTEIN STATISTICS and FERMI-DIRAC STATISTICS. Both lead to the classical MAXWELL-BOLTZMANN DISTRIBUTION LAW at sufficiently high temperatures, i.e. when a large number of energy levels are excited.

quantum theory. A theory based on the idea that the energy of a system cannot change continuously but only in discrete amounts, known as quanta. Planck first introduced the concept in order to explain the so-called *ultraviolet catastrophe*, i.e. the characteristic maximum found in the spectral distribution curve of a BLACK BODY. Other early applications of quantum theory were the BOHR THEORY of the atom and Einstein's explanation of the PHOTOELEC-TRIC EFFECT. QUANTUM MECHANICS was a later development of the theory.

quark. A fundamental constituent of matter. Every hadron, i.e. MESON and BARYON, is considered to have a quark substructure; a LEPTON does not have any substructure. There are several types of quark: *bottom quark, charmed quark, down quark, strange quark, top quark, up quark.* The bottom quark is the most massive quark currently known; like the strange and down quarks, its charge (taking the proton charge as 1) is $-1/3$. The least massive quark is the up quark, with charge $2/3$; it is the first GENERATION partner of the down quark, both of which are constituents of the proton and of the neutron. The top quark is a hypothetical particle of charge $2/3$ which would complete the third generation by partnering the bottom quark. The strange quark partners the charmed quark, of charge $2/3$, in the second generation.

quark confinement. The theory that there is some reason why quarks may exist inside elementary particles but not be observable in any experiment.

quarter wavelength line. A transmission line a quarter of a wavelength long, used as an impedance matching device. The main use is in the higher ratio frequency systems.

quarter wave plate. A thin mica or quartz sheet cut so that its interference colour in white light is pale grey, and of a thickness such that a quarter wavelength path difference is introduced between the unseparated ordinary and extraordinary rays into which normally incident unpolarized sodium light is split. When light which is plane polarized with its vibration plane at 45° to the principal plane of the plate is incident on the plate, circularly polarized light emerges; if the angle between the planes differs from 45°, elliptically polarized light emerges. *See also* POLARIZATION (electromagnetic); DOUBLE REFRACTION.

quartz. A colourless or white natural crystalline form of silica. It exhibits DOU-BLE REFRACTION and may be laevo- or dextrorotatory, the amount of rotation varying with wavelength. Its transmission range is 180 nanometre to 4000 nanometre. Quartz also exhibits the PIEZOELECTRIC EFFECT.

quartz clock. *See* CLOCKS.

quartz iodine lamp. A tungsten filament electric lamp moulded in a quartz envelope filled with iodine vapour. The working temperature is between 500° C and 600° C. Electric energy is converted to light much more efficiently than for the ordinary tungsten filament lamp.

quartz oscillator. A circuit whose resonant frequency of oscillation is stabilized by a QUARTZ crystal. *See* PIEZOELECTRIC OSCILLATOR.

quartz wedge. A very small angled wedge of quartz, by means of which any desired thickness (within limits) of quartz can be inserted in front of the objective of a polarizing MICROSCOPE. By adjusting the wedge, it is possible to cancel out the phase difference introduced by the specimen and to deduce the value of this phase difference from the thickness of quartz used.

quasar. A compact extragalactic source of very strong electromagnetic radiation, much of it in the infrared. Most quasars have a very large RED SHIFT, which if due to the DOPPLER EFFECT would indicate an enormous speed of recession and hence very great distance. There is much theoretical speculation on the nature of quasars.

quasi particle. A system of many interacting particles which has particle-like properties but does not exist as a free particle, an example being a PHONON.

quench. A capacitor or resistor or combination of the two placed across contacts, for example the make and break contacts of an INDUCTION COIL, to inhibit sparking when the current is cut off to an inductive circuit.

quenching. (1) Fast cooling of a hot metal, used for example in steel hardening.
(2) Sudden termination of discharge in a GEIGER COUNTER.
(3) A process in which the lifetime of excited atoms, ions etc. is reduced by treatment with a substance which deactivates them.

quick freezing. The preparation of food stuffs for cold storage by passing them through a temperature region of $0°$ C to $-4°$ C as quickly as possible. Maximum ice crystal formation results and so damage to plant and animal tissues, due to the formation of large ice crystals, is avoided.

quiescent current. The current flowing in a circuit when there is no applied signal.

quiet sun. The sun when it has a minimum of sunspots, flares or prominences.

Quincke's tube. A device used to demonstrate the interference of sound. As shown in fig. Q1, the single-frequency sound from

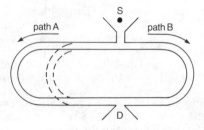

Q1 Quincke's tube

source S can reach detector D by either path A or path B. If the path difference is an odd number of half wavelengths, the vibrations arrive at D out of phase and minimum sound results. If, however, the path difference is an integral number of wavelengths, reinforcement occurs and a loud sound results. By moving the sliding tube a series of positions of maximum and minimum intensity can be obtained. By finding the average distance moved by the tube between consecutive minima; i.e. half a wavelength, the wavelength of the sound can be found and hence its speed.

quotient. The result of dividing one number by another.

Q value. The amount of energy produced in a nuclear reaction.

R

rad. A unit of absorbed DOSE of ionizing radiation, equal to 0.01 joule per kilogramme of material. *See also* RADIATION UNITS.

radar. A system for detecting distant objects by reflecting electromagnetic waves off them: the time taken by the waves to travel to the object and back enables the object distance to be calculated; the direction of the object is given by that of the receiving aerial. The waves may be transmitted either continuously or in pulses. The transmitting and receiving aerials can be rotated in order to scan an area. The return signals are fed to a cathode ray tube where they may be displayed in a variety of ways, depending on the type of scanning. A change in frequency of the reflected waves relative to the incident waves is produced by a moving object (*see* DOPPLER EFFECT). This frequency change is used to determine the velocity of the object; and the radar device is then known as *Doppler radar*. The name radar is an acronym for radio direction and ranging.

radial field. A field in which the lines of force diverge from or converge to a single point.

radian. The angle subtended at the centre of a circle by an arc of length equal to the radius of the circle; π radian is thus 180°.

radiance. Symbol L_e. The product of the RADIANT INTENSITY per unit area of a surface in a particular direction and the secant of the angle θ between the surface and that direction. Hence

$$L_e = \sec \theta \, dI_e/dA$$

where I_e is the radiant intensity and A the area of the surface. The unit is watt per steradian per square metre. *Compare* LUMINANCE.

radiant emittance. Another name for RADIANT EXITANCE.

radiant exitance. Symbol M_e. The RADIANT FLUX leaving a surface per unit area. The unit is watt per square metre. *Compare* LUMINOUS EXITANCE.

radiant flux. The rate of flow of radiant energy. The SI unit is the watt. *Compare* LUMINOUS FLUX.

radiant heat. *See* INFRARED RADIATION.

radiant intensity. Symbol I_e. The amount of radiation emitted from a point source per second in a given direction per unit solid angle. The unit is watt per steradian. *Compare* LUMINOUS INTENSITY.

radiation. (1) The emission of energy in the form of rays, waves or particle streams.
(2) The actual energy emitted in such a process.

radiation belts. Another name for VAN ALLEN BELTS.

radiation damage. Any harmful change produced by exposure to radiation. The damage may result from ionization, electronic excitation, transmutation or displacement of atoms. In living material such processes can alter the genetic structure of cells, interfere with their division or kill them, leading to long-term damage to the whole organism.

241

radiation laws. *See* KIRCHOFF'S LAW; STEFAN'S LAW; PLANCK RADIATION FORMULA.

radiation physics. The study of radiation, particularly IONIZING RADIATION and its physical effects on matter.

radiation pressure. The pressure exerted on a surface by the impact of radiation on it.

radiation pyrometer. *See* PYROMETER.

radiation sickness. Illness resulting from exposure of body tissue to large doses of IONIZING RADIATION, such as from a nuclear explosion.

radiation temperature. The surface temperature of a celestial body, calculated using STEFAN'S LAW on the assumption that the body behaves as a BLACK BODY.

radiation units. Units used to express the activity of a RADIONUCLIDE and ionizing radiation DOSE. The recommended SI units are the *becqueral*, symbol Bq, the *gray*, symbol Gy, the *sievert*, symbol Sv, and the coulomb per kilogramme, respectively replacing the CURIE, RAD, REM and RONTGEN. The temporary use of these old units is approved to allow time for familiarization with the new ones. The relationships between old and new are as follows:

1 curie = 3.7 × 10^{10} becquerel
1 rad = 10^{-2} gray
1 rem = 10^{-2} sievert
1 rontgen = 2.58 × 10^{-4} coulomb per kilogramme

radiative capture. See CAPTURE.

radiative collision. A collision between charged particles in which part of the kinetic energy is radiated as photons.

radiative transition. The emission of electromagnetic radiation due to the change of a quantum-mechanical system from one energy state to another.

radio. The communication of signals by means of RADIO FREQUENCY RADIATION. All forms of such communication, including television, are included in the term, which is however sometimes used specifically for sound broadcasting. A radio transmitter generates a CARRIER WAVE of fixed frequency, and it is by MODULATION of this wave that the information is transmitted. The modulated wave is projected from a transmitter aerial and carried either as a GROUND WAVE or a SKY WAVE to a receiver aerial where it is demodulated.

radioactive. Exhibiting RADIOACTIVITY.

radioactive age. The age of a geological or archaeological specimen determined by RADIOACTIVE DATING.

radioactive dating. Any method of determining the age of archaeological and fossil remains, rocks etc. by measuring the quantity of a specific radioisotope in a sample. *See* CARBON DATING; POTASSIUM-ARGON DATING; RUBIDIUM-STRONTIUM DATING; URANIUM DATING.

radioactive decay. Spontaneous nuclear DISINTEGRATION.

radioactive series. A series of radioisotopes such that each member of the series is formed by the decay of the preceding member. Four such series exist: the *uranium, thorium, actinium* and *neptunium series*, whose parents are respectively uranium-238, thorium-232, uranium-235 and plutonium-241. The first three parent radioisotopes have half lives in the region 10^9 year to 10^{10} year, in contrast to the much shorter half life of plutonium-241. The first three series all end in stable isotopes of lead whereas the neptunium series ends with thallium-81.

radioactive tracer. A definite quantity of radioisotope introduced into a mechanical or biological system to enable its route through the system and its regions of concentration to be monitored. This is achieved using a radiation detector outside the system.

radioactive waste. Any waste material containing radionuclides. It may arise from the mining of radioactive ores, from nuclear power stations, from the manufacture of nuclear weapons and from laboratories. The first stage in the disposal of such waste is to retrieve any reusable material; the residue is then buried either at sea or in stable deep cavities. Since much waste has a very long half life and is highly dangerous to any unprotected living matter it encounters, waste disposal is the subject of controversy.

radioactivity. The spontaneous disintegration of certain atomic nuclei with ALPHA PARTICLE, BETA PARTICLE or GAMMA RAY emission. If the nuclei are naturally occurring, the process is known as *natural radioactivity*. If the radioactivity is induced by irradiation with, for example, neutrons, the process is known as *artificial radioactivity*. Both natural and artificial radioactivity decrease exponentially with time. *See* HALF LIFE.

radio astronomy. The branch of astronomy which studies the radio frequency emissions of celestial bodies by means of RADIO TELESCOPES.

radiocardiography. A technique in which barium-137 of activity 10 millicurie is rapidly injected into the subclavian vein and so enters the right ventricle of the heart almost immediately. The flow is monitored by suitably placed radiation counters. Deviation of the RADIOACTIVITY from the standard pattern permits rapid diagnosis of the heart defect. The alternative technique of working a catheter from the bloodstream to the heart is much more risky and time consuming.

radiocarbon dating. Another name for CARBON DATING.

radio chemistry. The branch of chemistry concerned with radioactive materials.

radio direction and ranging. Original name for RADAR and from which the word radar is derived.

radio frequency. A frequency lying between 3 kilohertz and 300 gigahertz. *See also* FREQUENCY BAND.

radio frequency radiation. Electromagnetic radiation in the RADIO FREQUENCY range.

radiogenic. Resulting from or produced by radioactive DISINTEGRATION.

radiography. The use of radiation, generally X RAYS, to investigate internal structure. The X rays pass through parts of the object under investigation and form a shadow image, i.e. a *radiograph*, on a fluorescent screen or photographic film.

radio interferometry. *See* RADIO TELESCOPE.

radioisotope. An isotope that is radioactive.

radiology. The study of the science and applications of penetrating ionizing radiation such as X rays and gamma rays.

radiolucent. Denoting materials which partially transmit ionizing radiation. In contrast, materials passing most of the incident radiation are described as *radiotransparent* and those passing almost none of it are described as *radio-opaque.*

radioluminescence. The emission of visible electromagnetic radiation from a radioactive substance.

radiolysis. The process of decomposing a substance by bombarding it with radiation.

radiometer. An instrument for the measurement of radiant electromagnetic energy, especially that in the infrared region.

radionuclide. A NUCLIDE which exhibits RADIOACTIVITY.

radio-opaque. *See* RADIOLUCENT.

radio source. A source of radio waves in space. Examples include SUPERNOVAE, PULSARS, QUASARS, radio galaxies and also some stars, including the Sun.

radio telescope. An instrument for detecting radio waves from a particular direction in space. The simplest instrument consists of a steerable paraboloid dish aerial and ancillary amplifiers. In order that the dish may reflect the incoming radio waves to its focal point, it is necessary for its surface to be accurately constructed; thus for incoming radiation of 1 centimetre wavelength, a 100 metre diameter dish must be accurate to the nearest millimetre.

To circumvent the problem of constructing large dishes with high accuracy, the technique of *radio interferometry* has been developed. Two fixed or steerable small aerials at a suitable distance apart are connected to the same radio receiver. Since the aerials are separated, the path difference from a radio source to them will vary as the Earth rotates, i.e. as the source appears to move. The outputs of the aerials will therefore interfere with one another, giving a resultant signal of varying intensity. The resolution obtainable with the two small dishes is higher than that given by a single much larger dish. *See also* APERTURE SYNTHESIS.

radiotherapy. The use of ionizing radiation for medical treatment. The most commonly employed radiation is X RAYS. A beam of GAMMA RAYS is also frequently used and is generally obtained from cobalt-60. Electron, proton and neutron beams are not so much used but ALPHA RAYS are sometimes employed. Frequently the cross-fire technique is used: a deep-seated region is irradiated from several directions in order to reduce damage to healthy tissue. Alternatively the radiation beam may be moved around the region under irradiation.

radiotransparent. *See* RADIOLUCENT.

radio waves. Another name for RADIO FREQUENCY RADIATION.

radio window. *See* ATMOSPHERIC WINDOWS.

radium. Symbol Ra. A radioactive element whose most stable isotope, radium-226, has a half life of 1622 year. The element is used as a radioactive source in research and medicine.

radius. The distance between the centre of a circle or sphere and its circumference or surface.

radius of curvature. The reciprocal of the CURVATURE.

radius of gyration. Symbol k. The quantity $(I/m)^{1/2}$ where I is the moment of inertia of the body about the axis to which the radius refers, and m is the mass of the body.

radius vector. *See* POLAR CO-ORDINATES.

radix. The BASE of a number system. For example 10 is the radix of decimal notation.

rainbow. An arc of colour bands seen in the sky when sunlight is refracted, dispersed and totally internally reflected by raindrops, as illustrated in fig. R1. The rainbow is composed of spectral colours (*see* VISIBLE SPECTRUM) with the violet band on the inside of the bow, the mean altitude being 42°. If, as sometimes happens, a second total internal reflection occurs inside the raindrops, a second fainter rainbow with the red band on the inside of the bow can be seen at a mean altitude of 51°

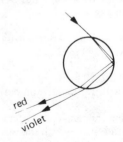

R1 Dispersion by rain drop

rales. The characteristic sound of respiratory disorder.

RAM. *See* MEMORY.

Raman effect. The decrease or increase of wavelength which occurs for a small proportion of monochromatic radiation passing through a substance. It is explained by assuming that after collision with incident photons, some of the excited molecules return to a higher or lower vibrational or rotational energy level in the ground electronic state. Information on molecular structure may thus be obtained from the study of Raman spectra, i.e. by *Raman spectroscopy*. For this purpose, LASER illumination is now generally used in order to obtain higher intensity.

Raman scattering. Another name for RAMAN EFFECT.

Raman spectroscopy. *See* RAMAN EFFECT.

Ramsden eyepiece. An eyepiece consisting of two identical planoconvex lenses with their convex surfaces facing each other and separated by a distance equal to two-thirds of the focal length of either lens.

random access memory. *See* MEMORY.

random error. *See* ERROR (def. 1).

random walk. A displacement followed by a given number of other displacements of equal magnitude but different directions. Calculation of the probability that the final displacement from the start will have a given value is relevant to many physical problems, for example those concerning diffusion and transport phenomena.

range of ionizing particle. The distance moved in a given material by an ionizing particle of particular energy before it just ceases to ionize.

range of projectile. *See* PROJECTILE TRAJECTORY.

ratio of specific heats. The ratio of the SPECIFIC HEAT CAPACITY of a substance at constant pressure to its specific heat capacity at constant volume.

rational number. A number which is an integer or can be expressed as a quotient of integers.

ray. A line drawn in the direction of travel of light or other radiation.

Rayleigh scattering. Radiation scattering by objects small compared with the wavelength of the radiation. The scattered intensity varies inversely as the fourth power of the wavelength, thereby explaining the colour change of the sky from reddish to blue, at sunrise and vice versa at sunset: at sunrise and sunset the light path is at its longest.

Rayleigh's criterion. A criterion concerned with RESOLVING POWER. For an optical magnifying system it states that resolution of two points is just possible when the cental maximum of the diffraction pattern due to one lies on the first dark ring due to the other, as shown in fig. R2. For a spectrometer it states that resolution of two wavelengths of equal intensity is just possible when the central maximum of one falls on the first minimum of the other.

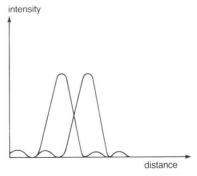

R2 Rayleigh criterion for resolution

rbe. The ratio of the REM to the RAD, the value giving an indication of the biological effectiveness of radiation: rbe is an abbreviation for relative biological effectiveness.

reactance. Symbol X. A property of an alternating current circuit containing inductance and/or capacitance. For a purely inductive circuit the reactance is $2\pi fL$, while for a purely capacitative circuit it is $1/(2\pi fC)$; f is the current frequency and L and C respectively represent inductance and capacitance. *See also* IMPEDANCE (electric).

reaction. An equal and opposite force which arises when any force acts on a body or system. *See* NEWTON'S LAWS OF MOTION.

reactor. (1) An apparatus for producing energy by nuclear FISSION or FUSION.
  (2) A component producing REACTANCE in an electric circuit.

read-only memory. *See* MEMORY.

real gas. A gas having molecules of finite size which exert non-negligible forces on each other. It is thus not an IDEAL GAS.

real image. *See* IMAGE.

real liquid. A liquid whose VISCOSITY is not negligible.

Réaumur scale. A temperature scale in which the ice point is taken as zero and the steam point as 80°. It is obsolescent.

reciprocal. A number divided into one; thus the reciprocal of 3 is 1/3.

reciprocal lattice. A geometric concept widely used in crystal diffraction problems.

reciprocal ohm. Former name for SIEMENS.

rectangle. A parallelogram whose angles are all right angles.

rectifier. An electronic component or circuit used for converting alternating current to direct current. The most usual type is now the semiconductor DIODE.

red giant. A type of GIANT STAR emitting red light. Red giants are one of the final phases in the evolution of a normal star, reached when the star's central energy source of hydrogen is exhausted. It has to obtain energy in other ways, causing its size, temperature and brightness to alter.

red shift. The displacement of the wavelengths of lines in the spectrum of a moving body towards longer wavelengths. The shift is due to the DOPPLER EFFECT and indicates that distant stars and galaxies which demonstrate the effect are receding from the Earth, the amount of shift increasing with the speed of recession. For objects beyond our galaxy the recession arises from the expansion of the universe: the greater the speed of recession, and hence red shift, the greater an object's distance.
  A high gravitational field produces a similar shift (*see* EINSTEIN SHIFT).

reduced equation. An EQUATION OF STATE of a gas in which pressure P, volume V and temperature T are replaced by the reduced values P/P_c, V/V_c and T/T_c respectively where P_c, V_c and T_c are respectively the critical pressure, volume and temperature of the substance

reduced mass. The quantity

$$m_1 m_2/(m_1 + m_2)$$

where m_1 and m_2 are respectively the masses of two particles which exert equal and opposite forces on each other and are not subject to any external forces. The motion of either particle with respect to the other as origin is the same, with respect to a fixed origin, as that of a particle whose mass is the reduced mass and which is acted on by the same force.

reflectance. Symbol ρ. The ratio of RADIANT FLUX or LUMINOUS FLUX reflected by a body to the amount of that flux incident on the body.

reflecting telescope. A TELESCOPE in which the object is first imaged by a curved mirror. This introduces no CHROMATIC ABERRATION and, provided a parabolic reflector is used, there is little SPHERICAL ABERRATION either. *See* NEWTONIAN TELESCOPE; CASSEGRAINIAN TELESCOPE; GREGORIAN TELESCOPE.

reflection. The process in which radiation incident on a surface is returned by that surface to the medium of incidence, rather than passing into the medium of the surface.

reflection density. Symbol D. The quantity $-\log_{10}\rho$ where ρ is the REFLECTANCE.

reflection factor. Another name for REFLECTANCE.

reflectivity. Symbol ρ_∞. The REFLECTANCE of a layer of material of such a thickness that no change of its reflectance would occur with increase of thickness.

reflector. Another name for REFLECTING TELESCOPE.

reflex angle. An angle lying between 180° and 360°.

refracting telescope. An optical TELESCOPE, such as the KEPLER TELESCOPE and the GALILEAN TELESCOPE, consisting of two lens systems, the focal length of the objective system being greater than that of the eyepiece system. Difficulty of grinding large lenses has resulted in the REFLECTING TELESCOPE being preferred for astronomical work.

refraction. The change in direction of radiation as it passes from one medium to another. The speed of the radiation also changes.

refractive index. Symbol n (or sometimes μ). The ratio of the sine of the angle of incidence of electromagnetic radiation on a medium to the sine of its angle of refraction in the medium. The refractive index also equals the ratio of the speed of the incident radiation to that of the refracted radiation. If the medium of incidence is a vacuum then the index is known as the *absolute refractive index*, otherwise it is known as the *relative refractive index* to the medium of incidence. Refractive index varies with wavelength, resulting in DISPERSION.

Refractive index may be measured by finding APPARENT DEPTH or, for material in the form of a prism, by SPECTROMETER: if the material is liquid, a hollow prism containing the liquid is used in the spectrometer. CRITICAL ANGLE measurement and interferometric methods may also be used for refractive index determination. Unless otherwise stated, refractive index is taken to mean its value for wavelength 589.3 nanometre for incidence in air, whose absolute refractive index is 1.000 29.

refractivity. A quantity defined as $n-1$ where n is the refractive index. Dividing refractivity by density gives the *specific refractivity*.

refractometer. Any of various instruments or pieces of apparatus used to measure REFRACTIVE INDEX.

refractor. Another name for REFRACTING TELESCOPE.

refrigerant. A substance, usually a volatile liquid or easily liquefied gas, used as a working fluid in a REFRIGERATOR.

refrigerator. A device for maintaining an enclosure at a lower temperature than its surroundings. It may be regarded as a HEAT ENGINE working backwards.

Regnault's hygrometer. A type of HYGROMETER, illustrated in fig. R3. It consists of two silvered tubes A and B mounted side by side, each with a thermometer. A contains ether through which air may be blown, causing cooling due to evaporation of the ether and hence condensation on A. The reading of the thermometer in A at which this occurs is noted and also its reading when, after standing, the condensation clears. From the mean of these two

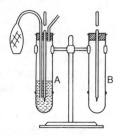

R3 Regnault's hygrometer

readings, i.e. the DEW POINT, and the room temperature as indicated by the thermometer in B, the realtive humidity at this room temperature may be obtained.

relative atomic mass. Symbol A_r. The average mass of the atoms of an element expressed in UNIFIED ATOMIC MASS UNITS. It replaced *atomic weight*, which was measured in atomic mass units.

relative biological effectiveness. *See* RBE.

relative density. For solids and liquids, the ratio of the density of a substance at a stated temperature to the density of water at 4° C. Thus the density expressed in gramme per cubic centimetre numerically equals the relative density. Any substance with a relative density less than 1 will float on water. The relative density of a gas is generally stated with reference to the density of air at standard temperature and presure.

relative density bottle. A small flask having a glass stopper with an axial capillary opening, which facilitates the filling of the bottle with the same amount of liquid on every occasion. By finding the mass m_0 of the empty bottle, then of the bottle full of water m_w and finally of the bottle full of liquid m_1 the relative density of the liquid can be found, being equal to

$$(m_1 - m_0)/(m_w m_0)$$

relative humidity. *See* HUMIDITY (def. 2).

relative molecular mass. Symbol M_r. The sum of the RELATIVE ATOMIC MASS over all the atoms contained in a molecule.

relative permeability. *See* PERMEABILITY (def. 2).

relative permittivity. *See* PERMITTIVITY (def. 2).

relative refractive index. *See* REFRACTIVE INDEX.

relative velocity. The velocity of one body A relative to another body B. At speeds small compared with that of light, A's velocity relative to B is found by compounding (i.e. vectorially adding) A's velocity with B's reversed velocity; in other words, A's velocity is compounded with a velocity which when applied to B would reduce B to rest. Thus for velocities in the same direction the relative velocity is obtained by simple subtraction, and for velocities in opposite directions it is obtained by simple addition.

relativistic mass. The total mass of a moving body, equal to

$$m_0 1(1 - v^2/c^2)^{-\frac{1}{2}}$$

m_0 is the REST MASS of the body under consideration, which is moving at speed v; c is the speed of light in a vacuum. When v is small the relativistic mass of a body thus differs inappreciably from its rest mass. *See also* RELATIVITY.

relativistic state. A state in which the laws of classical physics no longer apply.

relativistic velocity. A velocity whose magnitude is sufficiently close to the speed of light for effects predicted by RELATIVITY theory to be significant.

relativity. A theory put forward to account for the failure of NEWTON'S LAWS OF MOTION when applied to very high speed relative motion.

The *special theory of relativity* refers to nonaccelerated frames of reference. It assumes that physical laws are identical in all such frames and that the speed c of light in a vacuum is constant throughout the universe and independent of the speed of the observer. From these assumptions it follows that a object having length l_0 when at rest relative to an observer has length

$$l_0(1 - v^2/c^2)^{1/2}$$

when moving at speed v relative to the observer; this apparent shortening is consistent with the LORENTZ-FITZGERALD CONTRACTION. The RELATIVISTIC MASS formula also follows; this predicted increase in mass at high speed is supported by the behaviour of high speed particles in for example a CYCLOTRON. The increase of mass with velocity led to the idea of the equivalence of mass and energy as embodied in EINSTEIN'S LAW. Another prediction of the special theory is that of *time dilation*. This is the apparent reduction in the rate of a moving clock with respect to that of an identical clock at rest, when seen by a stationary observer; the reduction factor is

$$(1 - v^2/c^2)^{1/2}$$

Experimental verification of time dilation is provided by observations on the decay times of high velocity pions, which are found to be longer than those of pions at rest.

The special theory was formulated mathematically by Minkowski who considered an event to be specified by four co-ordinates: three of space and one of time. He also suggested that the motion of a particle could be described by a curve in this four-dimensional region, known as *Minkowski space time*. The LORENTZ TRANSFORMATION equations relate the co-ordinate axes of different observers.

In the *general theory of relativity* accelerated systems were included, leading to an analysis of gravitation. The presence of mass is considered to curve the space-time continuum in such a way that a gravitational field results. The theory predicts results in agreement with observation for the change of position with time of the PERIHELION of the planet MERCURY, for the RED SHIFT in an intense gravitational field (*see* EINSTEIN SHIFT) and for the BENDING OF LIGHT RAYS in the neighbourhood of massive bodies. The latter phenomenon is very precisely demonstrated by radio waves from spacecraft and from quasars.

relaxation phenomena. Any phenomena in which a system requires an observable amount of time to reach equilibrium following a sudden disturbance in the system.

relaxation time. A measure of the time taken to reach equilibrium in a relaxation phenomenon. It is the time required for an exponential function of time, t to reach $1/e$ of its initial value. Thus for the function $\exp(-at)$, where a is a constant, the relaxation time is $1/a$. This type of time development arises whenever the rate of change of a physical variable is proportional to the variable, as in RADIOACTIVE DECAY.

relay. An electric device which uses a current variation in one circuit to control the current in another circuit. For example the current in the first circuit may energize an electromagnet which operates a switch in the second circuit. Othe types of relay include devices such as the THYRATRON. Electronic switching using semiconductors is also employed.

reluctivity. The reciprocal of absolute PERMEABILITY.

rem. A unit of radiation. It produces an effect in man equivalent to 1 RAD of X rays. *See also* RADIATION UNITS.

remanence. The residual magnetization of a material when the external field has been reduced to zero in the HYSTERESIS LOOP.

renormalization group. A method of treating systems which involve all scales of length, from the atomic to the very large. It enables the UNIVERSALITY CLASS of a particular material and its CRITICAL EXPONENT values to be predicted. However the problem of critical phenomena in disordered systems remains.

research reactor. *See* NUCLEAR REACTOR TYPES.

resistance. Symbol R. The property of a material whereby it obstructs the flow of electric current through it by dissipating the energy of the current in some other form such as heat. The SI unit is the OHM. *See also* RESISTOR.

resistance thermometer. A type of thermometer whose action depends on changes of resistance with temperature. Usually a coil of fine platinum wire, wound on a mica former and enclosed in a porcelain sheath, forms one arm of a WHEATSTONE BRIDGE, and is inserted in the medium whose temperature is to be measured. The coil's resistances at this temperature and at FIXED POINT temperatures are found in the usual way, and so the unknown temperature can be calculated. The thermometer's range is from $-200°$ C to over $1200°$ C.

resistivity. Symbol ρ. The quantity RA/l where R is the resistance of a conductor of the material under consideration, of uniform cross sectional area A and length l. The SI unit of resistivity is the ohm metre.

resistor. A component used in an electric circuit in order to introduce RESISTANCE. Carbon resistors are widely used in electronic circuits, the value being indicated by the arrangement and colour of rings painted on each resistor. Coils of wire, usually manganin, are also used as resistors.

When resistors are connected *in parallel*, as shown in fig. R4, the potential difference between the terminals of each resistor is the same. The reciprocal of the effective

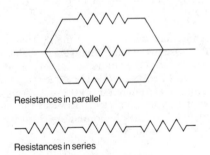

Resistances in parallel

Resistances in series

R4 Resistors in parallel and in series

resistance is then the sum of the reciprocals of the individual resistances. When resistors are connected *in series*, also shown in fig. R4, the potential difference between the first and last plate of the resistor bank is the sum of the potential differences across individual resistors. The effective resistance is then the sum of the individual resistances.

resolution. (1) The separation of a VECTOR into components.
(2) Another name for RESOLVING POWER.

resolving power. A measure of the ability of an optical instrument to produce detectably separate images of close objects, or to separate spectral lines whose wavelengths are close together. *See* RAYLEIGH'S CRITERION.

The *resolving power of the eye*. The angle subtended at the eye by the line joining the centres of two objects, for example two points or two lines, which can just be distinguished as two. The value is about 1 minute of arc.

The *resolving power of a microscope* is the minimum linear distance between the centres of two point objects when they are just seen as two. The value is

$$0.61 \ \lambda/(n \sin U)$$

where λ is the wavelength of the light used, n is the refractive index of the objective space and $2U$ is the angle subtended by the objective at the object.

The *resolving power of an optical spectrometer* is the ratio of wavelength to minimum difference in wavelength that can be detected. For a prism instrument the value of the resolving power if $t \ dn/d\lambda$, where t is the maximum thickness of prism traversed by the beam and $dn/d\lambda$ is the rate of change of refractive index n of the prism material with wavelength λ. For a diffraction grating instrument the resolving power is Nm, where N is the total number of grating lines illuminated and n is the order of the spectrum being used.

The *resolving power of a particle spectrometer* is a quantity which may be given as $e/\delta E$ or $P/\delta P$ or $M/\delta M$ where δE, δP and

δM are respectively the minimum difference in energy, momentum and mass which can be detected at energy E, momentum P and mass M. Sometimes the reciprocals of these quantities are quoted.

The *resolving power of a telescope* is the angle subtended by two objects, just resolved, at the objective or primary mirror. It equals $1.22\lambda/a$ where λ is the wavelength of the radiation used and a the diameter of the telescope objective or mirror.

resonance. (1) The state of a mechanical system, capable of vibration, when it is subjected to a periodic force of the same frequency as its own natural frequency of vibration.

(2) The state of an electric circuit when

$$2\pi fL = 1/(2\pi fC)$$

where f is the current frequency in the circuit and L and C respectively the circuit inductance and capacitance. The IMPEDANCE then has its minimum value, being determined only by the circuit resistance. The circuit itself is said to be *tuned*.

(3) The absorption of a photon with the correct energy to excite a nucleus, atom or molecule from a lower to a higher energy level.

(4) An elementary particle which has a mean life of about 10^{-23} second and which decays by STRONG INTERACTION.

resonance tube. A closed pipe of variable length, usually obtained by immersing more or less of a vertical open pipe in water. A measurement is made of the length of air column in the pipe which resonates with a tuning fork of known frequency. The speed of sound can then be calculated since the relationship between wavelength and resonant length is known (*see* VIBRATIONS IN PIPES). Two positions of resonance are found for the same fork in order to eliminate the end correction.

resonant frequency. The frequency at which RESONANCE occurs.

restitution coefficient. Symbol e. A measure of the ELASTICITY of colliding bodies. If two

spheres A and B moving in the same direction with speeds u_A and u_B collide and then continue to move in the same direction with speeds v_A and v_B, then

$$e(u_A - u_B) = v_B - v_A$$

For a perfectly elastic collision $e = 1$ and the kinetic energy is conserved; for inelastic collision $e < 1$ and the kinetic energy is partially converted to other forms such as heat.

rest energy. The energy corresponding to the REST MASS of a free particle. It has the value m_0c^2 where m_0 is the rest mass and c the speed of light.

rest mass. The mass of a body when it is at rest relative to the observer.

resultant. A VECTOR produced by the combination of two or more other vectors.

retentivity. Another name for REMANENCE.

retina. *See* EYE.

retinene. A small hydrocarbon group split from RHODOPSIN in the retina by the action of light.

reverse bias. A voltage applied in the direction in which an electronic or electric device has the greatest resistance.

reversibility of light rays. A general law of optics which states that a ray of light whose direction is reversed will retraverse its original path.

reversible process. A process whose effects can be counteracted so as to return the system involved to its original thermodynamic state.

Reynold's number. Symbol Re. The quantity $v\rho l/\eta$ where v is the speed of a fluid relative to a solid characterized by a linear measurement l: l could be the radius of a pipe through which the fluid flows; ρ and η are the fluid density and coefficient of VISCOSITY respectively. For a given value of Re, flow lines in any fluid take the same form. At a particular *Re* value, the nature of the flow changes from laminar to turbulent.

rheology. The study of the flow properties of fluids. *See also* VISCOSITY.

rheopexy. The acceleration of the gelation of some thixotropic soils by gentle mechanical agitation. Thus gypsum-water paste, allowed to stand, requires about ten minute to solidify but, when agitated, requires less than a minute.

rheostat. A *variable resistor* connected in series with a circuit in order to control the current through it. It usually consists of a coil of high resistance wire wound on a tube or ring, resistance variation being obtained by means of a sliding contact. The term is usually applied to physically large devices. A small rheostat suitable for electronic circuits is usually called a POTENTIOMETER.

rhodospin. A photosensitive pigment found in the retinal rods of the EYE. It is a complex protein with a relative molecular mass of about 40 000. *See also* COLOUR VISION.

rhombohedral system. *See* CRYSTAL SYSTEM.

rhombus. A parallelogram all of whose sides are equal.

rho meson. A collective name for vector meson resonances of total isospin 1, hypercharge 0, negative charge conjugation parity, and mass, relative to the proton as unit, of 0.82.

Richardson's equation. The basic equation of THERMIONIC EMISSION:

$$j = AT^2 e^{-b/T}$$

where j is the electric current density, T the thermodynamic temperature and A and b are constants. A depends on the nature of the emitting surface and $b = \phi/k$ where ϕ is the WORK. FUNCTION of the surface and k the BOLTZMANN CONSTANT.

right ascension. *See* CELESTIAL SPHERE.

right circular cone. *See* CONE (def. 1).

right-hand rule. *See* FLEMING'S RULES.

rigid body. A body such that the distance between any two points in it remains constant.

rigidity modulus. An ELASTIC MODULUS equal to the tangential force per unit area divided by the angular deformation in radian.

ripple. *See* GRAVITY WAVE.

ripple tank. A tank containing water into which a vibrating strip or spheres just dip. Disturbances then spread out from the vibrators, the water depth being greater than half the ripple wavelength. Observations can be made of straight waves from the strip or circular waves from a sphere, or the phenomena resulting when disturbances from the spheres meet.

RMS. Abbrev. for ROOT MEAN SQUARE.

rocket. A device propelled by the thrust produced by the ejection of high velocity matter, i.e. by JET PROPULSION. The rocket carries both its own fuel and oxidant. The rocket is thus independent of the Earth's atmosphere for lift, thrust and oxygen. It is the only known vehicle to achieve flight in space.

rods. *See* SCOTOPIC VISION.

ROM. *See* MEMORY.

rontgen. Symbol R. A unit of radiation exposure DOSE, equal to the dose of X or gamma radiation which would lose 8.34 × 10^{-3} joule in 1 kilogramme of pure dry air. *See* RADIATION UNITS.

Röntgen rays. Former name for X RAYS.

root. (1) One of two or more identical factors of a given number. The *n*th root of a number *y* is the number which multiplied by itself $n - 1$ times gives *y*; thus 2 is the *square root* of 4 and the *cube root* of 8.

(2) A value of a variable which satisfies a given equation. In general a quadratic equation has 2 roots, a cubic equation 3, and so on.

root mean square. The square root of the average of the squares of a number of values of a variable. For example if of *n* gas molecules n_1 have speed v_1, n_2 have speed v_2 etc., then the root mean square speed of all the molecules is

$$[(n_1 v_1^2 + n_2 v_2^2 + ...)/n]^{1/2}$$

For a periodically varying quantity, such as alternating current or voltage, the root mean square value is the square root of the average of the squares of the quantity over one cycle. For sinusoidal variation the root mean square value is given by the amplitude divided by $2^{1/2}$. The root mean square value is the *effective value*. Thus the power dissipated by an alternating current in a resistance *R* is RI^2 where *I* is the root mean square current.

rotary oil pump. A type of VACUUM PUMP. In one form a cylindrical rotor carrying two diametrically opposite scraping blades separated by a strong spring turns eccentrically inside a cylindrical casing. The casing is connected to the vessel to be evacuated and also, via a valve, to the atmosphere. As the rotor turns, the space between the vessel inlet and one blade increases thus causing gas to enter the space. The gas in the region between the other blade and the line of contact of rotor and casing wall is compressed, causing the valve to open and the gas to escape to the atmosphere; films of oil make the lines of contact gas-tight. The vessel is thus continuously evacuated. The lowest attainable pressure by a single pump is around 1 pascal.

rotational energy. (1) A quantity analogous to translational KINETIC ENERGY. For a body rotating with angular velocity ω about a fixed axis, the energy is given by $I\omega^2/2$ where *I* is the MOMENT OF INERTIA of the body about the axis of rotation. A rolling body has both rotational and translational movement. For example for a cylinder rolling without slipping down a slope, the kinetic energy is given by

$$I\omega^2/2 + Mv^2/2$$

I is the moment of inertia of the body about an axis through its centre of mass parallel to the instantaneous axis of rotation, which for the cylinder mentioned is its line of contact with the slope; ω is the angular velocity about this line, *M* is the mass of the body and *v* the speed of its centre of mass.

(2) The energy of a molecule attributable to its rotation. According to the EQUIPARTITION OF ENERGY principle it is $kT/2$ for each type of rotation of each molecule in a system, where *k* is the BOLTZMANN CONSTANT and *T* the thermodynamic temperature. *See also* ROTATIONAL ENERGY LEVELS.

rotational energy levels. The allowed energies of rotation about axes through the centre of mass of a molecule. By the quantum theory, change from one rotational energy level to a lower one is accompanied by the emission of a quantum of radiation. *See also* ENERGY LEVEL.

rotational motion. Motion which, when the angular acceleration is constant, is described by equations analogous to the four equations for LINEAR MOTION WITH CONSTANT ACCELERATION. Both sets of equations are given in the table,

linear	analogous rotational
$v = u + at$	$\omega_2 = \omega_1 + \alpha t$
$v^2 = u^2 + 2ax$	$\omega_2^2 = \omega_1^2 + 2\alpha\theta$
$x = ut + at^2/2$	$\theta = \omega_1 t + \alpha t^2/2$
$x = (u + v)t/2$	$\theta = (\omega_1 + \omega_2)t/2$

where ω_1 and ω_2 are the initial and final angular speeds, θ is the angle turned through in time *t* and α is the constant angular acceleration. The equation corresponding to $F = ma$, i.e. Newton's second law of motion, is $T = I\alpha$ where T is the TORQUE producing the angular acceleration and I is the MOMENT OF INERTIA of the body about the axis of rotation.

rotation spectrum. A spectrum resulting from a change of a molecule between rotational energy levels in the same vibrational energy level. Such spectra occur in the far infrared and microwave regions of the spectrum.

roton. A quantum of rotational motion in a SUPERFLUID.

rotor. The moving part of a GENERATOR or ELECTRIC MOTOR. It is either the field winding or the armature.

rotor diagram. A method of representing SIMPLE HARMONIC MOTION, shown in fig. R5. A point Q moves with constant speed round the circumference of a circle on a fixed diameter of the circle. The projection of Q is a point performing simple harmonic motion about the centre O of the circle.

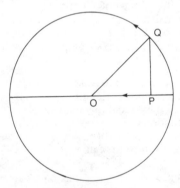

R5 Rotor diagram

rubidium-strontium dating. A RADIOACTIVE DATING technique for geological samples. Natural rubidium contains 27.85%
of the radioisotope rubidium-87, which decays with a half life of 5×10^{11} year into stable strontium-87. By measuring the ratio of rubidium-87 to strontium-87 present in a rock sample, an estimate of the age of the rock up to several thousand million year can be made.

ruby laser. *See* LASER.

Russell-Saunders coupling. Another name for LS coupling. *See* COUPLING.

Rutherford nuclear atom. The theory of atomic structure deduced by Rutherford from the results of GEIGER AND MARSDEN'S EXPERIMENT.

Rutherford scattering. The scattering of heavy charged particles by the COULOMB FIELD of an atomic nucleus.

Rydberg constant. Symbol R. A constant whose value is

$$me^4/(8h^3\varepsilon_0^2c)$$

where m and e are respectively the mass and charge of the electron, h the PLANCK CONSTANT, ε_0 the ELECTRIC CONSTANT and c the speed of light. The constant appears in equations for spectral series and has the value

$$1.096\,77 \times 10^7 \text{ per metre}$$

Rydberg series. A series of spectral lines in the absorption spectrum of a gas. As the wavelength decreases so does the separation between neighbouring lines. Eventually they merge into a continuum at a point corresponding to complete removal of the electron from the atom, i.e. to ionization. The ionization potential may therefore be deduced from the spectrum.

S

saccharimeter. A POLARIMETER especially designed for measuring the strength of sugar solutions. *Compare* SACCHARO-METER.

saccharometer. A HYDROMETER designed and calibrated for measuring the strength of sugar solutions. *Compare* SACCHARI-METER.

safety lamp. Another name for DAVY LAMP.

sampling. The selection and study of a small sample from a large collection. It is thereby possible to gain both qualitative and quantitative information about the collection as a whole.

satellite. (1) An object projected into space by man either to orbit the Earth or to visit and perhaps orbit other bodies in the solar system. The closer to Earth an Earth-orbiting satellite, the greater the drag it experiences due to the Earth's atmosphere. This causes the orbit to contract so that the satellite eventually burns up. The contraction effect is very small at heights greater than 1000 kilometre. The periodic time T of a satellite is given by

$$2\pi(r^3/(Gm))^{1/2}$$

where G is the GRAVITATIONAL CONSTANT, m the mass of the orbited body and r the distance of its centre of mass from the satellite; if T equals the rotation period of the Earth, the satellite is said to be synchronous. Artificial satellites are used in communications and also for information gathering about, for example, the Earth, other bodies and outer space; the information is radioed back to Earth.

(2) A relatively small naturally occurring body orbiting around a larger one. The Earth is thus a satellite of the Sun and the Moon is a satellite of Earth. The giant planets each have several satellites.

(3) A weak spectral line close to a strong one. It generally results from the presence of an isotope of low natural abundance.

saturated colours. The colours of visible monochromatic radiation. Mixing such radiation with white light yields radiation of an *unsaturated colour* but of the same HUE as the monochromatic radiation. The greater the proportion of white light, the more unsaturated the colour.

saturated mode. A mode of operation of a FIELD EFFECT TRANSISTOR. It is illustrated, together with *nonsaturated mode* of operation, in fig. S1.

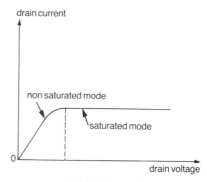

S1 FET modes

saturated vapour. A vapour in dynamic equilibrium with its liquid at a given temperature, i.e. at that temperature it can hold no more substance in the gaseous phase.

saturated vapour pressure. The maximum pressure that can be exerted by a vapour at a particular temperature.

saturation current. The maximum current, excluding breakdown current, which can be obtained from an electronic device, i.e. the current value which cannot be increased by a voltage increase.

saturation force. The maximum value of the EXCHANGE force of attraction between protons in a nucleus. Protons in a nucleus can partake in MESON exchange with a finite number of other protons. However, electrostatic repulsive forces act between all the protons in the nucleus. So, with increasing size of nucleus, the saturation force increases more slowly than the repulsive force.

Saturn. The sixth closest planet to the Sun. It is a GIANT PLANET and has about 95 times the mass and 9.5 times the diameter of Earth. Its axial and orbital periods are respectively 10.25 hour and 29.46 year. It has a spectacular ring system and possesses several satellites. It has been investigated by planetary probes.

sawtooth wave form. A wave form as illustrated in fig. S2. It repeatedly increases steadily to a maximum value and then drops suddenly to zero.

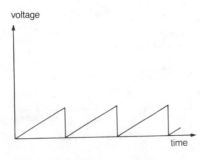

S2 Sawtooth wave form

scalar. A physical quantity characterized only by its magnitude. In contrast to a VECTOR there is thus no directional specification. Examples of scalars are mass and time.

scalar product. The product of two vectors A and B say; it is written $A.B$ and read as A dot B. It equals $AB \cos \theta$ where A and B are the magnitudes of the vectors A and B and θ is the angle between the vectors. For example WORK is the scalar produce of FORCE and DISPLACEMENT. If in CARTESIAN CO-ORDINATES the vectors A and B are (A_x, A_y, A_z) and (B_x, B_y, B_z) respectively, then

$$A.B = A_x B_x + A_y B_y + A_z B_z$$

Compare VECTOR PRODUCT.

scaler. An electronic device producing an output pulse when a specified number of input pulses have been received. The main application is in conjunction with a SCINTILLATION COUNTER in radioactivity measurement.

scanning. The methodical measurement or observation of some physical quantity at a series of positions on a surface or throughout a volume. It may be continuous, as in the movement of a radar aerial scanning a surrounding volume, or intermittent, as in the line by line movement of an electron beam across the fluorescent screen of a television receiver.

scanning electron microscope. *See* ELECTRON MICROSCOPE.

scattering. A process in which collision of a particle or radiation with another particle or discontinuity produces a change in direction or energy of participants in the collison.

In *inelastic scattering* the total kinetic energy changes, whereas in *elastic scattering* it is unchanged.

In *coherent scattering* a definite phase relation exists between scattered and incident waves. In *incoherent scattering* no definite phase relation exists.

See also COULOMB SCATTERING; RAYLEIGH SCATTERING; THOMSON SCATTERING; COMPTON EFFECT; RAMAN EFFECT.

scattering amplitude. A quantity specifying the wave function of particles scattered in a collision and whose modulus squared is proportional to the number of particles scattered in a given direction. In general, scattering amplitude depends on energy and scattering angle.

scattering length. A parameter used in analysing nuclear scattering at low energies.

schlieren photography. High speed photography using spark illumination, of inhomogeneities in a transparent medium. The inhomogeneties are due to density and refractive index variations in the medium. The variations show up as streaks on the photographs taken by transmitted light. Examples of the use of the technique are in the study of shock waves, sound waves, convection currents and flaws in glass.

Schmidt camera. A type of camera using a concave spherical mirror with a specially shaped lens in front of it to form the image. The lens is weakly converging near its centre and weakly diverging near the periphery, thus correcting the system for SPHERICAL ABERRATION. The resulting camera gives high light-gathering power combined with image sharpness and so is used in astronomical research and high altitude photography.

Schottky defect. *See* DEFECT.

Schottky diode. A metal-semiconductor contact which has similar rectifying characteristics to a P-N JUNCTION, but differs from the p-n junction diode in that its forward voltage is usually lower and it is more rapidly turned off by the application of reverse bias.

Schottky effect. The lowering of the WORK FUNCTION of a solid under the influence of an external electric field. As a consequence of the lowered work function, the thermionic emission is increased.

Schottky noise. The NOISE due to variations in current from an electronic device, produced by the random emission of electrons or holes from electrodes.

Schrödinger's equation. A wave mechanics equation describing the behaviour of a particle:

$$\nabla^2\psi + (8\pi^2m/h^2)(E - U)\psi = 0$$

∇^2 is the LAPLACE OPERATOR, m the particle mass, E and U total and potential energy respectively, h the PLANCK CONSTANT and ψ the WAVE FUNCTION of the particle.

Schwarzschild radius. The critical radius to which a collection of matter in space must be compressed in order to form a BLACK HOLE. For a nonrotating black hole its value is $2GM/c^2$, where G is the GRAVITATIONAL CONSTANT, M the mass and c the speed of light.

scientific method. The collection of observations followed by careful examination of the data obtained. Repeatable observations and regularities in the data are noted. By reasoning from the particular to the general, i.e. by inductive reasoning, rules followed by the phenomena under investigation are suggested. Deductions from these laws are then made and experiments performed to test the deductions. If the experiments fail the rules are modified; if they succeed the laws are accepted until evidence arises to refute them.

scintillation. The light flash which is produced by FLUORESCENCE and which occurs when a particle of suitable energy, known as a *scintillator*, falls on an appropriate substance.

scintillation counter. A device for counting SCINTILLATIONS. Sodium iodide activated with thallium is the most widely used phosphor in scintillation counters. Usually each scintillation produces an output pulse in a suitably mounted PHOTOMULTIPLIER; the pulses are counted by a SCALER. Since the magnitude of the pulse depends on the energy of the particle producing the scintillation, use of a PULSE HEIGHT ANALYSER permits a study of particle energies.

scintillator. *See* SCINTILLATION.

sclera. *See* EYE.

sclerometer. Any instrument for measuring the HARDNESS of materials. In one type of device a standard ball is dropped on the material from a prescribed height; the measured rebound height is an indication of the hardness.

scotopic vision. Vision at very low luminance levels, when the main light receptors in the retina are cells known as *rods*. Scotopic vision is devoid of colour sensation. *See also* LUMINOSITY CURVES; PHOTOPIC VISION.

scrambler. A circuit whose use in the transmission of signals renders them unintelligible on reception unless an appropriate circuit is used in the receiving equipment.

screen grid. A grid situated between the anode and control grid of a THERMIONIC VALVE and held at constant potential in order to reduce the capacitance between anode and control grid.

screening. The shielding of a piece of apparatus to protect it from electric and/or magnetic fields. Completely surrounding the apparatus by an earthed conductor provides electrical protection. For magnetic protection a surround of high permeability is used.

scruple. *See* Table 6D.

search coil. Another name for EXPLORING COIL.

Searle's method. A method of measuring the thermal conductivity λ of a good conductor in bar form. The horizontally mounted lagged bar is heated at one end by a steam bath. The other end of the bar is cooled by water flowing round it inside a copper spiral. In the steady state, the inlet and outlet water temperatures θ_3 and θ_4 respectively are measured as are also the temperatures θ_1 and θ_2 at two points on the bar a distance d apart. If A is the area of cross section of the bar and m the mass of

water flowing per second, then

$$\lambda = md(\theta_4 - \theta_3)/(A(\theta_2 - \theta_1))$$

λ can therefore be calculated.

sec. Abbrev. for secant. *See* TRIGONOMETRIC FUNCTIONS.

secant. (1) A line cutting off an arc of a curve.
(2) *See* TRIGONOMETRIC FUNCTIONS.

second. (1) Symbol s. The SI unit of time, defined as the duration of 9 192 631 770 periods of vibration of the radiation produced by transition between two hyperfine levels in the ground state of the caesium-133 atom. Previously the unit was defined by astronomical measurement. *See* TIME.
(2) Symbol ″. A unit of angle equal to 1/3600 degree.

secondary cell. *See* CELL.

secondary electron. An electron emitted from a material as a result of SECONDARY EMISSION.

secondary emission. The emission of electrons from a solid as a result of the impact of other electrons on the solid. If an incident electron is sufficiently energetic, it may result in the ejection of several secondary electrons; this effect is used in the PHOTOMULTIPLIER and the DYNATRON.

secondary standard. (1) A copy of a PRIMARY STANDARD and differing from it by a known amount.
(2) A quantity which is accurately known in terms of the primary standard and is used as a unit.

secondary waves. *See* HUYGHEN'S PRINCIPLE.

secondary winding. The winding on the output side of a TRANSFORMER.

second quantization. A process enabling a classical field to be considered as a collection of particles. *See also* QUANTUM FIELD THEORY.

sector. Part of a circle bounded by two radii and the arc contained between them. If rotated about one of the radii, a sector of a sphere of the same radius results.

sedimentation. A routine method of separating large particles from solution or suspension using a CENTRIFUGE. The particles end up in a hard pellet at the outer wall of the centrifuge and are said to have been sedimented.

sedimentation density gradient method. A method of measuring particle densities. A solution of, for example, caesium chloride is spun in a centrifuge. The solution density then increases radially outwards from the centrifuge axis and the variation can be accurately measured optically. If particles of unknown density are then added, they move until they reach a place where the density is the same as their own and so their density can be determined. A mixture of particles added to the solution will be separated thus enabling the densities of the individual components to be found.

sedimentation equilibrium method. A method of determining particle mass for particles which are too small for SEDIMEN-TATION. When a solution of such particles is centrifuged, an equilibrium gradient of concentration is eventually set up; the mass can be found from optical measurements of this gradient.

sedimentation velocity. Another name for TERMINAL VELOCITY.

Seebeck effect. *See* THERMOCOUPLE.

seed. A small crystal used to induce crystallization from a saturated or supersaturated solution or from a supercooled melt.

segment. (1) Part of a circle bounded by a chord and the arc it cuts off.
(2) Part of a line or curve lying between two points on it.

Segré chart. A plot of ATOMIC NUMBER (i.e. proton number) against NEUTRON NUMBER for the various nuclei. Points representing stable nuclei of nucleon number less than 40 lie on or close to a line of gradient 1, as shown in fig. S3. For nucleon numbers greater than 40, the stable nuclei line becomes steeper. Nuclei represented by points lying off the stable nuclei line decay to yield a nucleus lying nearer the line.

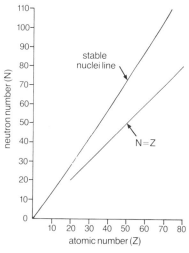

S3 Segré chart

seismograph. An instrument used for the detection and recording of ground movement. It consists essentially of a massive PENDULUM which is set in motion by the force produced at its suspension by the ground movement, which could for example be produced by an earthquake or a nuclear explosion.

seismology. The study of the Earth's structure from SEISMOGRAPH recordings at various points on the Earth's surface of the same event: the time taken by waves from the focus, i.e. origin of the disturbance, to reach a detector depends on the nature of the medium through which they travel.

selection rules. Rules which specify the permitted transitions, according to quantum mechanics, between different ENERGY LEVELS of a system. For example the vibrational quantum number specifying the energy of vibration of a molecule can only change by 1.

selenium cell. A type of PHOTOCELL in which a metal support carries a thin selenium layer covered with a transparent film of gold. When light falls on the film a potential difference due to the PHOTO-VOLTAIC EFFECT is set up between the selenium and the gold.

selenium rectifier. A type of METAL RECTIFIER consisting of alternate layers of iron and selenium.

self-absorption. The absorption of radiation by the substance emitting the radiation. for example radiation emitted by an incandescent gas is absorbed by the same gas at lower temperature.

self-energy. (1) The contribution to particle energy from interaction between different parts of the particle.

(2) In quantized field theory, the contribution to particle energy due to virtual emission and absorption of other particles, in particular mesons and photons.

self-inductance. *See* ELECTROMAGNETIC INDUCTION.

semicircular canals. A set of three structures forming one of the main parts of the inner ear. The canals act as the body's natural frame of reference and control balance.

semiconductor. A substance of negative temperature coefficient of resistance. Its electric conductivity lies in the range 10^{-2} ohm centimetre to 10^{-9} ohm centimetre at room temperature and is intermediate between that of insulators and that of metallic conductors.

A material which is naturally semi-conducting in the pure state is known as an *intrinsic semiconductor*. A semiconductor whose conductivity results from the presence of impurity centres or of imperfections is called an *extrinsic semiconductor* (*see* BAND THEORY); when the conductivity

depends only on the presence of impurities the material is sometimes called an *impurity semiconductor*. The process of DOPING allows impurities to be added to a semiconductor in order to produce a desired conductivity. When the density of CON-DUCTION ELECTRONS is greater than the density of mobile HOLES, then it is an *n-type semiconductor* (*see also* DONOR; ACCEPTOR).

The use of semiconductors is widespread in TRANSISTORS, rectifiers, amplifiers etc.

semiconductor detector. A device used for counting low-energy particles. It consists of a reverse-biased P-N JUNCTION so that any electrons or holes produced around the junction are rapidly swept away. Passage of a charged particle across the junction produces electron-hole pairs, which, if in sufficient numbers, result in a detectable pulse through the circuit. The semiconductor detector has about ten times the sensitivity of gaseous detectors. Moreover, because of its small size, it can be used in areas inaccessible to other types of counter. Its insensitivity to gamma rays and neutrons makes it extremely suitable for counting in the presence of such background radiation.

semiconductor diode. *See* DIODE.

semiconductor junction. A junction between a metal and an n-type SEMICONDUC-TOR, or between a metal and a p-type semiconductor. All three types conduct better in one direction than the other.

semiconductor laser. A LASER employing a P-N JUNCTION. It has the advantage of small size compared with other types of laser.

semiconductor rectifier. A device as illustrated in fig. S4. The effect of the external potential difference on the DEPLETION LAYER when opposing the junction's own potential difference is to restart the movement across the junction of electrons and holes as indicated. Thus current flows, its magnitude increasing with the voltage.

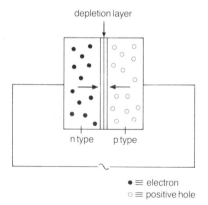

depletion layer

n type   p type

● ≡ electron
○ ≡ positive hole

S4 Semiconductor rectifier

Reversal of the external potential difference reinforces the junction's own potential difference and so no current flows. The P-N JUNCTION therefore acts as a rectifier. *See also* DIODE.

semipermeable membrane. A barrier which permits the passage of some substances in solution but which is impermeable to others. *See* OSMOSIS.

sensation level. Symbol L. The quantity $10 \log(I/I_0)$ where I and I_0 are respectively the sound intensity under consideration and the minimum audible sound intensity. I_0 is usually taken as 2.5 picowatt per square metre.

sensitivity of the ear. A property usually expressed as the minimum sound energy necessary for audibility, i.e. the *threshold of*

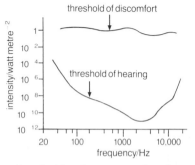

S5 Threshold v. frequency plots for ear

hearing. It is frequency dependent as indicated in fig. S5, unlike the *threshold of discomfort*, which is also shown.

sensor. Another name for TRANSDUCER.

separation energy. The energy required to remove a nucleon from the nucleus of a particular element.

sequence. A set of numbers or mathematical expressions in a fixed order, obtained by substituting the numbers 0, 1, 2, 3 ... in an expression for the general term. For example if the general term is $n^2 + n$, the sequence is

0, 2, 6, 12, 20 and so on

See also ARITHMETIC PROGRESSION; GEOMETRIC PROGRESSION; SERIES.

sequential access device. *See* STORAGE DEVICE.

series. (1) The summation of the terms of a SEQUENCE, written

$$\sum_{n=0}^{\infty} [\text{general term in } n]$$

The series may be finite or infinite and an infinite series may be convergent or divergent.
(2) Denoting a mode of connection of electrical apparatus for which the current is the same at all points. *See* CAPACITOR; RESISTOR; CELL. *Compare* PARALLEL.

series wound. Involving a type of ELECTRIC MOTOR winding in which the field coils are connected in series with the armature, resulting in a decrease in speed with increase in load.

servo mechanism. A mechanism in which a motion requiring a small amount of power is used to control a mechanical motion requiring a much large amount of power. The system may involve FEEDBACK.

sextant. An instrument for measuring the angle, up to 120°, between two directions. An important use is the determination of

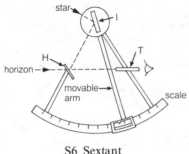

S6 Sextant

astronomical ALTITUDE. As illustrated in fig. S6, the horizon is viewed through telescope T and the upper unsilvered portion of H. The index glass I is rotated until an image of the star under observation is also seen in the telescope by reflection first at I and then by the lower silvered portion of H. The required angle is read off from the position on the scale of the pointer attached to I.

shade. *See* LIGHTNESS.

shadow. A dark region formed when an opaque body blocks part of the light falling on a surface. For a point source of light the shadow is completely dark. For an extended source the shadow consists of a completely dark part surrounded by a region of partial shade. The nature of shadows obeys the laws of geometric optics except for diffraction phenomena at the edges; these phenomena are more noticeable the smaller the opaque body.

shadow bands. Parallel bands of shadow moving rapidly across the Earth due to a refraction effect in the Earth's atmosphere. They are sometimes seen just before and after totality in a solar ECLIPSE.

shadow-free illumination. A state now readily attainable by taking FIBRE OPTICS light guides to a number of illuminating positions.

shear modulus. Another name for RIGIDITY MODULUS.

shear strain. *See* STRAIN.

shear stress. *See* STRESS.

shell. *See* ELECTRON SHELL; SHELL MODEL.

shell model. A theoretical model of the nucleus based on the assumption that the nucleons move in a central field, just as the electrons in the atom are assumed to move in the field of the nucleus. Furthermore, quantum mechanics predicts that a nucleon moves in a *shell* analogous to an electron shell of an atom. It is thought that if the number of nucleons is just sufficient to fill an exact number of shells, the nucleus will be particularly stable, i.e. it has MAGIC NUMBERS of protons and neutrons.

SHF. Abbrev. for SUPERHIGH FREQUENCY.

shield. A material, such as concrete, surrounding the CORE of a nuclear reactor in order to absorb dangerous radiation.

SHM. Abbrev. for SIMPLE HARMONIC MOTION.

shock wave. A very narrow region of high pressure and temperature formed in a fluid when there is relative supersonic motion between it and an obstacle (*see* SONIC BOOM). A shock wave may also be produced in a fluid by a violent disturbance of it, such as produced by a lightning stroke or a bomb blast.

shooting star. Another name for METEOR.

short circuit. A low-resistance electrical connection introduced between two points in a circuit in order to bypass the current through the part of the circuit between the points.

short sight. Another name for MYOPIA.

shot noise. Another name for SCHOTTKY NOISE.

shower. A group of elementary particles and photons arising from the collision of a single high-energy particle with an atom or molecule. *See also* COSMIC RAYS.

shunt. A low-value resistor connected in parallel with a piece of electrical equipment in order to reduce the current through the equipment by a known factor. Shunts are commonly used to increase the range of ammeters.

shunt wound. Involving a type of ELECTRIC MOTOR winding in which the field coils are connected in parallel with the armature, resulting in little speed variation with load.

sidereal. Involving or measured with respect to the stars.

sidereal time. Time measured with reference to the stars. The *sidereal year* is the interval between successive conjunctions of the Sun with a chosen fixed star; the *sidaereal month* is the interval between two successive conjunctions of the Moon with a fixed star; the *sidereal day* is the time between successive transits of a chosen star across the observer's meridian. The sidereal year equals 365.256 36 day; the sidereal month equals 27.321 66 day; the sidereal day equals 23 hour and 56.06 minute. The difference between sidereal time and SOLAR TIME arises from the imposition of the Earth's orbital motion on its rotational motion.

siemens. Symbol S. The SI unit of CONDUCTANCE, equal to the conductance of a conductor of resistance 1 ohm.

sievert *See* RADIATION UNITS.

sigma particle. *See* HYPERON.

sigma pile. A neutron source and MODERATOR without fissile material. Such piles are used in studies of moderator properties.

signal. Any agency carrying information, for example a changing electric current.

signal generator. A wide-range radio-frequency oscillator, generally with facility for audio or video frequency modulation. It is used for testing.

sign conventions. *See* OPTICS SIGN CONVENTIONS.

significance. A statistical estimate of the support that a particular measurement provides for a given theory.

significant figures. The number of reliable figures quoted for a numerical value of a physical quantity. It is to be noted that the number of significant figures quoted for a numerical value should never exceed the smallest number of such figures among any other variables used. Thus if a mass of 1.52 kilogramme is given an acceleration of 2.141 metre per second per second, the force necessary should be given as 3.25 newton not 3.254 32 netwon.

silicon cell. A photovoltaic barrier cell employing a P-N JUNCTION. It is used as a power source in spacecraft.

silicon chip. A small silicon slice to which appropriate impurities have been added. It is often used in INTEGRATED CIRCUITS.

silk screen process. A technique used in the manufacture of an INTEGRATED CIRCUIT on a ceramic substrate. The resistive properties of the various tracks are determined by the quantities of metals and other materials in the inks used.

silent discharge. An inaudible high voltage electric discharge involving a relatively high energy dissipation. Such a discharge readily occurs at a sharp pointed conductor.

similar triangles. Triangles whose corresponding angles are equal. In addition, lengths of corresponding sides are in the same ratio.

simple equivalent pendulum. A simple PENDULUM having the same period as a given compound pendulum.

simple fraction. Another name for common fraction. *See* FRACTION.

simple harmonic motion. The motion of a particle whose acceleration is always directed towards a fixed point and is directly proportional to the distance y of the particle from the point. The equation of motion is therefore

$$\ddot{y}(\equiv d^2y/dt^2) = -n^2y$$

where n is a constant. The solution of this equation is

$$y = A\sin(nt + B)$$

where A and B are constants. If $y = 0$ when $t = 0$, then

$$B = 0 \text{ and } y = A\sin(nt)$$

which is shown in fig. S7; the periodic time is $2\pi/n$ (*see also* ROTOR DIAGRAM). Common examples of simple harmonic motion are the small OSCILLATIONS of a pendulum, of a spring, of a rigid body and of a liquid.

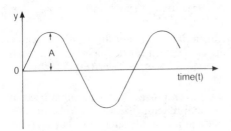

S7 Simple harmonic motion

simple microscope. *See* MICROSCOPE.

simple pendulum. *See* PENDULUM.

Simpson's rule. A rule used to estimate the area of an irregular figure. The area is divided into an even number of strips of the same width. The area then equals one-third of the strip width multiplied by the sum of the first and last boundary ordinates plus twice the sum of the odd ordinates plus four times the sum of the even ordinates. Examples of applications of the rule are the evaluation of definite integrals and the determination of pond areas.

simulator. A device that mimics the behaviour of an actual system but is made from components that are more easily, cheaply or conveniently manufactured.

simultaneous equations. Equations in one or more variables and whose solution is a value or set of values which satisfies all the equations.

sin. Abbrev for sine. *See* TRIGONOMETRIC FUNCTIONS.

sine. *See* TRIGONOMETRIC FUNCTIONS.

sine condition. For accurate imaging of a small object by a large-aperture optical system consisting of spherical refracting surfaces, where the object lies close to the axis,

$$ny\sin\alpha = n'y'\sin\alpha'$$

As illustrated in fig. S8, n and n' are the refractive indices of the media on either side of the optical system and y and y' are the object and image heights; α is the angle made with the axis by a ray leaving the object axial point and α' the angle it makes with the axis on arriving at the image axial point. For small-aperture systems the condition becomes

$$ny\alpha = n'y'\alpha'$$

S8 Sine condition

sine wave. Another name for SINUSOIDAL WAVE.

single crystal. A solid whose structure is regular throughout its volume. *See also* CRYSTAL.

singlet state. A state for which the MULTIPLICITY is 1.

single valued function. A function for which to every value of the independent variable there corresponds only one value of the DEPENDENT VARIABLE. Thus the straight line equation

$$y = mx + c$$

is a single valued function.

sink. *See* SOURCE.

sino-auricular node. The ACTION POTENTIAL source in the heart. It is situated in the right auricle of the heart and initiates the heart's pulsations. The action potentials travel in all directions stimulating the opening and closing the heart's valves.

sinusoidal wave. A wave of the same shape as that of a sine function. Thus $y = \sin x$ and $y = \cos x$ would both be described as sinusoidal waves.

siphon. A device, illustrated in fig. S9, used to transfer liquid from a higher to a lower level without pouring or using a tap. The siphon must be filled with liquid, i.e. primed, to start the flow. If P is atmospheric pressure, the pressure at A is $P - g\rho h$ and that at B is $P - g\rho H$ where h and H are as indicated, ρ is the liquid density and g is the acceleration due to gravity. Since H is greater than h, the pressure at A is greater than that at B and so the liquid is transferred. It is to be noted that some liquids will siphon in a vacuum, i.e. in the absence of atmospheric pressure. The

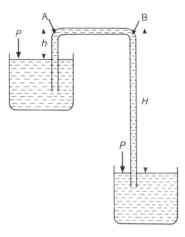

S9 Siphon

reason is that the greater mass of liquid in the longer column compared to that in the shorter column results in the liquid being pulled round due to cohesive forces between molecules, analagous to the motion of a chain on a pulley.

SI units. An internationally agreed system of COHERENT UNITS which is being increasingly used. The *base* and *supplementary* SI units are given in Table 1. *Derived* SI units with special names are in Table 2. Tables 3–5 are also concerned with SI units, while Table 6 deals with the relationship between other units and SI units.

Six's thermometer. Another name for MAXIMUM AND MINIMUM THERMOMETER.

size of atom. An atomic diameter is of the order of 10^{-10} metre.

size of nucleus. A nuclear diameter is of the order of 10^{-15} metre.

skin effect. The decrease in depth of penetration of an electric current in a conductor with increase in current frequency. At very high frequencies the current is restricted to a thin outer layer, and so hollow or stranded conductors are often used at such frequencies. The effect is due to the increase in internal SELF-INDUCTANCE of the conductor with depth below the surface.

skin friction. The drag experienced by a body in relative motion with a fluid.

sky wave. A radio wave transmitted between points on Earth by reflection from the IONOSPHERE rather than by direct transmission. *Compare* GROUND WAVE.

Slater determinant. An $n \times n$ DETERMINANT representing a quantum mechanical wave function for n fermions.

slide rule. An obsolescent mathematical instrument for simple calculations. It uses logarithmic scales so that numbers are multiplied or divided by adding or subtracting differences, using a moving scale.

slip. The motion of one plane of atoms in a crystal relative to another in a direction tangential to the planes. Slip occurs in the plastic deformation of solids.

slip ring. A copper ring connected to and rotating with an electric winding so that, by means of a brush or brushes resting on the ring surface, the winding may be connected to an external circuit. *See also* GENERATOR.

slow neutron. A neutron of kinetic energy not exceeding a few electron volt.

slug. An obsolescent unit of mass equal to *g* pound, where *g* is the acceleration due to gravity.

small angle scattering. The scattering at small angles of waves by particles or cavities in a medium, the wavelength being small compared with their dimensions. The most important type is probably the small angle scattering of X RAYS by discontinuities; it is used in molecular studies.

small circle. A plane section of a sphere not containing its centre. *Compare* GREAT CIRCLE.

smetic. A phase of matter intermediate between liquid and solid. *See* LIQUID CRYSTAL.

smoke. A suspension of small solid particles in a gas.

smooth. (1) Denoting the nature of contact between bodies in the absence of friction.
  (2) Denoting functions with no singularities or discontinuities.

smoothing circuit. A circuit as illustrated in fig. S10. The reactance of capacitors C_1

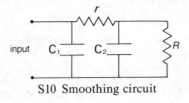

S10 Smoothing circuit

and C_2 is small compared with the values of the high resistance r and load R. The capacitors therefore charge quickly up to peak voltage and discharge only slowly. Thus the full wave rectified voltage has hardly dropped before the capacitors charge up again, giving a smoothed output voltage as shown in fig. S11.

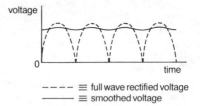

– – – – ≡ full wave rectified voltage
———— ≡ smoothed voltage

S11 Smoothed output

Snell's law. When light travelling in a medium of refractive index n strikes a medium of refractive index n' at an angle of incidence i, and enters it so that the angle of refraction is i', then

$$\sin i / \sin i' = n/n' = \text{a constant}$$

snow. (1) A finely divided hexagonal crystalline form of water produced in the atmosphere at temperatures below the freezing point of water.
  (2) A random pattern of white dots observed on the screen of a CATHODE RAY TUBE when there is little or no signal. It is due to thermal and shot noise.

sodium pump. A device in nerve cells whereby sodium ions are ejected across the axon membrane. It plays a part in the process of electrical transmission in neurons, although the pump mechanism is unknown.

sodium vapour lamp. A type of gas discharge lamp in which a discharge is started in an inert gas such as neon at room temperature. This causes the sodium present to vaporize and give rise to the emission of the characteristic intense orange-yellow light. Use of the neon gas explains the red colour of the sodium vapour lamp when first switched on.

soft radiation. IONIZING RADIATION with a low degree of penetration. The term is most commonly applied to long wavelength X RAYS.

soft vacuum tube. A vacuum tube whose vacuum is such that ionization of the residual gas has a significant effect on the electrical characteristics of the tube.

software. The PROGRAMS associated with a COMPUTER system. *Compare* HARDWARE.

solar battery. A SOLAR CELL array.

solar cell. Any device converting solar energy directly into electric energy. Examples are semiconductor devices operating by the PHOTOVOLTAIC EFFECT and THERMOPILE devices.

solar constant. The amount of solar energy passing normally through unit area at the Earth's averagae orbital distance from the Sun. In the absence of absorption by the Earth's atmosphere, the value would be 1390 watt per square metre.

solar day. *See* SOLAR TIME.

solar eclipse. *See* ECLIPSE.

solar energy. The energy emitted by the Sun. The rate of emission is about 4 × 10²⁶ watt.

solar flare. A violently explosive surge or outburst associated with electrical storms in the Sun's CHROMOSPHERE. The flares are very bright and may last for up to half an hour. Large amounts of short wave radiation and particles are emitted, which may affect the Earth's ionosphere and cause magnetic storms. The flares may be associated with sunspots.

solar heating. The utilization of SOLAR ENERGY in direct heating devices such as solar stills, solar furnaces and solar panels.

solarimeter. An instrument for measuring the radiation received from the Sun.

Thermopiles are often used for this purpose.

solar month. *See* SOLAR TIME.

solar neutrino. A NEUTRINO originating from one of several sources in the interior of the Sun. For detection, the number of chlorine-37 nuclei converted into argon-37 by neutrino interaction in a tank of tetrachloroethylene, buried deep in a Colorado mine, has been counted. However, the count is lower than anticipated and so suggestions for improved detection, using solar neutrino conversion of bromine-81 into krypton-81, have been made. Reliable detection should lead to clarification of energy generation mechanisms within the Sun.

solar time. TIME measured by reference to the Sun. The *solar day* is the time between successive transits of the Sun across the observer's meridian. The average of this time over a YEAR gives the *mean solar day*, which is used for general purposes: the mean solar day contains 24 mean solar hour, each one containing 60 mean solar second. The *solar month* is the time taken for the Moon to complete one revolution of the Earth, returning to the same longitude; it equals 27.321 58 mean solar day.

solar units. Units in which the mass, diameter, density and luminosity of the Sun are taken as unity; these properties for other stars are then expressed in solar units.

solar wind. A PLASMA of protons and electrons which streams away from the Sun at a speed between 300 kilometre per second and 800 kilometre per second. It is responsible for the formation of the VAN ALLEN BELTS and the AURORA. It also causes the Earth's magnetic field to extend further into space on the side of the Earth remote from the Sun. The solar wind is most intense during the occurrence of solar flares and sunspots.

solar year. *See* YEAR.

solder. An alloy used for joining metals together. Soft solder, melting point between 200° C and 300° C, consists of tin and lead in roughly equal amounts. For making electrical joints, a resin flux is used.

solenoid. A spiral coil of wire whose length is large compared with the diameter. Passing a current through the solenoid produces a magnetic field. The magnitude of the magnetic induction, B, at a position on the solenoid axis is

$$\mu_0 n I(\cos \theta_2 - \cos \theta_1)/2$$

where n is the number of turns per unit length, I the current, μ_0 the magnetic constant and θ_1 and θ_2 the half angles subtended by the ends of the solenoid at the axial point. If the solenoid is infinitely long then $\theta_2 = 0°$ and $\theta_1 = 180°$ and so

$$B = \mu_0 n I$$

If the axial point considered is at one end of the solenoid then $\theta_2 = 0°$ and $\theta_1 = 90°$ and so

$$B = \mu_0 n I/2$$

solid. (1) A state of matter in which it is sensibly incompressible by pressure and resists shear stress. The intermolecular forces in a solid are sufficiently large to keep the atoms or molecules in fixed positions about which they can vibrate. Amorphous solids have no regular structure whereas crystalline solids show a regular arrangement of atoms and molecules. *See* CRYSTAL. *See also* BAND THEORY.
(2) A three-dimensional geometric figure such as cube.

solid angle. Symbol ω or Ω. A quantity, illustrated in fig. S12, subtended

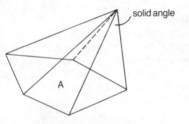
solid angle

A

S12 Solid angle

by any area A at a point not in its plane. The size of the solid angle is the area cut off on a sphere of unit radius, centre the point, by lines joining the perimeter of the area to the point. The unit of solid angle is the STERADIAN.

solid geometry. *See* GEOMETRY.

solid state. Involving or using semiconductor devices rather than thermionic valves in electronic apparatus.

solid state physics. The physics of the structure and properties of solids and of phenomena associated with solids, including ENERGY LEVEL studies and the study of the electric and magnetic properties of metals and semiconductors. *See also* FIELD EMISSION; PHOTOELECTRIC EFFECT; PHOTO CONDUCTIVITY.

soliton. A localized wave of permanent form which may interact strongly with other solitons in such a way that, on separating after the interaction, they regain their original forms.

solstice. Either of two points lying on the ECLIPTIC midway between the vernal and autumnal EQUINOXES. At the *summer solstice* the Sun appears at its maximum distance north of the CELESTIAL EQUATOR. At the *winter solstice* the Sun appears at its maximum distance south of the celestial equator. The Sun reaches these two points around June 21 and December 22, the actual times also being referred to as solstices.

solute. *See* SOLUTION.

solution. A homogeneous mixture of two or more substances; the atoms or molecules of the components are completely intermingled. When a solid or gas is dissolved in a liquid, the liquid is the *solvent* and the other substance(s) the *solute*. When liquids are dissolved in each other, the one of largest volume is the solvent and the other(s) the solute.

solvent. *See* SOLUTION.

sonar. A device for detecting an underwater object by reflecting ultrasonic pulses off it and measuring the time interval between the transmitted and reflected pulses; then knowing the ultrasound speed, the depth of the object can be calculated. The name sonar is an abbreviation for sound navigation ranging. *See also* ECHO SOUNDING.

sone. A unit of loudness equal to 40 decibel at 1000 hertz. It has been chosen so that a sound of x sone seems to the listener to be y times as loud as a sound of x/y sone; this has been experimentally confirmed for loudness from 0.25 sone to 250 sone.

sonic boom. The noise arising from SHOCK WAVES which are formed, as illustrated in fig. S13, when an aircraft travels at a speed greater than the local speed of sound. In level flight the intersection of the shock wave with the ground forms a hyberbola on which lie all points where the sonic boom is heard simultaneously.

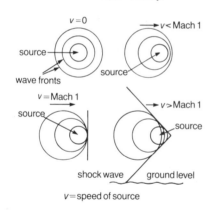

S13 Sonic boom

sonometer. Another name for MONOCHORD.

sorption pump. A type of vacuum pump in which gas is adsorbed on activated charcoal which is liquid-nitrogen cooled, or on some other efficient absorbent. Such pumps eventually become saturated and are then regenerated by heating.

sound. A mechanical longitudinal vibration of a medium by which energy is carried through the medium. The medium has elasticity and inertia; when acted upon by a vibrating body, it undergoes alternate compressions and rarefactions which travel through the medium with a velocity characteristic of it. Sound is not transmitted in a vacuum. The term sound is sometimes restricted to those vibrations audible to the human ear, i.e. vibrations in the frequency range of about 20 hertz to 20 000 hertz, but is sometimes used to describe any analogous wave motion. The velocity of sound in gas is independent of the gas pressure.

sound absorption coefficient. The ratio of the sound energy absorbed by a surface or material, at a particular frequency, to the incident energy.

sound flux. Symbol P. The rate of flow of sound energy. For a plane or spherical progressive wave of speed v in a medium of density ρ,

$$P = p^2 A/(\rho v)$$

where p is the SOUND PRESSURE and A is the area perpendicular to the flow direction through which the flow passes.

sounding balloons. Small balloons for lifting recording instruments into the Earth's atmosphere.

sound insulation. *See* DEAD ROOM.

sound intensity. The SOUND FLUX per unit area lying perpendicular to the flow direction.

sound pressure. The part of the instantaneous pressure at a point which is due to the sound wave. Over a period of time it has an average value of zero.

sound reflection coefficient. The ratio of the sound energy reflected by a surface or material, at a particular frequency, to the incident energy.

sound track. A recording of sound on film. The principle of the *variable area* method of doing this is illustrated in fig. S14. T is a triangular mask whose image is focused by lens L on slit S after reflection at mirror M of a moving coil galvanometer. Current produced in a microphone by the sound is passed through the galvanometer coil so that M follows the sound vibrations.

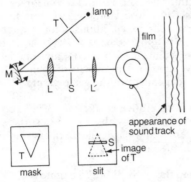

S14 Sound recording on film

The image of T therefore moves over S so that the illuminated length of S varies. The light from S is focused by lens L' on a strip of moving film, giving a sound track as indicated. To reproduce the sound, light passing through the moving sound track is focused on a photocell, connected via an amplifier to a loudspeaker. *Variable-density* sound tracks of constant track width but density varying with the sound signal are also available. *Compare* MAGNETIC RECORDING.

sound wave photography. *See* SCHLIEREN PHOTOGRAPHY.

source. (1) The point at which lines of force originate in a VECTOR FIELD, for example a magnetic field. The lines of force terminate at a *sink*.

(2) The point from which fluid emerges radially and uniformly. The quantity of fluid emitted in unit time is known as the *strength* of the source.

(3) The electrode in a FIELD EFFECT TRANSISTOR from which electrons or positive holes enter the interelectrode space.

(4) Any energy-producing device.

source impedance. The IMPEDANCE presented by any energy source to the input terminals of a circuit or device. Ideally a voltage source should have zero source impedance and a current source should have infinite source impedance.

space. (1) Any region outside the Earth's atmosphere.

(2) A geometric term; thus *Euclidean space* involves the problems of distance and angle.

space charge. A region in any device where the net charge density is greater than zero.

spacecraft. A vehicle using rocket motors to escape the Earth's gravitational field and thereafter using its propulsion system for course changes. For internal operations and radio communications, many spacecraft are SOLAR CELL powered.

space group. A group of symmetry operations which leave a crystal invariant; 230 have been identified. These have been developed from 14 basic lattices by rotation about an axis, reflection across a plane, screw rotation etc.

space quantization. *See* SPIN.

space-time. The reference system of three dimensions of space and one of time, used in RELATIVITY theory.

space-time diagram. A plot of distance travelled against time taken.

spallation. A nuclear reaction in which a high-energy photon or particle hits an atomic nucleus, causing it to emit several particles or fragments.

spark chamber. An apparatus for the detection and study of high-energy ionizing particles; it was developed from the SPARK COUNTER. Many parallel insulated metal wires separated by narrow gaps are mounted in a neon-filled chamber; alternate wires are at a much higher potential than their neighbours so that a passing particle triggers a succession of small sparks along its path. The track is photographed through the side of the chamber. The device has a very short recovery time.

spark counter. An apparatus for the detection and study of high-energy ionizing particles, especially ALPHA RAYS. It consists of an insulated metal wire close to an earthed metal base plate; the wire-plate potential difference is just less than that required to cause a discharge across the air gap between them. Approach of a particle causes a spark, after which the potential difference drops. The number of particles is recorded by a counting circuit designed to respond to a voltage change.

spark discharge. A visible disruptive discharge of electricity between two points whose potential difference is high. Ionization of the path precedes the discharge, which is generally accompanied by a sharp crackling noise due to rapid heating by the spark of the gas between the points. The maximum length of a spark depends on the shape of the electrodes and the potential difference between them.

spark gap. An arrangement of electrodes such that a spark occurs when the applied voltage reaches a predetermined value.

sparkover. Another name for FLASHOVER.

spark photography. Any form of photography for which a spark provides the illumination, as for example in SCHLIEREN PHOTOGRAPHY. Illumination by a spark of very brief but known duration is used in the photography of moving objects.

special theory of relativity. See RELATIVITY.

specific. A term usually denoting per unit mass; for instance specific volume is volume per unit mass. When a physical quantity has a capital letter symbol, the specific quantity is denoted by a lower case letter: for example L for latent heat, l for specific latent heat. Exceptions exist to the usual meaning of specific, an example being specific resistance.

specific activity. Symbol a. The ACTIVITY per unit mass of a radioisotope or radioactive material.

specific charge. The electric charge per unit mass. The term is generally used with reference to the charge to mass ratios of elementary particles. See also E/M.

specific gravity. Former name for RELATIVE DENSITY.

specific gravity bottle. Former name for RELATIVE DENSITY BOTTLE.

specific heat capacity. Symbol c. The HEAT CAPACITY of unit mass of a substance, i.e. the quantity of heat required to raise the temperature of unit mass of a substance by one degree.

The specific heat capacity of a gas has two values, depending on the conditions of temperature increase. If the volume is kept constant, the specific heat capacity at constant volume, c_V, is obtained. If the pressure is kept constant, the specific heat capacity at constant pressure, c_P, is obtained. The difference $c_P - c_V$, is equal to the work done in expansion, and for an ideal gas is equal to the UNIVERSAL GAS CONSTANT. According to the KINETIC THEORY OF MATTER, the molar heat capacity at constant volume equals $RF/2$ where R is the universal gas constant and F the number of degrees of freedom.

specific heat theories. Theories of the specific heats of solids, starting with DULONG AND PETIT's LAW, followed by EINSTEIN'S THEORY OF SPECIFIC HEATS and subsequent developments of it.

specific humidity. See HUMIDITY (def. 3).

specific inductive capacity. Former name for relative permittivity. See PERMITTIVITY (def. 2).

specific latent heat. (1) Symbol l_v. The quantity of heat required to change unit mass of liquid at its boiling point to vapour at the same temperature. In the reverse process an equal amount of heat is evolved. l_v decreases with increase of temperature, reaching zero at the CRITICAL TEMPERATURE T_c; empirically

$$l_v \propto (T_c - T)^{1/3}$$

where T is the temperature.

(2) *of fusion*. Symbol l_f. The quantity of heat required to change unit mass of solid at its melting point to liquid at the same temperature. In the reverse process an equal amount of heat is evolved.

(3) *of sublimination*. Symbol l_s. The quantity of heat required to change unit mass of solid to the vapour state at the same temperature. In the reverse process an equal amount of heat is evolved.

specific optical rotatory power. *See* OPTICAL ACTIVITY.

specific refractivity. *See* REFRACTIVITY.

specific reluctance. Former name for RELUCTIVITY.

specific resistance. Former name for RESISTIVITY.

specific volume. Symbol v. The reciprocal of DENSITY.

spectral line. An image of the entrance slit of a SPECTROMETER formed by radiation of a single wavelength. *See also* HYPERFINE STRUCTURE.

spectral luminous efficiency curve. Another name for LUMINOSITY CURVES.

spectral multiplet. *See* MULTIPLET (def. 2).

spectral satellite. *See* SATELLITE (def. 3).

spectral series. A series of wavelengths in an absorption or emission spectrum, all of which arise from electron transitions from or to the same energy level. For example for hydrogen the LYMAN SERIES corresponds to transitions from higher orbits back to

the innermost shell, while the BALMER SERIES corresponds to transitions from higher orbits to the second shell.

spectral types. *See* STELLAR SPECTRUM.

spectrograph. A SPECTROMETER designed to give a spectrogram, i.e. a photographic record of a spectrum.

spectrographic analysis. The analysis of substances by the spectra they produce. A routine method is to vaporize a small amount of sample in an electric arc of flame and to photograph its emission spectrum. From this the substance(s) may be identified; the densities of the lines give an indication of the amounts present.

spectrometer. Any of various instruments for producing, examining or recording spectra. In the simple prism spectrometer, illustrated in fig. S15, the light from the source passes through a collimator on to a prism which rests on a flat rotatable table. Refracted light is observed through the telescope which can be moved round the prism. The instrument is suitable for the observation of simple emission and absorption spectra and is also used for measuring refractive indices and prism angles.

spectrophotometer. An instrument for measuring the intensity of radiation at each wavelength in a spectrum. A prism or diffraction grating may be used to disperse the light. In the instrument shown in fig. S16, radiation from the source enters at slit S_1 and is reflected by the concave diffraction grating G on to slit S_2 and then strikes detector D. S_1 and S_2 are fixed and radiation of different wavelengths can be investigated by rotating G so that they fall in turn on D. The term spectrophotometer is often used synonomously with SPECTROMETER. *See also* MONOCHROMATOR; SPLIT BEAM SPECTROPHOTOMETER.

spectroscope. Another name for SPECTROMETER.

spectroscopic binary star. *See* BINARY STAR.

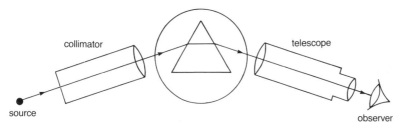

S15 Simple spectrometer

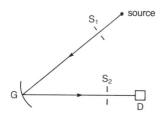

S16 Grating spectrophotometer

spectroscopy. The technique of producing spectra, analysing their constituent wavelengths and using the results either for chemical analysis or for the determination of energy levels and molecular structure. Spectroscopy is also used in astronomy. There are many and varied techniques, selection depending on the type of radiation under investigation.

spectrum. A plot, photographic record or visual display showing the variation of some property of a particular type of radiation (or system of particles) with some related quantity, for example the variation of intensity of electromagnetic radiation, or of sound, with frequency, or the variation of number of particles in a beam with kinetic energy. *See also* ABSORPTION SPECTRUM; BAND SPECTRUM; CONTINUOUS SPECTRUM; ELECTRONIC SPECTRUM; EMISSION SPECTRUM; INFRARED SPECTRUM; LINE SPECTRUM; MICROWAVE SPECTRUM; ROTATION SPECTRUM; ULTRAVIOLET SPECTRUM; VIBRATION-ROTATION SPECTRUM; X RAY SPECTRUM.

spectrum analyser. A circuit which splits an input wave form into its frequency components.

specular reflection. Reflection of radiation at a polished surface.

speculum. An alloy of copper and tin in a 2 : 1 ratio. When highly polished it is used for making mirrors and reflection diffraction gratings. It does not easily tarnish.

speed. (1) The ratio of distance travelled to time taken. Unlike VELOCITY, it is not a VECTOR quantity.
(2) A value specifying the sensitivity of photographic material to light.
(3) *See* F NUMBER.
(4) The rate of operation of a pump, i.e. the mass of pumped material transferred per second.

speed in circle. A quantity given by the product of the radius of the circle and the angular velocity of the object about the centre of the circle.

speed of light. Symbol c. A fixed constant equal to

$$2.997\ 924^{58} \times 10^8 \text{ metre per second}$$

It is the speed at which not only light but all electromagnetic radiation travels in a vacuum. The speed decreases when the radiation enters a material medium, the amount of decrease depending on the medium. The first calculation of the speed of light was made by Römer in 1676, using the observed times of the eclipses of a satellite of Jupiter. Since then numerous other methods have been evolved. The most recent make use of for example RADAR, CAVITY RESONATORS and the PIEZO-ELECTRIC EFFECT.

speed of sound. Symbol *c*. The value in dry air at standard temperature and pressure is 331.4 metre per second. In a vacuum the speed is zero. In any medium the speed of sound is given by

(appropriate elastic constant/density)$^{1/2}$

For solids the appropriate elastic constant is

$$K + 4n/3$$

where *K* is the BULK MODULUS and *n* the RIGIDITY MODULUS. For fluids the bulk modulus is required: at audio frequencies sound-type waves travel under adiabatic conditions and so the ADIABATIC BULK MODULUS is appropriate; at ultrasonic frequencies conditions are more isothermal than adiabatic and so the ISOTHERMAL BULK MODULUS is more suitable. The speed of sound in gas is independent of pressure but is proportional to the square root of the absolute temperature. A reasonably accurate measurement of the speed of sound in air was first made in the 18th century by timing the interval between the flash and sound of a gun about 20 miles away. Various other methods have been devised, using sound sources of known frequency and measuring the wavelength. *See* RESONANCE TUBE.

speed-time diagram. A graph of speed against time. Its slope at any point gives the magnitude of the acceleration at that point and the area under the diagram gives the distance travelled.

sphere. A surface or solid generated by rotating a circle about a diameter. If *r* is the radius of the circle, or sphere, the surface area of the sphere is $4\pi r^2$ and its volume is $4\pi r^3/3$.

sphere capacitance. *See* CAPACITANCE.

sphere gap. A spark gap having spherical electrodes. One use is the reliable measurement of extremely high voltages.

spherical aberration. An ABBERATION in an optical system arising from optical surfaces of spherical shape: as illustrated in fig. S17, rays entering near the periphery of

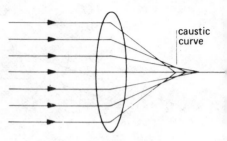

S17 Spherical aberration

the system are focused closer to the system than are rays entering close to the axis. Thus a point object is not imaged as a point but as a small disc, known as the *circle of least confusion*. Sharing the deviation of the light equally between the two faces of a spherical lens reduces spherical aberration. For mirrors it can be reduced by substituting a parabolic reflector for a spherical reflector when the object is distant and an ellipsoidal reflector for a spherical reflector when the object is close.

spherical co-ordinates. *See* POLAR CO-ORDINATES.

spherical harmonic. A function which satisfies the LAPLACE EQUATION in spherical co-ordinates.

spherical lens. A lens whose surfaces are portions of spheres.

spherical mirror. A reflecting surface which forms part of a sphere. Such surfaces are relatively easy to manufacture compared with other curved surfaces.

spherical refracting surface. A refracting surface which forms part of a sphere. Compared with other types of curved surface it is relatively easy to produce.

spherical triangle. A triangle drawn on a spherical surface and whose sides are arcs of GREAT CIRCLES.

spherical trigonometry. The branch of mathematics concerned with properties of SPHERICAL TRIANGLES.

spheroid.   Another name for ELLIPSOID.

spherometer.   An instrument for measuring the curvature of a spherical surface. It has three legs and a central movable screw which is adjusted until the spherometer has four point contact on the spherical surface. The radius may then be calculated from the reading of the movable screw position.

sphygmomanometer.   An instrument used for measuring arterial blood pressure. An inflatable bag is wrapped round the upper left arm and connected to a manometer. Pressure readings are taken when sounds in a suitably placed stethoscope indicate first maximum (i.e. *systolic) pressure and then minimum (i.e. diastolic*) pressure. The result is given as the ratio of systolic to diastolic pressure and is normally 120/80, the numbers corresponding to the pressure readings in millimetre of mercury.

spike potential.   Another name for ACTION POTENTIAL.

spin.   The intrinsic angular momentum of elementary particles, additional to any angular momentum due to orbital motion. The intrinsic angular momentum is quantized in values of

$$[s(s + 1)]^{1/2}h/(2\pi)$$

where s is the *spin quantum number* and h the Planck constant; for an electron $s = \pm \frac{1}{2}$. Due to spin there is a MAGNETIC DIPOLE MOMENT associated with each particle and so two close energy states result from the interaction of the magnetic moment due to orbital motion with that due to spin. Electrons jumping from such states to a lower energy one give rise to a DOUBLET in a spectrum.

Under the action of an external magnetic field, PRECESSION of the spin around the external field direction occurs. The allowed values of the component of the spin in the direction of the external field are

$$m_s h/(2\pi)$$

m_s is the magnetic quantum number and

has allowed values

$$s, s - 1, ..., - s$$

This phenomenon is called *space quantization*.

The resultant spin S of a number n of particles, each of spin s, can take the values

$$ns, (n - 1)s, ..., s, 0$$

Thus for two electrons S could be either 1 or 0. Often the spin of an elementary particle is denoted by J; for most elementary particles J is either an integer or a half integer. For a nucleus, the *nuclear spin quantum number* is the resultant of the spins of the constituent nucleons and is frequently denoted by I.

spin-orbit coupling.   The interaction between the spin and orbital angular momenta of a particle.

spin paired.   *See* PAULI EXCLUSION PRINCIPLE.

spin quantum number.   *See* SPIN.

spiral galaxy.   A flattened disc-shaped GALAXY containing prominent spiral arms of stars and interstellar gas and dust.

split beam spectrophotometer.   A SPECTROPHOTOMER, illustrated in fig. S18, designed for high accuracy. A *light chopper*

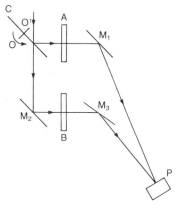

S18 Split beam spectrometer

C is incorporated in the form of two reflecting quadrants of a circle which rotates about the axis OO'. When the incident beam strikes a quadrant it is reflected through A, which contains the solution under investigation, and is then reflected at mirror M_1 on to photocell P. When the incident beam misses a quadrant it is reflected by M_2 through B, which contains pure solvent, and is then reflected by M_3 to P. Thus A and B are illuminated in turn for equal times. Ancillary equipment records only the difference of the photocell readings for light passing via A and that passing via B. Hence the effect of variations in the intensity of the incident light and of solvent absorption are eliminated.

split ring commutator. A device used to obtain unidirectional current from a GENERATOR. While the coil is rotating, the top lead runs over the top half of the ring and the bottom lead over the bottom half. When the leads arrive at the gap, the electromotive force reverses as also do the leads on the split ring. Hence the polarity of the upper half with respect to the lower half remains the same, resulting in a unidirectional but varying electromotive force.

spontaneous fission. Nuclear FISSION occurring independently of external factors.

spreading resistance. The part of a semiconductor's resistance not due to junction and contact resistance.

spring balance. *See* BALANCE (def. 2).

sputtering. A process in which atoms are ejected from a solid surface by bombardment of it with high-energy ions. It can occur in gas discharges due to ions hitting the cathode. The process is often used to coat a nonconductor with a thin adhesive metallic film.

sputter ion pump. *See* ION PUMP.

square. (1) A rectangle with all its sides equal.
(2) The result of multiplying a number by itself.

square root. *See* ROOT.

square wave. A rectangular PULSE train for which the MARK SPACE RATIO is one. It is usually illustrated as shown in fig. S19.

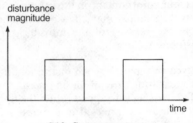

S19 Square wave

stable circuit. A circuit free from unwanted oscillations under any operating conditions.

stable equilibrium. *See* EQUILIBRIUM STABILITY.

staining. A technique of dyeing parts of an object, usually on a microscope slide, so that, provided the dye is selectively taken up, the intensity of the image varies from point to point much more markedly than without the staining. The visibility of the object is thus increased, but there is the disadvantage that most staining techniques kill cells.

standard atmosphere. Symbol atm. A unit of pressure equal to 101 325 pascal.

standard cell. Any electric cell of constant reproducible electromotive force which is used as a voltage standard, for example the CLARK CELL.

standard deviation. *See* DEVIATION (def. 1).

standard illuminants. Three sources defined for use in standard COLORIMETRY. They are of COLOUR TEMPERATURE 2848 K, 4800 K and 6600 K. The colour coordinates of a sample which is not self-luminous are found when it is illuminated at 45° to the normal by one of these sources, and viewed normally. The 2848 K source is the standard for artificial illumination. The other two are for daylight illumination: they differ because daylight itself differs with season, geographical location etc. Uniformity of practice is obviously desirable and so there are moves to have the equal energy source universally accepted.

standardization. The process of plotting a GAUSSIAN DISTRIBUTION so that the mean value coincides with the origin and the STANDARD DEVIATION is 1. A standardized normal frequency distribution is illustrated in fig. S20, the abscissa z being $(y - \mu)/\sigma$.

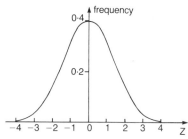

S20 Standardized normal frequency distribution

standards. Quantities defined in terms of physical situations which are believed to be unvarying. Some previously adopted standards suffered from the disadvantages of undergoing changes. For example corrosion, crystallization and other processes affected the length of the block of metal once chosen as the standard of length. *See* METRE.

standard temperature and pressure. Respectively 0° C, i.e. 273.15 K, and 101 325 pascal; the latter figure very nearly equals the former standard pressure of 760 millimetre of mercury. Standard temperature and pressure are standard conditions for the properties of gases.

standing wave. The wave formed when a PROGESSIVE WAVE travelling in one direction is superimposed on a similar progressive wave travelling in the opposite direction. If

$$y = a \sin 2\pi(t/T - x/\lambda)$$

represents a PLANE PROGRESSIVE WAVE incident on a rigid boundary, then the totally reflected wave at the boundary is represented by

$$y = a \sin 2\pi(t/T + x/\lambda)$$

and the combination of incident and reflected waves, i.e. the standing wave, is represented by

$$y = 2a \sin(2\pi t/T) \times \cos(2\pi x/\lambda)$$

This standing wave, in contrast to its parent waves, has permanent nodes at $\lambda/2$ apart and permanent antinodes, of amplitude $2a$, at $\lambda/4$ from each node. Other differences are that the amplitude varies from point to point along the standing wave and that the phase at every point between successive nodes is the same. In a progressive wave the amplitude at every point is constant and the phase changes progressively along the path. In practice total reflection may not occur, resulting in minimum amplitude at the previous standing wave node positions and smaller maximum amplitude at the previous antinode positions. At a rigid boundary the displacement is always zero; at an open boundary there is always an antinode, but otherwise rigid boundary treatment applies.

Using the apparatus shown in fig. S21, a standing wave can readily be obtained by suitably adjusting the mass in the scale pan. Standing waves, due to oscillating electrons, occur in aerials tuned to incoming radio waves and in aerials transmitting radio waves. Such waves are also formed in quantum mechancial stationary states. Standing waves can be set up in rods, in transmission lines and for light. *See also* VIBRATIONS IN PIPES; VIBRATIONS IN STRINGS.

star. Any of an enormous number of huge celestial objects which radiate electromagnetic radiation and are grouped into galaxies. Even within the galaxies, the stars, of which the Sun is one, are separated by vast distances and so appear from Earth only as points of light. *See also* HERTZSPRUNG-RUSSELL DIAGRAM.

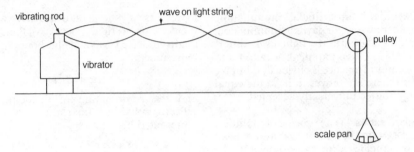

S21 Standing wave in string

Stark effect.   The alteration in the atomic spectrum of a source due to either an external electric field, or an intense field resulting from the presence of neighbouring atoms or ions. Each spectrum line is split into a number of sharply defined components, symmetrically arranged about the position of the original line. The displacement is proportional to field strength up to a strength of about 10^7 volt per metre..

Stark-Einstein equation.   The equation

$$E = hLf$$

where E is the energy of radiation of frequency f absorbed by a mole of substance in a photochemical reaction; h and L are respectively the PLANCK CONSTANT and the AVOGADRO CONSTANT.

stat-.   An obsolescent prefix used with practical electrical units to name the corresponding electrostatic units. *See* CGS UNITS.

state of matter.   Any of the three physical states, i.e. solid, liquid and gas, in which matter can exist. Changes in state occur at definite transition temperatures which are characteristic of each substance. Plasmas, liquid crystals, glasses, colloids and superfluids have all, at various times, been described as extra states of matter.

static electricity.   Another name for ELECTROSTATIC.

statics.   A branch of mechanics concerned with the properties of bodies in equilibrium, i.e. with systems of forces for which no motion occurs.

stationary orbit.   A circular SYN-CHRONOUS ORBIT in the same plane as the equator of a celestial body, usually the Earth. A satellite in such an orbit appears to remain stationary with respect to a point on Earth. Communication satellites in stationary orbit reflect radio waves and, by using several of them, world-wide coverage is obtained.

stationary point.   A point on a graph at which the tangent to the curve is parallel to one of the axes.

stationary state.   One of the time-independent quantized states of a system as described by QUANTUM MECHANICS.

stationary wave.   Another name for STANDING WAVE.

statistical mechanics.   The statistical prediction of the properties of matter in bulk from a knowledge of the properties of the constituent atoms or molecules and the forces between them. The term is sometimes restricted to the treatment of systems in equilibrium.

statistical weight.   Symbol g. A number equal to the number of times a particular value, or range of values, of a quantity occurs in a statistical investigation of the quantity.

statistics.   The systematic study of observations in order to determine the general behaviour of a system from a limited number of such observations. SIGNIFICANCE and PROBABILITY are involved.

stator. The stationary part of a GENERATOR or ELECTRIC MOTOR. It may be either the field winding or the armature.

steady state theory. The theory that the universe has always existed, its expansion being compensated for by the continuous creation of matter at a rate of about 10^{-43} kilogramme per cubic metre per second, thus keeping the average density of the universe constant. However, many observations tend to support the alternative BIG BANG THEORY.

steam. Water in gaseous state at a temperature of 100° C or above. Strictly steam is an invisible gas, but the term is also applied to steam carrying small liquid-water droplets and so appearing white.

steam calorimeter. A calorimeter in which the amount of heat supplied is calculated from the mass of steam condensed on the body under test. See JOLY'S STEAM CALORIMETER.

steam engine. A machine taking heat from a steam boiler, using it to perform external work and finally rejecting the unused heat to a condenser. See also CARNOT CYCLE.

steam point. The temperature at which liquid water and its vapour are in equilibrium at standard pressure. It is 100° C.

Stefan-Boltzmann law. Another name for STEFAN'S LAW.

Stefan's constant. See STEFAN'S LAW.

Stefan's law. The law

$$M = \sigma T^4$$

where M is the energy radiated per unit area per second from a black body at thermodynamic temperature T; σ is Stefan's constant and is given by

$$2\pi^5 k^4/(15h^3c^2)$$

where k is the BOLTZMANN CONSTANT, c the speed of light in vacuo and h the PLANCK CONSTANT. The value of σ is 5.6697×10^{-8} watt per square metre per kelvin to the fourth.

stellar evolution. The progressive series of changes occurring in a star as it ages. Stars similar in mass to the Sun eventually become a RED GIANT and finally a WHITE DWARF. Very large massive stars eventually become unstable, leading to a SUPERNOVA explosion, the core of the star collapsing to become a NEUTRON STAR or BLACK HOLE. See also HERTZSPRUNG-RUSSELL DIAGRAM.

stellar spectrum. The spectrum of the radiation given out by a star. It yields information about the star's temperature and composition. The shorter the wavelength

letter	region of maximum emission	temperature/kilokelvin
W	ultraviolet	80
O	blue	40
B	blue-white	20
A	white	9
F	yellow-white	7
G	yellow	6
H	orange	4.5
M	red	3
S R N	infrared	2.5–3

at which maximum emission occurs, the hotter the star and so stars have been grouped into *spectral types*; these are identified by letter, as shown in the table. For W stars, the spectra are of ionized gases. For lower temperature stars spectra of metallic elements, such as calcium appear. At the lowest temperatures elementary molecules survive. *See also* HERTZSPRUNG-RUSSELL DIAGRAM.

St Elmo's fire. Another name for CORONA DISCHARGE.

step down transformer. *See* TRANS-FORMER.

step up transformer. *See* TRANSFORMER.

step wedge. A sheet or block of material in the form of a series of bands of increasing opacity to radiation.

steradian. Symbol sr. A unit of SOLID ANGLE; it is the solid angle subtended at the centre of a sphere of unit radius by unit area of the surface of the sphere. So the solid angle subtended by the surface of a sphere at its centre is 4π steradian.

stere. A unit of volume equal to one cubic metre.

stereophonic reproduction. The reproduction of sound by two or more separate channels so as to give an impression of the spatial distribution of the original sound sources.

stereoscope. An instrument whereby two slightly different views of a scene are presented one to each eye of the observer. The retinal images are then combined by the brain to give an impression of depth. The usual method of obtaining the views is to photograph the scene from two slightly different positions.

stereoscopic microscope. *See* MICROSCOPE.

Stern-Gerlach experiment. An experiment which first demonstrated the quanti-zation of angular momentum of the electron. An atomic beam was passed through a nonuniform magnetic field and observed to divide into two separate components. *See also* SPIN.

stethoscope. A device for amplifying sounds, especially medical ones. It consists essentially of a horn, which is sited as close as possible to the sound under investigation, branching into two tubes each of which finishes in a small horn, i.e. one for each ear of the examiner. If the impedances of the several junctions are mismatched, serious distortion of the signal under investigation occurs. Although by trial and error reasonably efficient stethoscopes have been evolved, sound reproduction does vary from one instrument to another. This leads to a marked preference by a particular person for a particular instrument.

stilb. Symbol sb. A unit of LUMINANCE equal to 1 CANDELA per square centimetre.

stimulated emission. The process in which a photon colliding with an excited atom causes emission of a photon with the same energy as itself, moving in the same direction and having the same phase. The action of a LASER and of a MASER depends on this phenomenon.

stirrup. One of the ossicles housed in the MIDDLE EAR. It is articulated to the hammer by the anvil and has a foot plate attached to the oval window.

stochastic process. A process involving random behaviour and so describable in terms of probabilities.

stokes. Symbol st. The unit of kinematic viscosity in CGS UNITS. It equals 10^{-4} square metre per second.

Stoke's law. (1) The frictional force, i.e. drag, on a spherical body of radius r which is travelling at its terminal speed v through a viscous medium of coefficient of VIS-COSITY η is $6\pi\eta r v$. The law holds only for a restricted range of conditions.
(2) The wavelength of FLUORESCENCE

radiation is greater than that of the radiation that excites the fluorescence. The law is not always obeyed. *See also* RAMAN EFFECT.

stop.   An aperture used to limit the cross section of a beam of light in an optical system.

stop number.   Another name for F NUMBER.

stopping power.   A measure of the effect of a substance on the kinetic energy of a charged particle passing through it. Stopping power is often quoted relative to that of a standard substance, usually air or aluminium.

stopping potential.   The maximum value of the potential to which an electrically insulated plate rises when it emits electrons under the action of light.

storage battery.   Another name for ACCUMULATOR.

storage device.   A device used for holding information, usually in binary form in a COMPUTER. The devices include MAGNETIC TAPES and DISCS. Computer storage systems are classified into *direct access devices*, in which any part of a stored file can be retrieved directly, and *sequential access devices*, which are slower since the files are scanned in sequence until the information is found. Some digital computers have more than one type of store.

storage time.   (1) The time for which information can be stored, without significant loss, in any device.
(2) The time interval between application of reverse bias to a P-N JUNCTION and the stopping of forward current.

storage tube.   An electronic tube capable of storing information for a determined and controllable time and allowing extraction as required. In charge storage tubes, the information is stored in the form of electrostatic charge on a storage electrode. Storage tubes using photoconductivity effects are also used. Other types are also available.

STP.   Abbrev. for STANDARD TEMPERATURE AND PRESSURE.

strain.   The deformation, either temporary or permanent, produced in a body subjected to STRESS. The *tensile strain*, symbol e, is defined as the increase in length per unit length of a body. The *bulk strain*, symbol θ, is the increase in volume per unit volume of strained body. The *shear strain* is the angle in radian through which a body is twisted, the volume remaining constant: an example is a rectangular block strained so that two opposite faces become parallelograms, the other faces not changing shape.

strain gauge.   A device for measuring STRAIN. The term usually applies to small devices attached to the surface of a body in order to measure the deformation at different points. Devices depending on the change produced in electric resistance, or capacitance or inductance are available. Other devices monitor the piezoelectric or magnetostrictive effects produced by the strain.

strain hardening.   Hardening of a metal by subjecting it to STRAIN, by for example stretching or rolling. The hardening is due to dislocations produced in the crystal structure.

strangeness.   Symbol S. A QUANTUM NUMBER which is associated with elementary particles and is conserved in STRONG INTERACTION and in ELECTROMAGNETIC INTERACTION but not in WEAK INTERACTION. Strangeness was postulated in order to explain the longer than predicted lifetimes of some elementary particles; for example the observed and predicted lifetimes of the K meson were respectively about 10^{-9} second and 10^{-23} second. Strangeness is associated with the presence of the *strange quark* in the particle. S take integral values, given by HYPERCHARGE minus BARYON NUMBER.

strange quark. *See* QUARK; STRANGENESS.

stratopause. The upper boundary of the STRATOSPHERE.

stratosphere. The region of the Earth's atmosphere extending from a height of about 10 kilometre to about 80 kilometre above the Earth's surface. It is free of weather conditions and is approximately isothermal. *See* ATMOSPHERIC LAYERS.

stratospheric fallout. *See* FALLOUT.

stream function. A SCALAR function of position, describing the steady two-dimensional flow of an incompressible fluid.

streaming potential. The electric potential produced when a liquid is forced through a porous diaphragm or other permeable solid. *Compare* ELECTRO-OSMOSIS.

stream line. A line in a fluid such that at any point on the line the tangent to it is the direction of the velocity of the fluid at that point. If each stream line is continuous, then the flow is known as *stream line flow*.

stream line flow. *See* STREAM LINE.

stress. The force per unit area applied to a body. It produces or tends to produce STRAIN in the body. Several different types of stress have been identified: *tensile stress*, symbol σ, tends to stretch or compress a material in the direction of the applied force; *bulk stress* is the pressure on a body due to its immersion in a fluid; *shear stress*, symbol τ, is the tangential force per unit area and so tends to produce a twisting motion.

string galvanometer. Another name for EINTHOVEN GALVANOMETER.

stroboscope. An instrument for viewing a periodically moving object by an intense light which flashes on and off at a frequency equal to that of the object or to some multiple of that frequency; the object therefore appears stationary. The same effect may be obtained with steady illumination of the object, by viewing it through a slot in a disc which rotates at the appropriate speed. Slight variations from the flashing frequency, or disc speed, for a stationary appearing object cause the object to appear to move slowly. Stroboscopic methods make possible the study of phenomena which might otherwise not be observable.

An unfortunate stroboscopic effect may occur in cine and television pictures: the wheels of a moving vehicle can appear stationary. This occurs when a wheel makes one complete rotation between each frame or scan. Apparent rotation in the wrong direction is also possible.

strong interaction. A type of interaction occurring at short range, of about 10^{-15} metre, between HADRONS. The magnitude is about 100 times that of the ELECTROMAGNETIC INTERACTION. Strong interactions are the forces which hold nucleons together in the atomic nucleus and are unaffected by any charge on a nucleon, i.e. the force exhibits *charge independence*. In strong interaction, total angular momentum, charge, baryon number, isospin, strangeness, parity, charge conjugation parity and G parity are conserved. Strong interaction is now believed to be a remnant of a more powerful inter-QUARK force.

Student's t test. A statistical test of the significance of a series of observations.

subatomic. Smaller than an atom. Subatomic particles are particles found in atoms, i.e. electrons and nucleons.

subcritical. *See* CHAIN REACTION.

sublimation. The direct transition from solid to vapour or vice versa, without the formation of an intermediate liquid state.

subshell. A subdivision of an ELECTRON SHELL for which all the electrons have the same azimuthal quantum number. The subshells of a particular shell are denoted by s, p, d or f according to the value of their quantum number. *See* ATOMIC ORBITAL.

subsonic. Having or involving a speed less than Mach 1. *See* MACH NUMBER.

substandard. A standard of slightly less accuracy than a primary standard. The substandard is an intermediate link between the primary standard and the device being calibrated or checked.

substitution method. A method in which the effect of an unknown property, for example resistance, is recorded and then substituted by a known property, i.e. another resistance whose value is known, giving the same effect. The value of unknown property is thus determined.

subtractive process. The process in which COLOUR is produced by combining absorbing materials, such as pigments or dyes or filters, and presenting them to a beam of light. The colour of the reflected or transmitted light depends on the absorption spectrum of the mixture and on the spectral distribution of the illuminant. *See also* COLORIMETRY. *Compare* ADDITIVE PROCESS.

summer solstice. *See* SOLSTICE.

SU(N). A GROUP describing operations on *N* objects. For example SU(2) applies to the two quarks or two leptons in a GENERATION; SU(3) applies to the three colours of QUARK. Recently the three colours and two flavours have been described using SU(5).

Sun. The star around which orbit the Earth, the other planets and all other objects in the solar system. The Sun is a yellow main-sequence star of mass around 2×10^{30} kilogramme, situated about two-thirds of the radius of the Galaxy away from the central core of the Galaxy. The mean Sun-Earth distance is 149×10^6 kilometre and so solar radiation takes about 8 minute to reach Earth. The Sun is composed mainly of hydrogen and helium. Energy is produced in the central core, of about 400 000 kilometre diameter, by nuclear FUSION reactions: at the core temperature of 15×10^6 K, hydrogen is converted to deuterium and then to helium. At the

visible surface, i.e. the PHOTOSPHERE, the temperature is about 6000 K; it then increases again through the CHROMOSPHERE and CORONA.

sunspot. A region of the Sun's PHOTOSPHERE which is about 2000 K cooler, and therefore darker, than the surrounding area. Sunspot diameters range from a few hundred kilometre to several thousand kilometre. The spots last from a few hours to a few months. A typical sunspot has a dark central core surrounded by a lighter region. Sunspots tend to occur in groups, their number reaching a maximum value approximately every 11 years. Sunspots have intense magnetic fields and are associated with magnetic storms on Earth.

superconductivity. The disappearance of electric resistance which occurs in many metals and alloys at or below a transition temperature; this temperature is characteristic of the substance and close to absolute zero. The phenomenon results from a change in behaviour of the conduction electrons in the material at the transition temperature.

An important application of superconductivity is the production of ELECTROMAGNETS having a very high magnetic field: below the transition temperature, the current in a closed ring of superconductor continues to flow for a considerable time after the magnetic field which induced it has been removed and so the ring is itself an electromagnet. Alloys such as niobium-titanium, niobium-zirconium and niobium-tin maintained at about 4.5 K can produce magnetic flux densities of about 10 tesla, compared with a maximum value of 2 tesla for iron. The efficiency of electric generators using superconducting magnets is far higher than that of conventional machines; moreover the size is smaller. Superconducting magnets are being increasingly used in high energy PARTICLE PHYSICS research.

supercooling. The cooling of a system to a temperature below that at which a phase change normally occurs, without the change happening. Thus a pure liquid may sometimes be cooled slowly to a temperature a few degrees below its freezing point, without freezing. However, a slight mechanical disturbance, or the introduction of a very small amount of solid to a supercooled liquid, immediately starts solidification, which continues with the evolution of heat until the normal freezing point is attained. *See also* SUPERHEATING.

supercritical. *See* CHAIN REACTION.

superdense theory. Another name for BIG BANG THEORY.

superficial expansion. The increase in area of a body with temperature change.

superfluid. A fluid that flows without friction through fine channels and exhibits very high THERMAL CONDUCTIVITY at low temperatures. For wider channels, above a critical velocity there is a small amount of friction. Below 2.186 K helium is a superfluid, its thermal conductivity rising with decreasing temperature to a million times its value above that at 2.186 K.

supergiant. The largest and most luminous type of star. Supergiants are comparatively rare. *See also* HERTZSPRUNG-RUSSELL DIAGRAM.

superheating. The process of heating a liquid to a temperature above its normal boiling point without boiling occurring. It can only happen when there are no small particles of matter in the liquid or roughness on the vessel walls, i.e. no places where bubbles can form.

superheterodyne receiver. The most widely used radio receiver, in which the incoming signal is mixed with a locally generated signal. The HETERODYNE principle is thus used to produce a resultant signal of ultrasonic frequency, which contains all the original modulation. This signal is amplified and then passed to an audio frequency amplifier resulting in high gain amplification and great selectivity in the final reception.

superhigh frequency. A FREQUENCY in the range 3 gigahertz to 30 gigahertz.

superior planet. Any of the planets whose orbits are further away from the Sun than is Earth's, i.e. Mars, Jupiter, Saturn, Uranus, Neptune and Pluto. *Compare* INFERIOR PLANET.

superlattice. *See* SUPERSTRUCTURE.

supermultiplet. A set of quantum mechanical states each of which has some quantum numbers identical to, and some differing from, those of other states. The ranges of those which differ are determined by those which are the same.

supernova. A star which explodes due to instabilities following the exhaustion of its nuclear fuel. Much energy is released with the formation of a *supernova remnant*, i.e. an expanding shell of debris. Often a PULSAR remains at the shell centre.

superposition principle. When two or more waves traverse a medium, the resultant displacement at any point is the sum of the separate displacements due to the individual waves at that point.

supersaturated solution. A solution containing a higher concentration of solute than a saturated solution at the same temperature. It is formed by SUPERCOOLING.

supersaturated vapour. An unstable vapour whose pressure exceeds the SATURATED VAPOUR PRESSURE at that temperature. It is formed by SUPERCOOLING.

supersonic flow. The movement of a fluid at a speed exceeding the velocity of sound in the fluid. As the speed of an object in the fluid relative to the fluid increases from below to beyond the speed of sound in the fluid, the resistance to motion increases due to the formation of SHOCK WAVES. *See also* SONIC BOOM.

superstring theory. A quantum field theory modification in which the quanta are identified with the modes of vibration of either open or closed strings, rather than being considered as point-like objects.

superstructure. An ordered arrangement of atoms in a solid solution, forming a lattice superimposed on the normal solid solution lattice.

supersymmetry. A theoretical concept widely used in the modern approach to GRAND UNIFIED FIELD THEORIES. There is however no evidence that it plays any role in the description of nature.

supplementary angle. An angle related to a given angle by their sum being 180° Thus 120° is the supplementary angle of 60°.

supplementary SI units. *See* Table 1.

suppressor grid. The GRID between the screen grid and the anode in a PENTODE valve.

surd. An irrational root of a number. Thus $3^{1/2}$ and $5^{1/2}$ are surds but $4^{1/2} = 2$ is not.

surface active agent. *See* SURFACTANT.

surface barrier transistor. A TRANSISTOR in which the usual p-n junctions are replaced by Schottky barriers (*see* SCHOTTKY DIODE). Under saturation conditions there is no hole storage and so the transistors are suitable for very high speed switching applications.

surface charge transistor. A type of CHARGE TRANSFER DEVICE.

surface colour. The coloured light reflected by a surface rather than the more common body colour which includes light reflected after some penetration into the medium.

surface density of charge. Symbol σ. The amount of charge per unit area of surface. It is measured in coulomb per square metre.

surface energy. The energy per unit area of exposed surface. In general it exceeds the SURFACE TENSION which is the free surface energy.

surface forces. Forces between bodies in contact.

surface tension. Symbol γ. A phenomenon resulting from attractive forces between molecules in a liquid. A molecule in the interior of a liquid is attracted by other molecules on all sides, but one at the surface is only attracted inwards from the surface; this causes liquids to behave as if covered by an elastic skin. The phenomenon is responsible for the behaviour of liquids in vertical capillary tubes, the absorption of liquids by porous substances such as cloth and paper, and the tendency of a liquid surface to contract to the smallest possible area so that a free drop tends to be of spherical shape.

Surface tension is defined as the force tangential to the surface on one side of a line of unit length in the surface, and so is measured in newton per square metre. Alternatively it is the work required to produce isothermally unit increase in surface area, i.e. it is equal to the *free surface energy* but not the total surface energy, and is expressed in joule per square metre. The two definitions are numerically and dimensionally equivalent. Surface tension may be measured by observing the change of level of a liquid inside a vertical capillary tube introduced into the liquid (*see* JUVIN'S RULE), or by direct weighing of the pull of the liquid on a microscope slide, suspended from a balance arm so as to lie in the liquid surface, or by JAEGER'S METHOD.

surface wave. (1) Any wave propagated on the surface of an elastic medium.
(2) An electromagnetic wave propagated parallel to the Earth's surface.

surfactant. A substance used to increase the wetting or spreading power of a liquid. Many act by lowering the SURFACE TENSION.

surge. An abnormal transient electrical disturbance in a conductor, for example that produced in a TRANSMISSION LINE by lightning, sudden faults etc.

susceptance. Symbol B. The imaginary part of the ADMITTANCE, for which the real part is the CONDUCTANCE.

susceptibility. See MAGNETIC SUSCEPTIBILITY; ELECTRIC SUSCEPTIBILITY.

suspension. A mixture of a pure liquid and a substance in the form of very small particles, but sufficiently large to be distinguishable from the liquid.

switch. A device for opening or closing a circuit or for changing its operating conditions between specified levels. A mechanical device such as a circuit breaker or a solid state device such as a transistor may be used.

symbol. A letter representing a total quantity, i.e. both a number and a unit. When a symbol in a formula is replaced by a particular value in a particular problem, both the number and the unit should be inserted. This minimizes the danger of ending up with the wrong units in the final answer.

symmetry. A property of some geometric figures, mathematical expressions, patterns etc. such that various operations on them leave them unchanged. Rotation of 180° of a square about a diagonal leaves the square unchanged; the diagonal is said to be a two-fold axis of symmetry because two such rotations give a complete turn. The square also has a four-fold axis of symmetry perpendicular to its plane through its centre. The symmetry of shapes may be described in terms of mirror reflections. The properties of symmetry operators can be treated by the theory of GROUPS and are important in many branches of science, such as studies of crystal structures, of shapes and energy levels of molecules, of properties of elementary particles.

synapse. The junction between two neurons.

synchrocyclotron. A type of CYCLOTRON in which the frequency of the accelerating electric field is slowly decreased so that the particles stay in phase with the field as their speed increases. The decrease in frequency compensates for the relativistic increase in mass. Particle energies in the range 400 mega-electronvolt to 500 mega-electronvolt can be produced.

synchronous clock. A clock in which a SYNCHRONOUS MOTOR drives the mechanism controlling the hands. The time keeping is therefore entirely dependent on the frequency of the alternating current passing through the motor.

synchronous motor. An alternating current electric motor which runs at a speed proportional to the frequency of the power supply.

synchronous orbit. A satellite orbit about a celestial body, such as the Earth, in which the satellite makes one complete traverse of the orbit in the time during which the body rotates once on its axis.

synchronous speed. The speed of rotation of the magnetic flux in an alternating current machine. It equals the ratio of the frequency of the alternating current supply to the number of pairs of magnetic poles for which the alternating current winding has been designed. It is measured in revolution per second.

synchrotron. A cyclic ACCELERATOR based on the BETATRON but employing a constant frequency electric field in addition to a changing magnetic field. It is illustrated in fig. S22. Inside the circular vacuum tube is a metallic cavity with a gap across which a high frequency electric field is applied in synchronism with the angular frequency of the electrons. The electrons are accelerated while in the cavity. The relativistic increase in electron mass with speed may be compensated for by increasing the magnetic field strength. Electron energies in the giga-electronvolt range have been achieved. See also PROTON SYNCHROTRON.

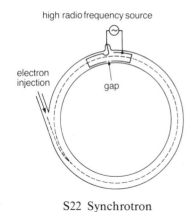

S22 Synchrotron

synchrotron radiation. Electromagnetic radiation produced when electrons move at high speed in a magnetic field. The electrons travel in circular paths. The radiation is emitted in the direction of motion of the electron and is plane polarized perpendicular to the magnetic field direction. The frequency of the radiation depends on the electron speed and can vary from that of radio waves to that of X rays. Such radiation is produced in a SYNCHROTRON, hence the name. It is also generated in many regions of the universe which are associated with very high magnetic fields.

synodic period. For any object in the solar system, the mean time elapsing between successive returns of the object to the same position relative to the Sun, as seen from Earth.

synoptic chart. A map used in weather forecasting, showing wind speed, barometric pressure etc. at a particular time.

synovial fluid. A liquid which helps in the lubrication of human joints. It is a NON-NEWTONIAN FLUID.

systematic error. See ERROR (def. 2).

Système Internationale d'Unites. Known as SI UNITS. See also Tables 1–6.

T

tachometer. An instrument for measuring speed of rotation.

tachyon. A hypothetical elementary particle whose speed exceeds that of light so that, according to RELATIVITY theory, either its rest mass or its energy is imaginary. Such a particle should emit CERENKOV RADIATION in a vacuum. Tachyons still await detection.

Tacoma bridge disaster. A disaster due to RESONANCE. The bridge absorbed excessive energy, resulting in vibrations of such large amplitude that the bridge collapsed.

Talbot's law. The apparent brightness of an object, viewed through a rotating disc with a sector removed, is proportional to the ratio of the angle α of this sector to $2\pi - \alpha$ for rotation speeds sufficiently high for persistence of vision to give the sensation of uniform brightness.

tan. Abbrev. for tangent. *See* TRIGONOMETRIC FUNCTIONS.

tandem generator. An electrostatic generator consisting of a pair of VAN DE GRAAFF GENERATORS in series, thus doubling the voltage. The machine is used in accelerators.

tangent. (1) *See* TRIGONOMETRIC FUNCTIONS.
(2) A line or plane touching a curve or surface at a point. The derivative at the point gives the slope of the tangent at the point.

tangent galvanometer. A type of galvanometer in which the current to be measured is passed through a vertical coil of wire of n turns, whose plane is parallel to the horizontal component H of the Earth's magnetic field. A small magnet is pivoted at the centre of the coil so that it can move in a horizontal plane. It is thus deflected through an angle θ due to the magnetic action between the fields of coil and Earth. The current is given by

$$2Hr \tan \theta / n$$

where r is the coil radius. The instrument was used to compare currents but is now mainly employed in finding H, by passing a known current through the coil.

tangential acceleration. The quantity $r\,d\omega/dt$, often written $r\dot\omega$, where r is the perpendicular distance of the point in question from the axis of rotation and ω is the angular speed of rotation.

tangential speed. The quantity $r\omega$ where r is the perpendicular distance of the point in question from the axis of rotation and ω is the angular speed of rotation.

tangent law. The relationship

$$ny \tan \alpha = n'y' \tan \alpha'$$

n and n' are respectively the object and image space refractive indices; y and y' are respectively the object and image heights; α is the angle made with the axis by a ray leaving the object axial point and α' the angle it makes with the axis on arriving at the image axial point. Except within the paraxial region where sines and tangents may be equated, the tangent law and the SINE CONDITION are contradictory. This implies that geometrically perfect imagery by an optical system is only possible for one pair of conjugates.

tape recorder. *See* MAGNETIC RECORDING.

tau lepton. A negatively charged LEPTON in the third GENERATION. It is a heavier analogue of the electron and MUON.

telecommunications. The transference of information by any kind of electromagnetic system, for example telephone, radio, television or telegraph.

telegraph. An apparatus for transmitting and receiving messages by sending electric pulses along cables.

telemeter. Any instrument for transmitting measured data from its source to a distant observer. Thus measurements of conditions within a nuclear reactor core are transmitted along wires to the controller.

telephone lens. A photographic compound lens consisting of a converging lens system followed by a diverging lens system such that the combination gives a real image of greater magnification then the normal camera lens which it replaces when required.

telescope. (1) An optical instrument for the production of a magnified image of a distant object. A REFRACTING TELESCOPE uses a combination of lenses, whereas a REFLECTING TELESCOPE has at least one reflecting surface and may also incorporate lenses. A telescope producing an erect final image is known as a *terrestrial telescope*, although it may be suitable for astronomical work. A telescope producing an inverted final image is known as an *astronomical telescope*. A telescope is said to be in *normal adjustment* when both the object and the image of it formed by the telescope are at infinity, i.e. parallel light both strikes and emerges from the instrument.
(2) *See* RADIO TELESCOPE.

television. The electrical transmission of visual scenes and images by cable or by radio waves. The scene to be transmitted is optically focused on a photoelectric screen consisting of a mosaic of light-sensitive particles in a *camera tube*. The optical image is thereby converted to an electrical image. The electrical image is scanned horizontal line by horizontal line by an electron beam whose jump to the next line is controlled by a signal of sawtooth wave form. By this means an electric current, proportional at each instant to the intensity of the part of the electrical image being scanned, is obtained. In Europe the scanning rate is 625 lines in 1/25 second, whereas in the USA it is 525 lines in 1/30 second.

The signal so produced modulates a VHF or UHF CARRIER WAVE and is transmitted with an independent sound signal. The picture signals arriving at the receiving aerial are demodulated in the television receiver. They can then control the electron beam intensity arriving at the receiver's CATHODE RAY TUBE, which thus shows the reconstructed picture. The audio and visual signals are synchronized in the receiver. *See also* COLOUR TELEVISION.

temperature. Symbol T. A measure of the hotness of a body. The temperature determines the direction of heat flow, which is always from a higher temperature to a lower temperature. The temperature of a body is a measure of the average KINETIC ENERGY of its atoms or molecules.

temperature coefficient of resistance. Symbol α. A coefficient defined by the equation

$$R_t = R_0(1 + \alpha t - \beta t^2)$$

where R_0 and R_t are the resistances of a conductor at $0°$ C and t^0 C respectively; in general β is negligible and α is generally positive for conductors and negative for semiconductors.

temperature gradient. The rate of change of temperature with length along a substance through which heat is being conducted. For a well lagged bar of uniform cross section, the temperature gradient is constant along the length of the bar.

temperature scales. Scales developed in order to quantify temperature. In the empirical approach, two or more repro-ducible temperature dependent events were selected and assigned fixed points on a scale; an example of such a scale is the CELSIUS SCALE. An alternative is to measure temperature as a function of the energy possessed by matter; this leads to the con-cept of THERMODYNAMIC TEMPERATURE.

temporal bone. A bone at the side of the head. It is the hardest bone in the body and provides protection for the middle ear.

tensile strain. *See* STRAIN.

tensile strength. The tensile STRESS neces-sary to break a material under tension.

tensile stress. *See* STRESS.

tensimeter. An apparatus for measuring the VAPOUR PRESSURE of a liquid by com-paring it with that of water at the same temperature.

tensor. An abstract mathematical entity used in transformations from one co-ordinate system to another. *Tensor analysis* is a generalization of VECTOR ANALYSIS.

tera-. Symbol T. A prefix meaning 10^{12}.

term diagram. An energy-level diagram for an atom. Possible transitions between the levels correspond to spectral lines from the infrared to X ray wavelength range.

terminal. (1) A point on an electric circuit, battery etc. at which a connection can be made.
(2) A piece of apparatus connected to a computer for the input or output of information.

terminal velocity. The velocity of a body relative to a fluid in which it is moving, when the resultant force on the body is zero. For a sphere of radius r and density ρ, falling under gravity in a fluid of density σ, the terminal velocity is given by

$$2(\rho - \sigma)r^2 g/(9\eta)$$

where g is the acceleration due to gravity

and η the coefficient of VISCOSITY. See STOKES' LAW (def. 1).

terminal voltage. The voltage between the terminals of a battery when it is supply-ing current.

termination. A load impedance placed at the end of a transmission line to ensure IMPEDANCE MATCHING and to prevent reflection.

terminator. The line separating the bright and dark parts of a celestial body.

terrestrial abundance. *See* ABUNDANCE.

terrestrial magnetism. The Earth's mag-netism. The Earth behaves as if it con-tained a large bar magnet making a small angle with the Earth's axis of rotation. The field is most intense at the magnetic North and South Poles, whose positions do not coincide with those of the geographic poles. Moreover, the positions of the magnetic poles change slowly with time. The magnetic field at points on the Earth's surface is characterized by its horizontal and vertical components and by the angles of DIP and of DECLINATION. The Earth's magnetic field is not fully understood but may arise from an iron core. The field is influenced by solar phenomena and the ionosphere. In addition to a slow change with time, the field shows periodic annual and daily variations. *See also* AGONIC LINE; MAGNETIC EQUATOR; ISODYNAMIC LINE; ISOGONAL LINE; ISOCLINE; PALAEOMAG-NETISM.

terrestrial planet. A planet whose mass does not exceed that of Earth.

terrestrial telescope. *See* TELESCOPE.

tesla. Symbol T. The SI unit of MAGNETIC INDUCTION, equal to one weber per square metre.

Tesla coil. An apparatus for producing high voltages of high frequency. It consists of an INDUCTION COIL in which the inter-rupter in the primary circuit is replaced by a high frequency spark gap. A Tesla coil is often used to excite a luminous discharge in glass vacuum apparatus in order to test for leaks.

tetragonal system. *See* CRYSTAL SYSTEM.

tetrahedron. A polyhedron with four faces each of which is a triangle. For a *regular* tetrahedron, the triangles are equilateral and congruent.

tetrode. A THERMIONIC VALVE with four electrodes. It is thus a triode with an extra grid, known as a SCREEN GRID, between anode and control grid, thereby reducing the capacitance between them and so improving the valve's performance as a high frequency amplifier.

TeV. Symbol for tera-electronvolt, i.e. 10^{12} electronvolt.

theodolite. An instrument in which a small telescope can move over angular horizontal and vertical scales. It is used in surveying for measuring angles.

theorem. A proposition that can be proved logically from a set of basic assumptions. The PARALLEL AXES THEOREM is an example.

theory. (1) A proposition suggested in order to explain observed facts, but not yet verified.
(2) A collection of principles, methods etc. used to explain a wide set of phenomena. QUANTUM THEORY is an example.

therm. A unit of heat energy used in the gas industry. It equals 10^5 BRITISH THERMAL UNIT, i.e. $1.055\,056 \times 10^8$ joule.

thermal agitation. The random motion of the molecules of a substance. According to the KINETIC THEORY OF MATTER, the energy of the motion is the heat content of the substance.

thermal analysis. The use of heating and cooling curves to detect endothermal or exothermal reactions associated with phase transitions. It is used in metallurgy.

thermal capacity. Another name for HEAT CAPACITY.

thermal conductance. Symbol K. The rate of heat flow across a substance for a temperature difference of 1 K between its faces.

thermal conductivity. Symbol λ. The quantity of heat flowing per second across unit area of a substance for unit temperature gradient, the plane of the area being perpendicular to the temperature gradient.

thermal diffusion. A phenomenon which occurs when a mixture of gases is subjected to a temperature gradient; the heavier atoms or molecules concentrate in the cooler region, and the lighter ones in the hotter region. The phenomenon is used in the separation of some isotopes. *See also* CLUSIUS COLUMN.

thermal diffusivity. Symbol α. A quantity defined as $\lambda/(\rho c)$ where λ is the THERMAL CONDUCTIVITY, ρ the density and c the specific heat capacity of the substance under consideration.

thermal expansion. The increase in the length, area or volume of a substance as a result of raising its temperature. *See* COEFFICIENT OF EXPANSION.

thermal insulator. A substance of very low THERMAL CONDUCTIVITY.

thermal ionization. IONIZATION occurring in a gas due to collisions between atoms or molecules moving with thermal energy.

thermalize. To reduce the energies of neutrons with a MODERATOR so that they become THERMAL NEUTRONS.

thermal neutrons. Neutrons of kinetic energy of the same order of magnitude as the kinetic energies of atoms or molecules, which is about 0.025 electronvolt at room temperature.

thermal power station. A power station producing electric energy by burning fossil fuel such as coal, coke or oil.

thermal radiation. The radiation given out by a body due to its temperature. It is excited by the thermal agitation of molecules or atoms and has a continuous spectrum. *See also* BLACK BODY.

thermal radiator. A body emitting THERMAL RADIATION.

thermal reactor. *See* NUCLEAR REACTOR TYPES.

thermionic cathode. A heated CATHODE which gives off electrons by THERMIONIC EMISSION. A tungsten wire cathode is usually directly heated by passing current through it, while a cathode coated with barium and strontium oxides is generally indirectly heated using a separate heating coil.

thermionic diode. *See* DIODE.

thermionic emission. The release of electrons obtained by heating a solid. The effect occurs when significant numbers of electrons have sufficient kinetic energy to overcome the solid's WORK FUNCTION. The effect increases with temperature. *See* RICHARDSON'S EQUATION. *See also* SCHOTTKY EFFECT.

thermionics. The study and design of devices based on thermionic emission, for example thermionic valves or electron guns.

thermionic valve. A multielectrode evacuated ELECTRON TUBE using a THERMIONIC CATHODE as electron source. Since current only flows when a positive potential is applied to the anode, the valve has rectifying properties. With three or more electrodes, applying a signal to one of the extra electrodes causes modulation of the anode current and hence amplification of the signal. For most applications, valves have now been superseded by their solid state equivalents: semiconductor DIODES, FIELD EFFECT TRANSISTORS etc. However, some special-purpose valves are still important.

thermistor. A piece of semiconducting material whose resistance falls rapidly with temperature. By using a thermistor as one arm of a suitable electrical BRIDGE, temperature changes as small as 0.005° C can be measured. Thermistors are also used to compensate for the increased resistance of ordinary resistors on heating. Other uses are in vacuum gauges, time delay switches and voltage regulators.

thermocouple. A thermoelectric device used for measuring temperature. Two dissimilar metal wires or semiconducting rods are joined at both ends. A temperature difference between the junctions results in the production of an electromotive force; the phenomenon is known as the *Seebeck effect*. The electromotive force may be measured by a millivoltmeter or potentiometer incorporated in the circuit. For temperature measurement, one junction is maintained at a constant known temperature and the other at the temperature to be measured, which can be obtained from a calibration curve of voltage against junction temperature difference.

Suitable pairs of metals are ones showing a large increase in electromotive force for a small temperature difference. For temperatures up to about 500° C copper/constantan or iron/constantan junctions are suitable; up to 1500° C chromel/alumel or platinum/10% rhodium-platinum alloy are used and at still higher temperatures, iridium/rhodium-iridium alloy.

thermocouple pyrometer. *See* PYROMETER.

thermodynamic equilibrium. A state said to exist when all parts of a system are at the same temperature.

thermodynamics. The branch of science concerned with the relationship between HEAT, WORK and other forms of ENERGY. It is based on the following basic laws. *First law of thermodynamics*. The law

$$\delta Q = \delta W + \delta U$$

where δQ is the heat absorbed by any system δW the work done by the system and δU the internal energy change. The law represents the application of CONSERVATION LAW to heat energy. An alternative statement of the first law sometimes quoted is that it is impossible to construct a continuously operating machine that does work, without obtaining energy from an external source.

Second law of thermodynamics. A law formulated in various ways. One of the most frequently used formulations is that heat cannot pass from a colder to a hotter body without any other external effect taking place, i.e. there can be no spontaneous self-sustaining heat flow against a temperature gradient. Another way of stating the law is that an isolated system, free of external influence, always passes from states of relative order to states of relative disorder, until eventually it reaches the state of maximum disorder: in other words, the ENTROPY of a closed system increases with time.

Third law of thermodynamics. The entropy of a substance in a perfect crystalline state is zero at the temperature of absolute zero. This law thus enables absolute values to be stated for entropies, in contrast to the second law which is concerned with entropy changes.

Zeroth law of thermodynamics. If two bodies are each in thermal equilibrium with a third body, then all three bodies are in thermal equilibrium with each other.

thermodynamic temperature. A quantity defined by the equation

$$T_1/T_2 = q_1/q_2$$

where q_1 is the quantity of heat taken up by a reversible engine working on a CARNOT CYCLE at thermodynamic temperature T_1 and q_2 is the quantity of heat rejected by the engine at thermodynamic temperature T_2. Thermodynamic temperature is thus a physical quantity, measured in KELVIN, which is defined in such a way that the TRIPLE POINT of water is 273.16 K. In practice, measurements of thermodynamic temperature are made using the INTERNATIONAL TEMPERATURE SCALE.

thermoelectric effects. Effects in which a potential difference arises as a result of a temperature gradient and vice versa. They include PELTIER EFFECT the Seebeck effect (*see* THERMOCOUPLE) and the THOMSON EFFECT.

thermoelectricity. Electricity produced by THERMOELECTRIC EFFECTS.

thermoelectric series. A list of metals:
bismuth, nickel, cobalt, palladium, platinum copper, manganese, titanium, mercury, lead, tin, chromium, molybdenum, rhodium, iridium, gold, silver, zinc, tungsten, cadmium, iron and antinomy
such that when any two of them are used to form a THERMOCOUPLE, the direction of the current at the hot junction is from the metal coming earlier in the series to the one coming later. In general, the further apart the two metals in the series, the greater the electromotive force of the thermocouple they form.

thermoelectric thermometer. Another name for THERMOCOUPLE.

thermogram. An image produced by the thermal radiation from a body. It is a valuable diagnostic aid since human skin, independent of its colour, is a good approximation to a BLACK BODY; marked changes in the radiation temperature are produced by circulatory and cellular abnormalities and are thus shown up on the thermogram. The method has been particularly successful in breast cancer screening.

thermography. (1) Continuous recording of temperature change over a period of time.
(2) The production of THERMOGRAMS.

thermoluminescence. LUMINESCENCE produced when certain solid materials, containing electrons trapped at defects, are heated. The effect of heating is to free the electrons, which then fall to lower energy states with the emission of photons. The phenomenon is used for example in pottery dating: it is assumed that ionizing radiation caused the trapping of the electrons in the pottery; the older the pot, the more ionizing radiation it has received and hence the greater should be the thermoluminescent intensity, which is therefore measured.

thermometer. Any instrument for measuring temperature. There are many different types, each depending on the easy measurement of some physical property that varies with temperature. Examples are MERCURY IN GLASS THERMOMETER: ALCOHOL THERMOMETER; CONSTANT PRESSURE GAS THERMOMETER; CONSTANT VOLUME GAS THERMOMETER; RESISTANCE THERMOMETER; THERMOCOUPLE; THERMISTOR; PYROMETER.

thermometric property. The physical property whose temperature change is made use of in a THERMOMETER.

thermonuclear bomb. Another name for fusion bomb. *See* NUCLEAR WEAPON.

thermonuclear reaction. Another name for FUSION (def. 2).

thermophone. A thin metal strip of small heat capacity mounted between two terminal blocks. On passing alternating current through the strip, periodic temperature variations are produced in it; these cause expansions and contractions of the surrounding air, i.e. sound waves. The sound output requires amplification before use.

thermopile. An instrument for measuring radiant heat. It consists of a THERMOCOUPLE collection connected in series: alternate junctions, blackened to increase absorption, are exposed to the radiation, the other junctions being shielded. A measurable electromotive force, proportional to the radiation intensity, is thus produced.

Thermos flask. A commercial type of DEWAR FLASK.

thermosphere. The layer of the atmosphere extending from about 80 kilometre to about 400 kilometre above the Earth's surface. *See* ATMOSPHERIC LAYERS.

thermostat. A device for maintaining a constant temperature, usually in an enclosure. It consists of a temperature sensing instrument connected to a switching device. The sensing device is frequently a BIMETALLIC STRIP which triggers a simple electric switch. The term is sometimes used to describe the constant temperature enclosure.

thick lens. A lens in which the axial separation between its two faces is not negligible.

thin film. A thin layer of substance deposited on the surface of another substance, for example a solid deposited by vacuum evaporation on another solid. Such thin solid films are used in many semiconductor devices.

thin lens. A lens in which the axial separation between its two faces is negligible.

thin prism. A prism whose refracting edge angle is not more than 10°. The deviation produced by such a prism is independent of the angle of incidence for small angles of incidence.

thixotropy. The phenomenon whereby the viscosity of certain media decreases with increasing velocity of flow; such media are known as *thixotropic*. A useful application is in nondrip paints, which are firm on the brush but flow when applied.

Thomson effect. The phenomenon in which, when a temperature gradient is maintained along a conductor through which a current is passed, heat is evolved at a rate roughly proportional to the product of the current and the temperature gradient. If either the gradient or the current is reversed, heat is absorbed rather than evolved. The phenomenon is superimposed on the normal heating effect of a current.

Thomson scattering. The scattering of electromagnetic radiation by free or loosely bound electrons. Provided that the incident photon energy is much less than the rest energy of the electron, the total intensity of the scattered radiation is given by $8\pi r^2 I/3$ where r is the classical electron radius and I the intensity per unit area of the incident radiation.

thoron. A radioisotope of radon, radon-220, produced by the decay of thorium.

thou. One thousandth of an INCH.

three-body problem. The problem of the motions of three particles moving under the influence of their mutual interactions. In general no exact solution is possible, although the motions can be obtained by numerical methods. The problem is a special case of the MANY-BODY PROBLEM.

three-colour theory. *See* COLOUR VISION.

three-dimensional television. At present a subject under investigation, using HOLOGRAM transmission.

threshold frequency. The minimum frequency giving rise to the PHOTOELECTRIC EFFECT for a particular substance.

threshold of discomfort. *See* SENSITIVITY OF THE EAR.

threshold of hearing. *See* SENSITIVITY OF THE EAR.

threshold voltage. The voltage at which a particular characteristic of an electronic device first occurs.

thrust. The propelling force generated by a jet engine or rocket. It is usually measured by the product of the rate of mass discharge and the speed of the exhaust gases relative to the vehicle.

thunderstorm. An atmospheric phenomenon originating when intense heating causes a parcel of moist air to rise, leading

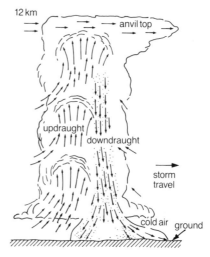

T1 Thunderstorm cell cross section

to the formation of the cloud type illustrated in fig. T1. The exact mechanism remains to be elucidated.

thyratron. A type of thermionic gas-filled TRIODE used in particle counters and relays. Current flow is initiated by a small increase in grid potential; it then becomes independent of grid potential and can only be stopped by a suitable reduction of anode potential. The thyratron is now being superseded by solid state devices.

thyristor. A silicon controlled rectifier which performs much as a THYRATRON. It usually consists of a four layer, three P-N JUNCTION chip.

tides. The regular rise and fall of the level of water in the Earth's oceans due to the gravitational forces of the Sun and Moon on the Earth; the effect of the Moon is approximately twice that of the Sun. Since the vector sum of the forces acting varies according to the Moon's position relative to the Earth, high water has different heights at different times. Tidal power stations using specially designed tidal water operated turbines to drive generators and so produce electricity have been built at several sites.

tight binding approximation. A method for finding energy states and wave functions of electrons in a solid by summing the product of a phase factor and a pure atomic wave function for an individual lattice atom, for all lattice atoms.

timbre. *See* QUALITY.

time. The duration between two events as determined by comparison with some agreed process. Usually a periodic process is chosen: examples are the swing of a pendulum (*see* CLOCKS), the vibration of electromagnetic radiation (*see* SECOND) or various astronomical processes (*see* SOLAR TIME, SIDEREAL TIME, LUNAR TIME). The *calendar year* has an average length equal to that of the solar YEAR. So that the calendar year may contain an exact number of days, each year has 365 days with a *leap year* containing 366 days every fourth year: only century years yielding an exact integer when divided by 400 are leap years, thus 2000 will be a leap year but 1900 was not.

time base. A circuit that repeatedly generates a voltage which increases linearly with time and then quickly drops to zero, i.e. of SAWTOOTH WAVE FORM. Such a circuit is used in a CATHODE RAY OSCILLOSCOPE.

time constant. The time taken by the current in a reactive component to which a direct electromotive force is applied to reach $(e - 1)/e$ of its final value, where e is the irrational number 2.718... For an inductance L in series with a resistance R, the time constant is L/R; for a capacitance C in series with a resistance R, the time constant is RC.

time dilation. *See* RELATIVITY.

time lapse photography. A technique for obtaining a speeded up record of a slow process such as the opening of a flower. Single exposures are made at regular intervals on cine film, keeping the same camera position throughout. The developed film is then projected at normal speed.

time reversal. The multiplication of the time co-ordinate in the equation of motion of a dynamical system by -1.

time sharing. The technique whereby the processing time of a COMPUTER can be shared by two or more separate terminals.

time switch. A switch incorporating a type of clock mechanism so that an electric circuit can be made or broken at times predetermined by the setting of the mechanism.

tint. Another name for DESATURATED COLOUR.

Tokamak. A toroidal-shaped FUSION REACTOR.

tone. (1) An audible note with no PARTIALS, such as a sound of a single frequency and sinusoidal wave form.
(2) The QUALITY of a music note.

tonne. A mass of 1000 kilogramme.

topology. A branch of geometry concerned with those properties of solids unaffected by continuous deformation such as twisting and stretching. It is used in the analysis of complicated electrical networks.

top quark. *See* QUARK.

toric lens. Another name for TOROIDAL LENS.

tornado. A very violent vortex-like wind distribution with a central funnel up which air rushes at high speed; the radius of a horizontal section is up to 100 metre. Tornados frequently occur in groups, and have been responsible for much destruction in America and Australia.

toroidal. Having a TORUS shape.

toroidal coil. A coil in which the turns of wire are uniformly distributed round a closed circular ring. When current is passed through such a coil a uniform magnetic field, which can be made very large, is set up inside the windings.

toroidal lens. A lens with a TOROIDAL surface. Such lenses are generally used for the correction of eyes which have another defect in addition to ASTIGMATISM.

torque. Symbol *T*. A force or system of forces producing a turning effect, measured by its MOMENT. The torque on a rigid body is the product of the MOMENT OF INERTIA of the body about the axis of rotation and the angular acceleration about that axis.

torque meter. A device for measuring the TORQUE exerted by a rotating body.

torr. A unit of pressure used in high vacuum technology and equal to 133.322 pascal.

Torricellian vacuum. The space above the mercury in a mercury BAROMETER. It contains mercury vapour which exerts a vapour pressure of about 1.7×10^{-5} pascal.

torsion. A twisting deformation produced by a TORQUE.

torsional constant. The TORQUE necessary to produce a twist of one radian in a suspension device.

torsional hysteresis. *See* HYSTERESIS (def. 3).

torsional vibration. A vibration associated with a twist occurring first in one direction and then the other. For example if a horizontal disc is suspended from its centre by a vertical wire which is firmly clamped at the other end, and the disc is rotated in its own plane and then released, it performs torsional vibrations due to the twisting of the wire first in one direction and then the other. By timing the oscillations, the RIGIDITY MODULUS of the wire can be found.

torsion balance. A device in which a TORQUE is measured by the amount of twist it produces in a vertical wire or fibre. The angle of twist is usually obtained from the deflection of a beam of light by a small mirror attached to the wire or fibre; the angle is proportional to the torque. The instrument is used to measure small forces, such as those associated with SURFACE TENSION and static charges.

torsion pendulum. A PENDULUM which performs TORSIONAL VIBRATION.

torus. A surface or solid generated by rotating a circle about an external line in its plane. It thus has the same shape as a doughnut or anchor ring.

total heat. Another name for ENTHALPY.

total internal reflection. The reflection of an incident ray of light at the interface between the medium of incidence and one of lower refractive index. For total internal reflection to occur the angle of incidence, *i*, must exceed the CRITICAL ANGLE, *c*, as illustrated in fig. T2. For angles of incidence less than the critical angle, the ray is refracted in the normal way at an angle *r*.

totality. The condition said to occur when the Sun's disc is completely obscured during a total solar ECLIPSE.

total radiation pyrometer. *See* FÉRY TOTAL RADIATION PYROMETER.

total reflecting prism. A right-angled crown glass isosceles prism. When light is normally incident on it, as shown in fig. T3(a), the light strikes the hypotenuse face at an angle of 45°; this is greater than the CRITICAL ANGLE, whose value for crown glass is 41° 30′. The light is thus totally internally reflected and emerges as shown. The advantages of such a prism over a mirror are that multiple images are avoided as is also the 10% intensity loss occurring at a silvered surface. For light incident as in fig. T3(b), an inverted image is formed after two total internal reflections.

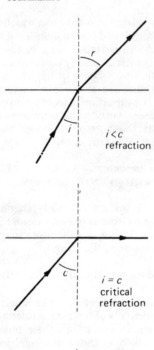

i < c
refraction

i = c
critical
refraction

i > c
total internal
reflection

T2 Refraction and total internal
reflection

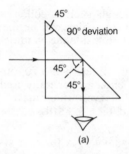

(a)

T3 (a) 90° deviation by total reflection
prism

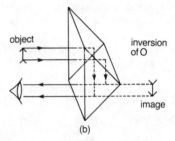

(b)

T3 (b) Inversion by total reflection prism

tourmaline. A mineral crystal exhibiting both DICHROISM and the PIEZOELECTRIC EFFECT.

Townsend discharge. A type of electric discharge occurring between two electrodes in a gas at low pressure. In contrast to the GLOW DISCHARGE, the voltage gradient along the tube is uniform as is also the luminosity. The current of a few microampere characteristic of the Townsend discharge is obtained by using a high external resistance in the tube circuit.

tracer. An isotope used for investigating chemical or physical changes. For example, a layer of radioactive metal isotope deposited on the surface of a normal metal will slowly diffuse into the normal metal; the extent of diffusion can be found by investigating the radioactivity of thin slices cut from the material.

track. The illuminated path of an ionizing particle as seen in a CLOUD CHAMBER, BUBBLE CHAMBER etc.

tracking. The formation of electrically conducting paths on the surfaces of solid dielectrics and insulators under the influence of a strong electric field.

trajectory. *See* PROJECTILE TRAJECTORY.

transcendental. (1) Denoting a number such as π which cannot be written as a fraction.
(2) Denoting a mathematical function which cannot be written as a finite number of terms, for example $\sin y$, e^y, $\log y$.

transducer. A device for converting energy of any type into electric energy and vice versa, so that the output is a function of the input. Familiar transducer examples are microphones, loudspeakers, pick-ups etc. For most transducers the output is a continuous function of the input, but some give a digital output.

transfer of heat. *See* HEAT TRANSFER.

transformation. (1) A change in the form of a mathematical expression by replacing the original variables in it with ones related to them by specified equations. Thus

$$y = x^2$$

may be transformed by substituting

$$y = y_1 - 3 \text{ and } x = x_1 + 2$$

to give the equation

$$y_1 = x_1^2 + 4x_1 + 7$$

In this case the transformation corresponds to a translation of axes in Cartesian co-ordinates.
(2) A change of one nuclide into another, for example by the emission of an alpha particle.

transformer. A device for changing alternating-current power from one voltage or current level to another without frequency change by using ELECTROMAGNETIC INDUCTION. The transformer consists essentially of two coils of wire round a ferromagnetic core. Power fed into the PRIMARY WINDING produces a changing magnetic field which induces power in the SECONDARY WINDING. The magnetic coupling is strengthened by the presence of the core, which is generally laminated to reduce EDDY CURRENT losses; 98% efficiency is achievable. If losses are negligible, the ratio of the primary to the secondary voltage equals the ratio of the number of primary turns to the number of secondary turns. If the secondary voltage is greater than the primary, the transformer is said to be a *step up transformer*. If the secondary voltage is less than the primary, the transformer is said to be a *step down transformer*. Transformers with equal numbers of primary and secondary turns, i.e. equal primary and secondary voltages, are used to isolate a piece of equipment from its power supply: the lack of direct electrical connection enhances safety.

transient. A brief disturbance or oscillation in a circuit resulting from a sudden rise in current or voltage.

transistor. A device composed of SEMICONDUCTOR material in which the flow of current between two electrodes is controlled by the potential applied to a third electrode. It is thus capable of amplification in addition to rectification. It is the basic unit of radio, television and computer circuits and has almost completely replaced the THERMIONIC VALVE. There are various types of transistor.

A *bipolar junction transistor* consists either of a thin N-TYPE SEMICONDUCTOR flanked on either side by P-TYPE SEMICONDUCTOR regions, forming a *p-n-p transistor*, or a thin p-type semiconductor flanked on either side by n-type semiconductor regions, forming an *n-p-n transistor*. These two forms are shown in fig. T4; they are biased

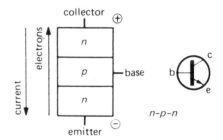

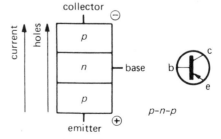

T4 Junction transistors

as illustrated. In the n-p-n transistor, electrons flow from the EMITTER into the BASE and holes from the base flow to the emitter. Most electrons entering the base reach the COLLECTOR. A small current input to the base results in a large current drawn from the collector, i.e. amplification occurs. The p-n-p transistor works similarly except that the emitter is positive with respect to the collector and the main carriers are holes. Since holes have a lower mobility than electrons, p-n-p transistors operate more slowly than do n-p-n transistors.

See also FIELD EFFECT TRANSISTOR.

transition. (1) Any change accompanied by a marked alteration of physical properties, for example a PHASE change, a conversion of one crystal structure to another or the onset of SUPERCONDUCTIVITY.

(2) A sudden change in ENERGY LEVEL in an atom, ion, molecule or nucleus.

transition temperature. (1) The temperature at which a change of PHASE occurs.

(2) The temperature at which SUPERCONDUCTIVITY commences.

transit time. The time taken for a charged particle to travel between two specified points in an electric field.

translation. The movement of a body or system in such a way that all points move through equal distances in parallel directions.

translucent. Denoting material which allows only partial transmission of radiation incident on it.

transmission coefficient. Another name for TRANSMITTANCE.

transmission density. Symbol D. The logarithm to base 10 of the reciprocal of the TRANSMITTANCE.

transmission electron microscope. See ELECTRON MICROSCOPE.

transmission factor. Another name for TRANSMITTANCE.

transmission line. A system of conductors forming a continuous path from one place to another in order to conduct electromagnetic energy along this path.

transmittance. Symbol τ. The ratio of the electromagnetic radiation flux transmitted by a body to that incident upon it.

transmitter. The telecommunication equipment in a system whereby the signal is transmitted to the receiving parts of the system.

transmutation. The change of one element into another, either occurring naturally or resulting from particle bombardment.

transparent. Denoting a material which transmits all the radiation incident upon it.

transport number. Symbol t. The fraction of the total current carried by a particular type of ion when an electrolyte conducts electricity.

transverse wave. A wave in which the vibrations of the transmitting medium lie in a plane perpendicular to the direction of travel of the wave. Examples are light waves and water waves.

trapezium. A quadrilateral with one pair of opposite sides parallel. Its area is the product of half the sum of the parallel sides and their distance apart.

travelling wave. Another name for PROGRESSIVE WAVE.

travelling wave tube. See VELOCITY MODULATION.

triangle. A three-sided plane geometric figure. Its area is half the product of its base and height.

triangle law of vector addition. If lines representing two vectors are laid tip to tail as shown in fig. T5, the line drawn from the free tail to the free tip represents the sum of

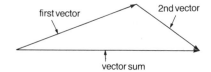

T5 Triangle law of vector addition

the two vectors. A similar method can be applied to sum any number of vectors.

triboelectricity. Static electricity produced by friction. It is thought to be caused by a transfer of electrons between a rubbed body and the rubbing body so that one is left with an excess of electrons (i.e. negatively charged) and the other with an electron deficit (i.e. positively charged). Thus when a glass rod is rubbed with silk, the silk becomes positively and the glass negatively charged.

tribology. The study of friction between solid surfaces. It includes the study of the origin of the frictional forces and the practical problems of the lubrication and wear of moving parts.

triboluminescence. Light emission produced by friction between certain solids.

trichromatic theory. *See* COLOUR VISION.

triclinic system. *See* CRYSTAL SYSTEM.

trigger. Any stimulus that initiates a particular response from an electronic circuit or device.

trigonal system. *See* CRYSTAL SYSTEM.

trigonometric functions. Functions of an angle, defined as ratios of the lengths of the sides of a right-angled triangle which contain the angle. For example, for the right-angled triangle illustrated in fig. T6

$$\text{sine } \alpha = a/c$$

$$\text{cosine } \alpha = b/c$$

$$\text{tangent } \alpha = a/b$$

$$\text{cosecant } \alpha = c/a$$

$$\text{secant } \alpha = c/b$$

$$\text{cotangent } \alpha = b/a$$

The accepted abbreviations are respectively

sin α, cos α, tan α, cosec α, sec α, cot α

The following relations hold:

$$\sin^2\alpha + \cos^2\alpha = 1$$

$$\sec^2\alpha = 1 + \tan^2\alpha$$

$$\csc^2\alpha = 1 + \cot^2\alpha$$

Moreover,

$$\sin \alpha = \alpha - \alpha^3/3! + \alpha^5/5! \ldots$$

$$\cos\alpha = 1 - \alpha^2/2! + \alpha^4/4! \ldots$$

(*see* FACTORIAL).

trigonometry. The study of the properties of triangles with particular reference to the TRIGONOMETRIC FUNCTIONS. The subject is much used in surveying, navigation and astronomy.

trimmer. A variable capacitor connected in parallel with a much larger fixed capacitor so that the capacitance of the combination is finely adjustable.

trimming capacitor. Another name for TRIMMER.

Trinitron. *See* COLOUR PICTURE TUBE.

triode. Any electronic device with three electrodes, such as a bipolar junction TRANSISTOR, a three-electrode THERMIONIC VALVE or a THYRATRON.

triple point. The temperature and pressure at which the solid, liquid and gaseous

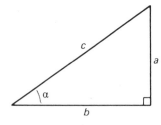

T6 Right-angled triangle

phases of a pure substance can coexist. The triple point for water is 273.16 K and 101 325 pascal.

triplet state. A state for which the MULTIPICITY is three.

tristimulus values. *See* CHROMATICITY CO-ORDINATES.

tritium. Symbol T. A hydrogen isotope with one proton and two neutrons in its nucleus. It is produced by the irradiation of lithium in a nuclear reactor. It is radioactive and is used in tracer studies.

triton. The nucleus of the TRITIUM atom.

trochoid. The locus of a point on the radius of a circle as the circle rolls along a straight line.

tropical year. *See* YEAR.

tropopause. The upper limit of the TROPOSPHERE.

troposphere. The part of the Earth's atmosphere extending from the surface of the Earth to about 10 kilometre above it. The temperature progressively falls throughout it with increasing distance from Earth (*see* ATMOSPHERIC LAYERS). The troposphere is the most turbulent part of the atmosphere and most meteorological phenomena, such as cloud precipitation, occur there.

tropospheric fallout. *See* FALLOUT.

troy units of mass. A system of units used for gems and precious metals. *See* Table 6D.

true anomaly. *See* ANOMALY.

tube. *See* ELECTRON TUBE.

tuned circuit. An alternating current circuit which has been adjusted for RESONANCE at a particular frequency. The adjustment process is usually accomplished by altering the capacitance or inductance of the circuit.

tuning fork. A steel fork with two prongs and a central handle. When struck, the fork produces a pure note of known pitch and so is used as a standard of pitch for musical instruments.

tunnel diode. A highly doped P-N JUNCTION diode for which, over part of its operating range, the current decreases with increasing voltage as illustrated in fig. T7. The explanation of the phenomenon lies in the TUNNEL EFFECT: electrons tunnel from the valence ENERGY BAND to the conduction band; the effect is larger the more negative the voltage.

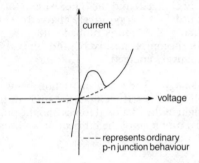

T7 Characteristic of tunnel diode

tunnel effect. The movement of particles through barriers which, on classical theory, would demand more energy than the particles possess. However, according to WAVE MECHANICS, there is a definite probability of a particle penetrating such a barrier.

turbine. An engine in which rotational energy is obtained from the motion of a fluid. For example the fluid may emerge through nozzles to strike the blades of a wheel, thus causing it to rotate.

turbulent flow. Fluid flow in which the motion is irregular. The velocity at any point may vary with time in both magnitude and direction. *See also* REYNOLD'S NUMBER.

TV tube. The CATHODE RAY TUBE or a TELEVISION receiver.

T wave. The potential variation on the trace in ELECTROCARDIOGRAPHY which occurs at the end of ventricular contraction.

tweeter. A small loudspeaker designed to reproduce sounds of comparatively high frequency. *Compare* WOOFER.

two-dimensional grating. A diffraction grating with lines ruled on it in two directions at right angles. It gives rise to an intensity pattern consisting of discrete dots since maxima will only be produced where the grating equations for the two directions are satisfied simultaneously.

tympanic canal. One of the canals inside the cochlea. It ends in the ear's round window.

Tyndall effect. The SCATTERING of light by small particles in its path, thus causing the path of a light beam in, for example, a dusty atmosphere to become visible.

U

UHF. Abbrev. for ULTRAHIGH FREQUENCY.

ultimate strength. The limiting stress, expressed as force per unit area of original cross section, at which a material completely breaks down.

ultracentrifuge. A CENTRIFUGE designed to work at very high angular speed, so that the centrifugal force produced is sufficiently large to cause SEDIMENTATION of colloids. Since the rate of sedimentation depends on particle size, this can be investigated. Relative molecular masses of proteins and other macromolecules can also be measured. In quantitative instruments, the formation of sediment is photographically recorded.

ultrahigh frequency. A frequency in the range 0.3 gigahertz to 3 gigahertz.

ultrahigh vacuum. *See* VACUUM.

ultramicrobalance. A MICROBALANCE capable of weighing to 10^{-11} kilogramme.

ultramicroscope. A MICROSCOPE suitable for observing particles smaller than those resolvable by ordinary microscopy. DARK FIELD ILLUMINATION is usually employed.

ultrasonic. Denoting or involving sound-type vibrations of frequency greater than 20 kilohertz, i.e. inaudible frequencies to humans.

ultrasonics. The study of the properties and applications of ULTRASOUND.

ultrasound. Vibrations of ULTRASONIC frequency. They can be generated using either the PIEZOELECTRIC EFFECT or MAGNETOSTRICTION, or by GALTON WHISTLE or HARTMANN GENERATOR. The vibrations may be detected using a piezoelectric oscillator. Alternatively the diffraction of light, produced in a liquid traversed by the ultrasonic vibration, may be used: for a narrow beam of light, wavelength λ, entering the liquid at right angles to the ultrasonic beam, wavelength λ_u,

$$\sin \theta_k = k\lambda/\lambda_u$$

where θ_k is the angle of diffraction for the kth order. Hence λ_u can be very accurately measured. There are numerous applications of ultrasound, such as flaw detection in metals, depth finding, cleaning, dispersion of one liquid in another coagulation of some dispersions, medical diagnosis: in pregnancy, for example, foetal defects can be diagnosed by ULTRASOUND SCANNING of the mother's abdomen.

ultrasound scanning. The process of passing an ultrasonic beam over an area of interest and studying the reflections.

ultraviolet catastrophe. *See* QUANTUM THEORY.

ultraviolet microscope. *See* MICROSCOPE.

ultraviolet radiation. Electromagnetic radiation in the wavelength range of about 10 nanometre to 380 nanometre, i.e. lying between X ray and visible light wavelengths. The ultraviolet region is arbitrarily divided into the *near ultraviolet*, wavelength range 200 nanometre to 380 nanometre, and the *far ultraviolet*, also known as the VACUUM ULTRAVIOLET.

The main sources of ultraviolet radiation are the Sun, arc, spark and gas discharge tubes. The radiation is produced by emission from excited atoms or ions.

Ultraviolet radiation may be detected by either its photographic action, or by its fluorescence, or photoelectric or photochemical effects. Practically important properties of the radiation are its production of vitamin D from ergosterol in the human body, and its induction of photosynthesis in plants. Substances transparent to light absorb increasingly strongly in the ultraviolet as the wavelength decreases. In the far ultraviolet most substances are opaque or show selective absorption. Absorption by ozone in the STRATOSPHERE prevents damaging solar radiation of wavelength less than 290 nanometre from reaching Earth.

ultraviolet spectrum. An emission or absorption spectrum in the ultraviolet region. The spectra yield information on energy levels and molecular structure. Optical components of glass cannot be used in ultraviolet spectroscopy because of their strong absorption. In the near ultraviolet quartz components are suitable; at shorter wavelengths vacuum mounted reflection gratings are employed.

umbra. (1) A region of complete shadow. (2) The central darker region of a SUNSPOT.

uncertainty principle. *See* HEISENBERG UNCERTAINTY PRINCIPLE.

undercurrent release. A tripping device which operates when the current falls below a predetermined value.

underdamping. *See* DAMPING

undervoltage release. A tripping device which operates when the voltage falls below a predetermined value.

uniaxial crystal. A doubly refracting crystal, except for light passing in the direction of the principal crystallographic axis, for which it is singly refracting. *See also* DOUBLE REFRACTION.

unified atomic mass unit. Symbol u. A unit of mass equal to 1/12 of the mass of a

nuclide of carbon-12. It equals

$$1.6605 \times 10^{-27} \text{ kilogramme}$$

or approximately 931 mega-electronvolt. It replaces the *atomic mass unit*, which was related to oxygen. *See also* RELATIVE ATOMIC MASS.

unified field theory. *See* FIELD THEORY.

unifilar suspension. A suspension used in some electrical instruments. It employs a single wire, thread or strip to suspend a moving part; twisting of the suspension provides the restoring TORQUE.

uniform acceleration. Acceleration of constant magnitude and direction.

uniform circular motion. Motion at constant speed in a circular path.

uniform temperature enclosure. An enclosure whose walls are maintained at constant temperature. The radiation inside such an enclosure depends solely on the walls' temperature and is thus BLACK BODY radiation.

uniform velocity. Velocity of constant magnitude and direction.

unijunction transistor. A TRANSISTOR usually of the form shown in fig. U1. The n-part is lightly doped and the p-part is heavily doped. Contacts are made as shown; the BASE is the n-part and the EMITTER is the p-part.

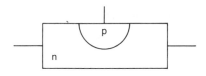

U1 Unijunction transistor

unipolar transistor. A TRANSISTOR in which current flow is due to majority carriers only. The FIELD EFFECT TRANSISTOR is such a device.

unimolecular. Involving one molecule. Thus a unimolecular layer of substance is a layer whose thickness is one molecule.

unit. An amount of a physical quantity which is arbitrarily defined as being the unit amount of that quantity. For example a particular piece of platinum has been assigned a mass of one kilogramme and is the unit of mass in SI UNITS. Any measurement is effectively a comparison of what is being measured with the appropriate standard, which may be a PRIMARY STANDARD or a SECONDARY STANDARD depending on the accuracy required and on convenience.

unitary symmetry. A theory which is concerned with SU(3) (*see* SU(N)) and predicts that, with respect to STRONG INTERACTION, elementary particles can be grouped into multiplets containing 1, 8, 10 or 27 particles and that the particles in each multiplet can be regarded as different states of the same particle. The theory is usually illustrated graphically by plotting HYPERCHARGE against ISOSPIN QUANTUM NUMBER. The BARYON octet plot is shown in fig. U2. The theory predicted the existence of the OMEGA MINUS PARTICLE prior to its discovery, but has not been completely successful.

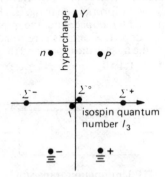

U2 Octet of baryons

unit cell. The smallest repeated unit in a crystal lattice. *See also* FACE-CENTRED; BODY-CENTRED.

unit vectors. Vectors of unit magnitude. Unit vectors lying along the x, y and z axes are usually denoted by i, j and k respectively. A vector function F may then be written

$$F = ix + jy + kz$$

universal gas constant. Symbol R. The constant occurring in the EQUATION OF STATE for one MOLE of an IDEAL GAS. It is equal to

8.314 35 joule per kelvin per mole

Combining the equation of state with the expression for the pressure of an ideal gas predicted by the KINETIC THEORY OF MATTER leads to the result that R is two-thirds of the total translational energy of the molecules in 1 mole of an ideal gas at a temperature of 1 K.

universality class. A collection of different substances having the same CRITICAL EXPONENT values.

universal motor. An electric motor suitable for use with either direct of alternating current.

universal shunt. A galvanometer shunt such that either 0.1 or 0.01 or 0.001 of the main circuit current passes through the galvanometer, whose range is thus increased.

universe. The whole of space and all the matter in it. The mass of matter is estimated at 10^{41} kilogramme and the age in the range 5×10^9 year to 50×10^9 year. The origin is a matter for speculation although the BIG BANG THEORY is now widely accepted. *See also* EXPANDING UNIVERSE.

unsaturated colour. *See* SATURATED COLOUR.

unsaturated vapour. A vapour which at the prevailing temperature contains less than the maximum amount of the vaporized substance.

unstable equilibrium. *See* EQUILIBRIUM STABILITY.

upper atmosphere. The atmosphere above a height of about 30 kilometre; it includes part of the STRATOSPHERE. *See* ATMOSPHERIC LAYERS.

up quark *See* QUARK.

upsilon. A MESON of mass 9 giga-electronvolt, built from a bottom QUARK and bottom antiquark. It is a member of the most massive family of particles at present known.

upthrust. The upward force experienced by an object, wholly or partially immersed in a fluid, due to the fluid pressure on it.

uranium. Symbol U. A metallic radioactive element of which 99% is uranium-238, 0.7% uranium-235 and the remainder uranium-234. Uranium-238, the most stable isotope, is used in BREEDER REACTORS as a source of the plutonium-239 isotope. Uranium enriched with uranium-235 is used as a NUCLEAR REACTOR fuel.

uranium dating. A type of RADIOACTIVE DATING mainly used for rocks. It is thought that when rocks were formed only uranium-238 was present and that the lead-206 now found is the product of the radioactive decay of the uranium during the intervening period. Hence by determining the amounts of uranium-238 and lead-206 now present in a rock sample, the age of the sample can be calculated.

uranium series. A RADIOACTIVE SERIES beginning with uranium-238, of half life 4.5×10^9 year, and ending with lead.

Uranus. The seventh closest planet to the Sun. It is a GIANT PLANET with a dense atmosphere of hydrogen and methane and was the first to be discovered telescopically. Its orbital period around the Sun is 84 year and its axial rotation period is about 10.75 hour. The temperature is probably about $-200°$ C. It has several satellites and a dark ring system. It has been investigated by a planetary probe.

useful angular magnification. That angular magnification i.e. MAGNIFYING POWER, of an optical instrument for which both eye and instrument are working at the LIMIT OF RESOLUTION. Larger values of the angular magnification will make objects appear larger but will not increase their definition. For example the limit of resolution of a telescope for light of wavelength 6×10^{-7} metre is

$$1.22 \times 6 \times 10^{-7}/D \text{ radian}$$

Where D is the diameter of the telescope objective. The limit of resolution of the eye is

$$\pi/(180 \times 60) \text{ radian}$$

The useful angular magnification of the telescope is therefore

$$\pi D/(180 \times 60 \times 1.22 \times 6 \times 10^{-7}) = 400D$$

In practice a somewhat larger value than this is used in order to reduce eye strain when examining very small image detail.

uv radiation. Short for ULTRAVIOLET RADIATION.

V

vacancy. An unoccupied crystal lattice position that would normally house an atom, ion or molecule.

vacuum. A space in which there are relatively few atoms or molecules. A perfect vacuum would contain none but is unattainable in practice because of the finite vapour pressure of the enclosing walls. For pressures down to 10^{-2} pascal, the vacuum is said to be *soft* or *low*; from 10^{-2} pascal to 10^{-7} pascal, the vacuum is said to be *hard*; for pressures less than 10^{-7} pascal, it is said to be an *ultra high vacuum*.

To obtain a vacuum, a *vacuum pump* is connected to a sealed apparatus; the lower limit of the pressure that can be achieved depends on pumping speed, on the presence of adsorbed gas, volatile materials, leaks etc. Vacuum technology is important in many branches of research and industry such as light bulb and cathode ray tube manufacture, freeze drying and vacuum evaporation.

vacuum evaporation. A process for depositing a thin film of solid on a surface by evaporating the material at high temperature in a vacuum.

vacuum flask. Another name for DEWAR FLASK.

vacuum pump. *See* ROTARY OIL PUMP; GAEDE MOLECULAR PUMP; DIFFUSION PUMP; ION PUMP; SORPTION PUMP.

vacuum ultraviolet. The part of the ultraviolet region of the spectrum, of wavelengths below about 200 nanometre, in which the radiation is absorbed by air and so experiments using such radiation are performed in an evacuated chamber.

valence band. The band of energies of the VALENCE ELECTRONS in a solid. *See* BAND THEORY.

valence bond. A single linkage between two atoms in a molecule. It is assumed to comprise a pair of electrons shared between the atoms.

valence electrons. Outer atomic shell electrons which participate in the chemical bonding when the atom forms compounds. *See also* BAND THEORY.

valve. A glass envelope, either evacuated or gas filled, containing two or more electrodes. One electrode is a primary source of electrons, usually provided by THERMIONIC EMISSION. The name arises from the fact that current flows in one direction only.

valve voltmeter. An amplifier incorporating VALVE components. The output current, which is proportional to the input voltage, is read on a meter. The instrument uses very little input current because of its high input impedance. It is suitable for either alternating current or direct current input. The valve voltmeter is being superseded by a voltmeter with DIGITAL DISPLAY.

Van Allen belts. Two regions within the Earth's MAGNETOSPHERE in which charged particles become trapped and oscillate to and fro as they spiral round magnetic field lines. The lower belt, situated from 1000 kilometre to 5000 kilometre above the equator, contains protons and electrons; the particles are captured from the SOLAR WIND or produced by collisions between upper atmosphere atoms and COSMIC RAYS. The upper belt, situated from 15 000 kilometre to 25 000 kilometre above the equator, contains mainly electrons from the solar wind.

Van de Graaff accelerator. An ACCELERATOR in which a VAN DE GRAAFF GENERATOR is used to provide a high-voltage source.

Van de Graaff generator. An electrostatic generator capable of producing potentials of millions of volt. As shown in fig. V1, needle points P apply up to 100 kilovolt potential from an external source to a continuous vertically moving insulated fabric belt. The belt transfers the charge to a large hollow metal sphere S, the collector points C removing it from the belt. The potential of the sphere continuously increases to a limit of some millions of volt which is determined by the leakage rate of the supporting insulators. In recent models a series of metal bands connected by insulating string replace the fabric belt. Two such generators in series will develop 30 million volt.

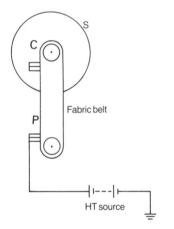

V1 Van de Graaff generator

Van der Waals equation. The EQUATION OF STATE for a mole of real gas:

$$(P + a/V^2)(V - b) = RT$$

where a and b are constants for a particular gas. The term a/V^2 allows for molecular attraction and b allows for the finite size of real molecules.

Van der Waals forces. Weak interatomic or intermolecular forces arising mainly from interaction between dipoles. Some of the DIPOLE forces are repulsive but the average force is attractive. For molecules the dipoles may be either permanent or induced. The fluctuating electronic distribution in atoms results in small instantaneous dipole moments in the atoms themselves. The potential energy of each interaction is given by $-A/r^6$ where A is a constant for a particular atom or molecule and r is the atomic or molecular separation.

Van't Hoff factor. Symbol i. The ratio of the total number of entities, i.e. ions, molecules etc., of solute in a solution to the number that would be present if the solute consisted of undissociated molecules. The factor occurs in the equation for OSMOTIC PRESSURE of electrolytes:

$$\Pi V = iRT$$

it thus represents the fact that the osmotic pressure Π depends on the number of entities present in solution.

vaporization coefficient. Symbol α. A coefficient defined by the equation

$$dm/dt = \alpha p [M/(2\pi RT)]^{1/2}$$

where dm/dt is the rate, per unit surface area, of vaporization, p the vapour pressure of the substance, M the RELATIVE MOLECULAR MASS of the vapour, R the universal gas constant and T the absolute temperature.

vapour. A gas at a temperature below the CRITICAL TEMPERATURE. It could therefore be liquefied by pressure alone.

vapour density. The density of a vapour under specified conditions, or sometimes the density relative to hydrogen under the same conditions. Vapour density measurement provides a method of determining RELATIVE MOLECULAR MASS since the latter is twice the vapour. *See also* SATURATED VAPOUR PRESSURE.

vapour pressure. The pressure exerted by a vapour. *See also* SATURATED VAPOUR PRESSURE.

vapour pressure thermometer. A thermometer which depends on the fact that the SATURATED VAPOUR PRESSURE of a liquid is a function of temperature only. Such thermometers are most reliable at temperatures below 5 K, which is the boiling point of helium.

variable. A quantity which can occur in a mathematical expression and can take a range of values.

variable capacitor. *See* CAPACITOR.

variable focus condenser. A modified form of ABBÉ CONDENSER in which the illuminated field area is increased.

variable resistor. *See* RHEOSTAT.

variable star. A star whose brightness varies with time. An example is a PULSATING STAR.

variac. A device for producing a variable alternating voltage from a fixed input voltage. It works on the same principle as the RHEOSTAT but has a heavy-duty toroidal winding on a core.

variance. The square of the standard DEVIATION of a statistic.

variation method. The most important general method of finding approximate solutions to SCHRÖDINGER'S EQUATION. The method is based on the proposition that an approximate energy, found by integrating the HAMILTONIAN function over any approximate wave function, is never less than the true ground state energy. The problem of solving a differential equation is thus replaced by that of finding a maximum or minimum value of an integral.

variometer. A variable INDUCTOR. It generally consists of two coils in series and arranged so that one may be rotated inside the other, thereby changing the self-inductance of the pair.

varistor. A resistor which does not obey OHM'S LAW. Usually the resistance decreases with voltage increase for both current directions.

vector. A quantity which has both magnitude and direction: displacement, velocity and acceleration are all examples of vectors. A vector can be represented pictorially by a line of length proportional to the magnitude of the quantity and drawn in its direction. The symbol of a vector can be represented in text in bold italic typeface or with a bar over it, i.e. as say a or $\bar{a}$. *See also* COMPONENT. *Compare* SCALAR.

vector addition. If vectors A and B are completely represented by two sides of a parallelogram drawn from one corner, then the vector $A + B$ is completely represented by the diagonal of that parallelogram drawn from the same corner. *See also* TRIANGLE LAW OF VECTOR ADDITION.

vector analysis. The mathematical treatment of physical problems using vector notation.

vector boson. Another name for VIRTUAL PARTICLE.

vector coupling coefficient. One of the coefficients expressing an eigenfunction of the sum of two angular momenta in terms of the products of eigenfunctions of these angular momenta.

vector field. A region of space for which a particular property – magnetic, electrostatic etc. – at any point is completely described by a VECTOR. Such a field can be mapped by lines whose density at any point (i.e. the number of lines per unit area crossing an infinitesimally small area perpendicular to the lines) is proportional to the magnitude of the vector at that point and whose direction is that of the vector.

vector meson. A MESON possessing the same spin as the photon. Annihilation of an electron and positron may directly produce a vector meson.

vector product. A vector whose magnitude is the product of the magnitudes of two vectors and the sine of the angle between their directions; the direction of the vector product is perpendicular to the plane of the two vectors. If *a* and *b* represent the vectors, the vector product is written *a* × *b* or *a* ∧ *b* and read as a cross b. The direction of *a* × *b* is such that a right-handed screw turning from *a* towards *b* moves in the direction of *a* × *b*; *b* × *a* would thus have the opposite direction. An example of a vector product is the force *F* on a particle charge *Q* moving at velocity *v* in a magnetic field *B*:

$$F = Qv \times B$$

vector resolution. *See* RESOLUTION (def. 1).

vector subtraction. A process accomplished for vectors *A* and *B*, say, by applying the VECTOR ADDITION process to the vectors *A* and −*B*; −*B* is a vector of the same magnitude but of opposite direction to *B*

velocity. Symbol *v*. The rate of increase of distance traversed by a body in a particular direction. Velocity is thus a VECTOR quantity. The velocity of a particle travelling in a curved path has the direction of the tangent to the path at the point considered.

velocity modulation. The production of pulses of charged particles from a continuous stream of them. It is accomplished by passing the stream into a sharply defined region such as a CAVITY RESONATOR and there subjecting it to a rapidly fluctuating alternating electric field. The field will either accelerate or retard the particles according to the nature of the field half cycle when they enter. For the particles immediately following, the field is reversed and so the effect on them is the opposite. The particles are thus bunched and proceed in pulses. Velocity modulation is used for the amplification and generation of MICROWAVE frequencies in ELECTRON TUBES such as the *klystron* and the *travelling wave tube.*

velocity potential. A scalar function of the position of a moving particle such that its gradient gives the particle velocity at any point. It has applications for example to fluid flow and to the movement of electric charge.

velocity ratio. *See* MACHINE.

velocity selector. A device in which an electric and a magnetic field act at right angles to each other. By varying the strengths of the fields it can be arranged that all charged particles, except those of a chosen velocity, can be deflected from the initial path of a stream of mixed velocity particles. *See also* MASS SPECTROMETER.

vena contracta. The section of minimum cross sectional area in a jet of fluid discharged from an orifice.

ventricle. Any small cavity in the human body.

Venturi meter. A flow meter for use in closed pipes. It is shown in fig. V2. It consists of a constriction, known as a *Venturi tube,* inserted in the line of piping, and some means for measuring the excess pressure *p* at inlet over that at throat. The inlet speed is then given by

$$(2p/[\rho(a_i/a_t)^2 - 1])^{1/2}$$

where a_i and a_t are the inlet and throat cross sectional areas respectively and ρ is the fluid density. *See also* CANNULA; INJECTOR; PITOT TUBE.

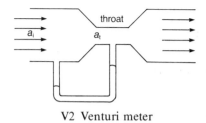

V2 Venturi meter

Venturi tube. *See* VENTURI METER.

Venus. The second closest planet to the Sun and the nearest planet to Earth. It is slightly smaller than Earth and orbits the Sun every 225 day. The planet has been extensively studied using space probes. Its atmosphere has been shown to extend to about 250 kilometre above the surface and to consist mainly of carbon dioxide. The surface temperature is about 470° C and the pressure about 90 times Earth's. Venus has no satellite and is observable as a very bright morning or evening star.

vernal equinox. *See* EQUINOX.

vernier. A short scale sliding on the main scale of a length- or angle-measuring instrument, and used to obtain a more accurate value of the measurement. The vernier scale is constructed as shown in fig. V3. Ten of its divisions occupy the same space as nine divisions on the main scale. The zero on the vernier scale is the instrument's pointer. When this falls between two main scale graduations, the next decimal place in the reading is given by the number of the graduation on the vernier that coincides with a graduation on the main scale.

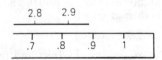

V3 Vernier scale

versed sine. One minus the cosine of the angle in question.

vertex. The point of intersection of two sides of a plane figure.

vertex focal length. The distance from the last surface of a lens system to the PRINCIPAL FOCUS.

vertex power. The reciprocal of the VERTEX FOCAL LENGTH.

very high frequency. A frequency in the range 30 megahertz to 300 megahertz.

very low frequency. A frequency in the range 3 kilohertz to 30 kilohertz.

vestibular canal. A passage ending in the COCHLEA of the ear.

VHF. Abbrev. for VERY HIGH FREQUENCY.

vibrating reed electrometer. An ELECTROMETER in which a low direct potential difference is converted to an alternating potential difference by connecting it to a capacitor, one of whose plates is mechanically vibrated. The alternating potential is then amplified and recorded.

vibration. Originally an elastic oscillation of a body. The word is now used synonymously with OSCILLATION. In *longitudinal vibration* the particles move to and fro along the line of travel of the wave. In *transverse vibration* the particles move to and fro in a direction perpendicular to the line of travel of the wave. *Resonant vibrations* occur when a periodic force of the same frequency as the natural frequency of a system is applied to the system. The amplitude of the vibrations may become very large (*see* RESONANCE).

vibrational energy. The sum of the KINETIC and POTENTIAL ENERGIES of an oscillating body. For undamped SIMPLE HARMONIC MOTION, the sum is constant and equal to $m\omega^2a^2/2$; m is the mass of the body, a its amplitude of swing (assumed small) and $2\pi/\omega$ its period. At maximum displacement all the energy is potential; for zero displacement it is all kinetic.

vibrational energy level. The value of the vibrational energy of an atom about its equilibrium position. According to QUANTUM THEORY only certain vibrational energy levels are allowed and changes in vibrational energy only occur from one allowed level to another. *See also* ENERGY LEVEL.

vibration-rotation spectrum. A spectrum resulting from a change of a molecule from a rotational energy level in one vibrational energy level to a rotational energy level in a different vibrational energy level. Such spectra occur in the near infrared region.

vibrations in pipes. The vibrations which may be set up of the air contained in pipes. Study is usually confined to straight tubes for which the length is large compared with the width. Two cases are considered: *open pipes*, meaning pipes open at both ends, and *closed pipes*, meaning pipes open at one end but closed at the other. By suitable excitation, such as blowing across the top of a pipe or sounding a tuning fork of suitable frequency just above the pipe, the air in a pipe can be caused to emit a loud sound. This results from the setting up of a longitudinal STANDING WAVE; there is an antinode just beyond an open end and a node at a closed end for the FUNDAMENTAL, as shown in fig. V4.

In a closed pipe of length l, allowing for END CORRECTION, the wavelengths λ, λ_1, λ_2, λ_3,.... of possible vibrations of the air are given by

$$l = \lambda/4 = 3\lambda_1/4 = 5\lambda_2/4 = 7\lambda_3/4 \text{ etc.}$$

For an open pipe of the same corrected length, the wavelengths λ', λ'_1 λ'_2, λ'_3,.... are given by

$$l = \lambda'/2 = \lambda_1' = 3\lambda_2'/2 = 2\lambda_3' \text{ etc.}$$

The first PARTIAL, i.e. second HARMONIC, is also shown for open and closed pipes.

Since the product of frequency and wavelength is constant, if f is the fundamental frequency for the closed pipe then the pipe's other possible vibration frequencies are $3f$, $5f$, $7f$ etc. In the case of the open pipe the fundamental frequency is $2f$ and its other possible vibration frequencies are $4f$, $6f$, $8f$ etc. Thus to obtain the same fundamental from open and closed pipes, the open pipe needs to be twice as long as the closed pipe. Even so, the sounds would be different because of the different partials for the two pipes. A given pipe can sound several harmonics simultaneously since the air column divides itself into sections, each of which vibrates as if it were a whole.

vibrations in plates. A source of sound, first investigated by Chladni. *See* CHLADNI'S FIGURES.

vibrations in rods. Vibrations which may be longitudinal, transverse or torsional; some of the analysis is complicated. Use is made of rod vibrations in KUNDT'S TUBE.

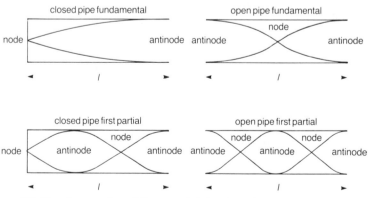

V4 Fundamental and first partial for closed and open pipes

vibrations in strings. The vibrations which may be produced in a stretched string. Study is usually concerned with the vibrations of a cord or wire stretched between two fixed points with a tension T sufficient for disturbance of the string to result in a sound. It is assumed that there is always zero disturbance at the fixed points.

Most attention has been given to the disturbance of a string by displacing it at its mid-point perpendicular to its length. This can result in a transverse STANDING WAVE in the string. For a string of length l the wavelengths $\lambda, \lambda_1, \lambda_2, \lambda_3,...$ of possible vibrations are given by

$$l = \lambda/2 = 3\lambda_1/2 = 5\lambda_2/2 = 7\lambda_3/2 \text{ etc}$$

The first two are illustrated in fig. V5. It can be shown that the frequency, f, of the FUNDAMENTAL is given by

$$(T/m)^{1/2}/(2l)$$

where m is the mass per unit length of the string. The string's other possible vibration frequencies are then $3f, 5f, 7f, ...$, since the product of wavelength and frequency is constant. Stopping the string at its mid-point and plucking at a point a quarter of the way along it from the support produces a node at the fixed point and a vibration of frequency $2f$; by plucking at a point an eighth of the way along a vibration of frequency $4f$ is produced; other possibilities also exist.

For a similar reason to that given in the entry VIBRATIONS IN PIPES, a string, like a pipe, can sound several harmonics simultaneously. Although the standing wave in the string is transverse, the string oscillations set the air in contact with the string into longitudinal VIBRATION: on reaching the ear these give rise to the sound sensation in the usual way. A longitudinal wave can also be set up in a string by stroking along the length of the string: a high-pitched sound results. A string can also be set into TORSIONAL VIBRATION by bowing hard perpendicular to its length.

videofrequency. A frequency in the range 10 hertz to 2 megahertz, used for transmitting TELEVISION signals.

vidicon. A type of television camera in which the image is projected on to a photoconducting mosaic coated on an insulating screen, which is scanned by a low-velocity electron beam; a positive potential is applied to the mosaic. The amount of charge on each mosaic element depends on the amount of illumination it receives. This charge is released by the electron beam, thus producing a current which depends on the amount of charge and therefore on the amount of incident illumination.

vignetting. The reduction in the cross sectional area of a beam of light passing through an optical system, due to obstruction by apertures, lens mounts etc.

virgin neutron. A neutron that has not experienced a collision.

virial coefficients. *See* VIRIAL EQUATION.

virial equation. An EQUATION OF STATE for real gases of the form

$$PV = RT + BP + CP^2 + DP^3 + ...$$

P and V are respectively the pressure and volume of a mole of gas at absolute temperature T, R is the UNIVERSAL GAS CONSTANT and $A, B, C, D, ...$ are empirical constants, known as the *virial coefficients* of the gas.

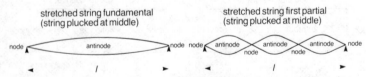

stretched string fundamental (string plucked at middle)

stretched string first partial (string plucked at middle)

V5 Fundamental and first partial for string plucked at mid-point

virtual cathode. The surface in the region of a space charge inside a THERMIONIC VALVE at which the electric force is zero. The virtual cathode acts as if it were the electron source.

virtual image. *See* IMAGE.

virtual object. *See* OBJECT.

virtual particle. A particle existing for a very short time and whose continued existence would violate the CONSERVATION LAW for mass and energy: very short term violation is allowable because of the HEISENBERG UNCERTAINTY PRINCIPLE. An example is the emission and absorption of virtual photons during an electromagnetic interaction between two charged particles.

virtual work. The work which would be done if a system subjected to a set of forces was given an infinitesimal displacement. For such a system in equilibrium, the virtual work is zero.

viscoelasticity. The phenomenon shown by a solid or liquid which can both store and dissipate energy during mechanical deformation.

viscometer. An instrument or apparatus for measuring fluid VISCOSITY. The main types are based on
(a) the flow of fluids through capillary tubes (*see* POISEUILLE'S EQUATION; OSTWALD VISCOMETER);
(b) the time of fall of a sphere through a liquid (*see* STOKES' LAW);
(c) the TORQUE required to keep two concentric cylinders rotating when the space between them is occupied by the fluid under test;
(d) the rate of DAMPING of a vibrating body by the fluid.

viscosity. The resistance of fluids to flow caused by intermolecular forces in the fluid. *Newton's law of viscosity* for STREAM LINE FLOW in a pipe is

$$F = \eta A \; dv/dx$$

F is the tangential force acting on an area A parallel to the fluid flow direction and dv/dx is the rate of change of speed of the fluid with perpendicular distance x of A from the containing pipe wall; η is a constant of the fluid known as the *coefficient of viscosity* or *dynamic viscosity*. *See also* ANOMALOUS VISCOSITY.

According to kinetic theory, the coefficient of viscosity of a gas is $\rho c L/3$; ρ is the gas density and c and L respectively the mean molecular speed and mean molecular free path. The coefficient should thus be independent of pressure; however at low gas pressure the coefficient is proportional to pressure, a property used in the design of low pressure manometers.

viscous. Having a high VISCOSITY.

viscous damping. DAMPING in which the opposing force is proportional to the speed.

viscous force. A frictional force in a fluid. For STREAM LINE FLOW the fluid can be considered to consist of parallel layers which move at different rates; the viscous force is the tangential force opposing the motion of the layer.

visible spectrum. Electromagnetic radiation in the wavelength range 380 nanometre to 780 nanometre. It is represented by points on the horseshoe part of the CHROMATICITY CHART boundary. This boundary radiation broadly appears to divide into seven colours: red, orange, yellow, green, blue, indigo and violet (memonic: Richard of York Gained Battles in Vain). The wavelength ranges in nanometre for these colours are respectively 780–620, 620–585, 585–575, 575–500, 500–445, 445–425, 425–380. In reality there is a continuous gradation of colour through the spectrum. The line joining the extremes of the horseshoe represents purples, i.e. not spectral colours.

visual acuity. The minimum angular separation of two points of light at which they can just be resolved by the eye. If the points subtend an angle smaller than this at the eye, the eye perceives them as a single point source.

visual axis. The direction in which the eye sees most clearly. It is inclined at an angle of 5° to 6° to the OPTICAL AXIS of the eye.

visual binary star. *See* BINARY STAR.

visual display unit. A device in which COMPUTER output is fed to a CATHODE RAY TUBE to produce figures, letters, diagrams etc. on a screen. It is normally used in conjunction with a keyboard for feeding information into the computer. A light pen may also be fitted.

visual purple. Another name for RHODOPSIN.

vitreous. Having the appearance or structure of glass.

vitreous humour. *See* EYE.

VLF. Abbrev. for VERY LOW FREQUENCY.

voice frequency. A frequency in the range 300 hertz to 3400 hertz. If communication systems reproduce frequencies in this range, speech is intelligibly rendered.

Voigt efffect. The DOUBLE REFRACTION produced by passing light through a vapour subjected to a magnetic field, the field being in a direction perpendicular to the direction of incidence of the light.

volt. Symbol V. The SI unit of electrostatic POTENIAL, of POTENTIAL DIFFERENCE and of ELECTROMOTIVE FORCE. It is defined as the potential difference between two points on a conductor carrying a steady current of one ampere, when the power dissipated between the points is one watt.

voltage. Symbol *V*. The potential difference or electromotive force, measured in VOLT.

voltage amplifier. *See* AMPLIFIER.

voltage between lines. The voltage between two lines of a single-phase electric power system, or between any two of the lines of a symmetrical three-phase system.

voltage divider. Another name for POTENTIAL DIVIDER.

voltage doubler. An arrangement of two rectifiers, as shown in fig. V6, in order to give double the voltage produced by a single rectifier.

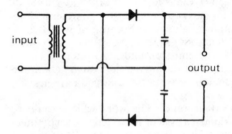

V6 Voltage doubler circuit

voltage drop. The POTENTIAL DIFFERENCE between two points of a circuit.

voltage multiplication. The ratio of the voltage developed across a reactive component in a circuit at RESONANCE to the impressed voltage.

voltage stabilizer. A device for maintaining a constant output voltage in spite of variations in input voltage and load current. A ZENER DIODE is mainly used.

voltage transformer. An instrument TRANSFORMER whose primary and secondary windings are connected to the main circuit and to the instrument respectively. Such transformers are used to extend the range of alternating current instruments and to isolate instruments from high voltage circuits.

voltaic cell. Another name for primary cell. *See* CELL.

voltameter. An electrolytic cell used to measure current by ELECTRODEPOSITION of a metal, usually copper or silver. If m is the mass of metal deposited at the cathode in time t, then the current is given by $m/(Zt)$ where Z is the ELECTROCHEMICAL EQUIVALENT of the metal. *See also* FARADAY'S LAWS OF ELECTROLYSIS.

Volta's pile. A pile of pairs of silver and zinc discs, each disc being separated from its neighbour by a brine-soaked cardboard disc. It is the forerunner of modern batteries.

voltmeter. A device for measuring VOLTAGE. One of the common forms is the permanent magnet MOVING COIL INSTRUMENT, which is suitable for direct voltages only. The DIGITAL DISPLAY voltmeter, CATHODE RAY OSCILLOSCOPE and ELECTROSTATIC VOLTMETER are also used. All voltmeters are required to take little current from the circuit and so a series resistance is used with the moving coil instrument; cathode ray oscilloscopes take no current from the circuit.

volume. A measure of the amount of space occupied by a body. The SI unit is the cubic metre. *See* Table 6B.

volume compressor. A device which automatically decreases the circuit amplification for a large-amplitude input signal and increases it for a small one. A device producing the opposite effect is known as a *volume expander*. Use of a compressor at the transmitter and an expander at the receiver improves the signal-to-noise ratio of a transmission system.

volume density of charge. Symbol ρ. The electric charge per unit volume of a medium or body; it is measured in coulomb per cubit metre.

volume expander. *See* VOLUME COMPRESSOR.

volume expansivity. *See* COEFFICIENT OF EXPANSION.

vortex. An intense spiral motion of a fluid in a limited region.

vulgar fraction. Another name for common fraction. *See* FRACTION.

W

wall effect. Any appreciable effect of the inside wall of a container or reaction vessel on the behaviour of the contents, for example a contribution to the current in an IONIZATION CHAMBER by electrons liberated from its walls.

wall energy. The energy per unit area of boundary between domains in a ferromagnetic material.

wandering star. *see* FIXED STAR.

warble tone. A sound whose frequency varies cyclically between two limits whose separation is small compared to the average frequency; a warble occurs several times a second.

water. A colourless, odourless, tasteless liquid which covers about three-quarters of the Earth's surface and occurs in all living matter and many minerals. It is used as a standard for several physical quantities (*see* for example CELSIUS SCALE; RELATIVE DENSITY). Water has some unusual properties: thus ICE is less dense than the liquid at 0° C, the respective values being 916.8 kilogramme per cubic metre and 999.8 kilogramme per cubic metre (*see also* ANOMALOUS EXPANSION OF WATER). Water is an excellent solvent for many compounds.

water equivalent. The mass of water that would have the same HEAT CAPACITY as a given object.

water mattress. A mattress used to reduce bed sores. It works because the body is evenly supported everywhere.

watt. Symbol W. The SI unit of POWER whether mechanical, thermal or electrical. *See* Table 2.

wattage. Symbol *W*. The POWER as measured in watt.

wattmeter. An instrument for measuring power directly in watt.

wave. A periodic disturbance in a medium or in space. *See* PROGRESSIVE WAVE; STANDING WAVE; TRANSVERSE WAVE; LONGITUDINAL WAVE.

wave equation. The partial differential equation

$$\nabla^2 U = (\partial^2 U/\partial t^2)/c^2$$

∇^2 being the LAPLACE OPERATOR, i.e.

$$\nabla^2 U = \partial^2 U/\partial x^2 + \partial^2 U/\partial y^2 + \partial^2 U/\partial z^2$$

where U is the displacement produced by the wave at time t and c is the wave speed. *See also* SCHRÖDINGER'S EQUATION.

wave form. The graph obtained by plotting displacement produced by a wave against time.

wave front. A surface which is the locus of points having the same phase in a wave motion.

wave function. Symbol ψ. A mathematical function appearing in the SCHRÖDINGER EQUATION and representing the amplitude of a wave associated with a particle. Its physical significance is that the square of the value of ψ at any point is proportional to the probability of finding the particle at that point. *See also* HEISENBERG UNCERTAINTY PRINCIPLE.

wave group. A travelling disturbance intermediate between a PULSE and a pure SINUSOIDAL WAVE.

waveguide. A DIELECTRIC inside either a hollow conducting tube or a dielectric tube of different permittivity. An ultrahigh frequency electromagnetic wave is reflected from the internal surfaces of the guide as it passes through it and thus energy dissipation is reduced.

wavelength. Symbol λ. The distance in a wave between adjacent points of the same phase, for example the distance between adjacent peaks in the same direction.

wavelength of light. The range of wavelengths which constitute the VISIBLE SPECTRUM; the limits may vary for individual observers.

wavelength of sound. The range of wavelengths from 0.016 metre to 16 metre in air; the limits may vary for individual observers.

wavelength response of the eye. *See* LUMINOSITY CURVES.

wave mechanics. One of the forms of QUANTUM MECHANICS arising from the theory that a particle can be regarded as a DE BROGLIE WAVE, so that any system of particles, for example electrons orbiting atomic nuclei, can be described by SCHRÖDINGER'S EQUATION. The basic problems in wave mechanics are thus those of formulating and solving Schrödinger's equation for the system under consideration. In general it is found that a particle can have only certain allowed energies (*see* CHARACTERISTIC VALUE) and associated wave functions (*see* CHARACTERISTIC FUNCTION); the latter give the probabilities of finding the particle at different points in space.

wave meter. An instrument for measuring the frequency of radio waves. It comprises a circuit which can be tuned to different frequencies, usually by means of a variable capacitor calibrated in frequency, and a device which indicates correct tuning.

wave number. Symbol σ. The number of single waves in unit distance, i.e. the reciprocal of wavelength.

wave packet. A pulse resulting from the superposition of waves of different wavelengths. Its amplitude is finite over only a limited region.

wave particle duality. A fundamental characteristic of all atomic species. Radiation acts sometimes like particles and sometimes like waves; similarly particles on occasion behave like radiation. *See also* DE BROGLIE WAVE.

wave power. The use of the energy transmitted by water waves. Various devices have been tried to enable the energy to be used in electric power generation. A great advantage is that it is a renewable energy source.

wave properties. Those common to every PROGRESSIVE WAVE are reflection, refraction, diffraction and interference. Polarization is confined to TRANSVERSE WAVES.

wave train. A succession of waves, especially a group of waves of limited duration.

wave trap. A tuned circuit incorporated in a radio circuit to reduce interference at a particular frequency.

W boson. Another name for W PARTICLE.

weak interaction. An interaction between elementary particles of magnitude about 10^{-12} times that of STRONG INTERACTION. The weak interaction force is considered to have a very short range. If both strong and weak interaction are permissible, only the strong will occur. Beta decay and the decay of some mesons and hyperons are examples of weak interaction. Weak interaction proceeds slowly on the nuclear time scale.

weber. Symbol Wb. The SI unit of magnetic flux, defined as that flux which, when linking a circuit of one turn, produces an electromotive force of 1 volt when, over a period of 1 second, the flux is reduced to zero at a uniform rate.

Weber-Fechner law. A general law of human sensation stating that the change in stimulus necessary to produce a just perceptible change in sensation is a constant fraction of the whole stimulus.

Wehnelt interrupter. An INTERRUPTER consisting of a large lead plate immersed in 30% sulphuric acid into which also dips the tip of a platinum wire. When this electrolytic cell is connected in series with the primary of an INDUCTION COIL, the platinum being anode, intermittent current results; make and break occurs at the platinum point. Very high interruption frequencies are possible.

weight. (1) Symbol W. The force exerted on a body due to the gravitational attraction of the Earth. Its direction is from the body to the Earth's centre and its magnitude is mg where m is the mass of the body and g is the magnitude of the ACCELERATION DUE TO GRAVITY.

(2) A standard mass.

weight thermometer. A small glass or silica bulb with a narrow tube attached, used to determine the COEFFICIENT OF EXPANSION of a liquid. The thermometer is filled with the liquid under investigation; the liquid temperature is then raised by a known amount t and the mass m of expelled liquid and mass m' of remaining liquid determined. The apparent coefficient of cubical expansion of the liquid is then given by $m/(m't)$.

Weiss magneton. The magnetic moment of a single molecule.

Weston cell. A cell constructed in a H-shaped glass vessel as shown in fig. W1. The cell has a very low temperature coefficient of electromotive force. Its voltage at $t°$ C is given by

$$1.018\,58 - 4.06 \times 10^{-5}(t - 20) -$$
$$9.5 \times 10^{-7}(t - 20)^2 + 10^{-8}(t - 20)^3$$

wet and dry bulb hygrometer. A type of hygrometer in which two mercury in glass thermometers are placed side by side, one having its bulb encased in muslin whose other end dips into water. The wet muslin

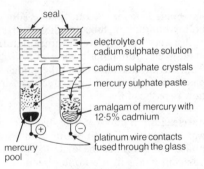

W1 Weston cell

is cooled by evaporation and so the thermometer whose bulb it surrounds indicates a lower temperature than the other thermometer. Hence using special tables, the relative HUMIDITY can be found from the difference in readings of the two thermometers and the dry thermometer reading.

wetting agent. A substance which causes a liquid to spread when in contact with a solid. A wetting agent acts by reducing SURFACE TENSION.

Wheatstone bridge. A network of resistors arranged as shown in fig. W2 and used for

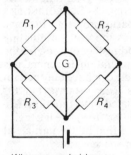

Wheatstone bridge

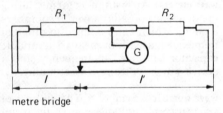

metre bridge

W2 Wheatstone bridge and metre bridge

the measurement of resistance. When there is no deflection of the galvanometer G, the bridge is said to be *balanced* and resistances R_1, R_2, R_3 and R_4 satisfy the relationship

$$R_1/R_2 = R_3/R_4$$

If three of the resistances are known then the fourth one can be found. Generally R_3 and R_4 are lengths of resistance wire of constant resistance per unit length so that $R_3 \propto l$ and $R_4 \propto l'$. G is connected to a contact which can slide over the wire. *See* METRE BRIDGE.

wheel and axle. A MACHINE as indicated in fig. W3. The axle is rigidly attached to the wheel. The effort is applied by a rope attached to the wheel rim and the load is carried by a rope attached to the axle. The mechanical advantage equals the ratio of the radius of the wheel to that of the axle.

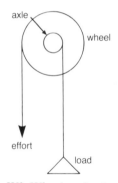

W3 Wheel and axle

whistler. A noise of descending pitch interfering with radio reception. It results from the reflection off the IONOSPHERE of radio waves produced by lightning flashes.

white dwarf. A very faint star of density in the range of 10^7 kilogramme per cubic metre to 10^{11} kilogramme per cubic metre and of mass less than about 1.4 times that of the Sun. Such stars are thought to represent a final stage of stellar evolution and to result from gravitational collapse following

the exhaustion of nuclear fuel. *See also* HERTZSPRUNG-RUSSELL DIAGRAM.

white light. Light which produces a colour sensation of white. Such a sensation is produced by radiation having an equi-energy spectrum, but can also result from radiation of many other spectral distributions, for example a mixture of red, green and blue monochromatic radiation. The definition of white light in terms of any particular spectral distribution is therefore to be deprecated.

white noise. *See* NOISE.

Wiedemann effects. Effects associated with the twisting of a rod produced by appropriate magnetic fields.

Wiedemann-Franz law. The ratio of the thermal and electric conductivities of all pure metals is approximately constant at a given temperature and is proportional to the thermodynamic temperature. Except for very low temperatures the law is fairly well obeyed.

Wien effect. The increase in conductivity of an electrolyte when subjected to a voltage gradient of around 2 megavolt per metre.

Wien displacement law. For BLACK BODY radiation, the product of the wavelength at which maximum emission occurs and the thermodynamic temperature is constant.

Wigner coefficient. Another name for VECTOR COUPLING COEFFICIENT.

Wigner effect. The change in the crystal structure of a solid due to irradiation. An example is neutron bombardment of graphite which produces a size change in the graphite by displacing the carbon atoms from their lattice positions.

Wigner force. A nonexchange force, acting over small distances, between the nucleons in an atom.

Wigner nuclides. An ISOBAR pair of nuclides. Each nuclide has an odd mass number and a neutron number differing by 1 from its proton number. An example is tritium and helium-3.

Wigner-Seitz method. A method of calculating the energy levels of electrons in solids, assuming that each electron occupies a cell in which its potential energy has spherical symmetry.

will o' the wisp. A pale flickering moving flame sometimes seen at night over marshy ground. It is caused by the combustion of methane produced by decaying vegetation.

Wilson cloud chamber. Another name for CLOUD CHAMBER.

Wilson effect. The DIELECTRIC POLARIZATION set up in an insulating material when moved through a magnetic field.

Wimshurst machine. An electrostatic generator consisting of two parallel insulating plates rotating in opposite directions about a common axis. Attached to each plate are radial conducting strips, shown in fig. W4 for one plate; these strips pass in turn past stationary brushes BB (two for each plate). Initially one radial strip is charged; then rotation is started and charges of opposite sign are induced on strips of the other plate as they pass. Charges on both plates are collected by pointed combs at PP where a large potential difference can be built up.

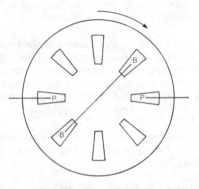

W4 Wimshurst machine

windage loss. The power loss in an electrical machine due to motion produced by the machine in the surrounding medium.

winding. A group of insulated conductors which is designed to produce or to be acted upon by a magnetic field.

wind instrument. A musical instrument in which sound is produced by the vibrations of an air column. *See* VIBRATIONS IN PIPES.

window. (1) The range of wavelengths over which a medium, opaque to electromagnetic radiation of most wavelengths, transmits the radiation. *See* ATMOSPHERIC WINDOWS.
(2) A sheet of material allowing the passage of electromagnetic or particle radiation, for example a thin sheet of mica through which radiation enters a *Geiger counter*.

wind power. The utilization of wind energy. Various types of windmill are being tried for converting wind energy to electric power. Like SOLAR HEATING and WAVE POWER it is a renewable source.

wind tunnel. A hollow tube through which passes a uniform flow of air. A scale model of the object under test, such as a car, aircraft or missile, is suspended in the tunnel. Deductions about the actual behaviour of the object can be made from that of the model.

winter solstice. *See* SOLSTICE.

wobbulator. A signal generator whose output frequency varies periodically through a definite range of values. It is used to investigate the frequency response of circuits.

Wollaston prism. A prism made by cementing together two triangular prisms of quartz or calcite, as shown in fig. W5. The optical axis of prism P_1 is parallel to AB; that of prism P_2 is perpendicular to the plane of the paper. The prism separates the ordinary and extraordinary components

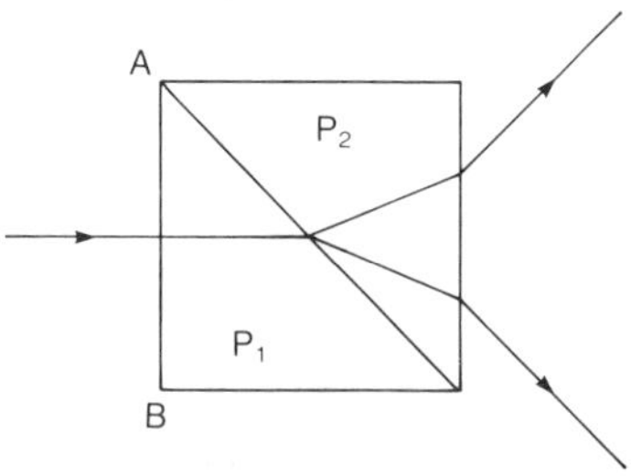

W5 Wollaston prism

of unpolarized light and thus yields plane polarized light. *See also* NICOL PRISM.

Wollaston wire. A thin platinum wire whose diameter has a lower limit of one micrometre. It is made by drawing out a silver-encased platinum wire of larger diameter and then dissolving away the silver with acid.

Wood's glass. A glass of high transmission factor for ultraviolet radiation, but of very low transmission factor for the visible spectrum.

woofer. A loudspeaker of large size for use at low audio frequencies.

word. A number of bits processed as a single unit in a COMPUTER.

work. Symbol W. The product of the magnitude of a force and the distance moved by the point of application of the force in the direction of the force. Work done on a system increases the energy of the system. Work done by a system decreases the system's energy. *See* Table 6G.

work done by couple. The product of the TORQUE due to the couple and the angle of rotation measured in radian.

work done in stretching wire. Half the product of the stretching force and the extension.

work-energy principle. For any displacement, the work done by the resultant force is equal to the increase in kinetic energy in the displacement.

work function. Symbol ϕ. The minimum energy required to remove an electron from a solid at the temperature of absolute zero. *See also* BAND THEORY; PHOTOELECTRIC EFFECT.

work hardening. Another name for STRAIN HARDENING.

W particle. Either of two massive particles, one positively charged the other negatively charged, which participate in WEAK INTERACTION.

wrench. A couple together with a force which acts along the axis of the couple. The moment of the couple divided by the magnitude of the force is the *pitch* of the wrench and has the dimension of length. Any system of forces can, in general, be reduced to a wrench.

write. To enter information into the store of a COMPUTER.

X

xenon. Symbol Xe. A gas obtained from the atmosphere as a byproduct in the liquefaction of air. It is used in fluorescent lamps, lasers and light bulbs.

xerography. An electrostatic method of document copying. After striking the document to be copied, ultraviolet radiation falls on an electrostatically charged plate, usually selenium coated. The loss of charge of any plate area increases with the amount of ultraviolet energy falling on it. A dark powder, consisting of thermosetting resin and graphite mixture and bearing a charge opposite to that of the plate, is then applied to the plate. Since like charges attract, the greater the charge on any plate area, the greater the amount of powder adhering there; hence a positive image of the document results. It is transferred from the plate to a charged paper on which it is fixed by heat treatment.

xi particle. Symbol Ξ. Any elementary particle of STRANGENESS −2. It may have zero charge or a charge equal to that of the electron. The mass is about 1530 mega-electron volt. A xi particle is a HYPERON.

X radiation. Electromagnetic radiation composed of X RAYS.

X ray absorption. A process involving usually either the PHOTOELECTRIC EFFECT or COMPTON EFFECT. For the former, an X ray photon is completely annihilated, part of its energy being used to free an electron from an atom or molecule and the remainder giving the freed electron kinetic energy. In the Compton effect, the X ray photon encounters an effectively free electron and loses part of its energy to it. The photoelectric effect is predominant for low energy photons and rapidly increases with atomic number. The Compton effect shows little dependence on photon energy and is only directly proportional to atomic number. In biological materials, the Compton effect predominates for photon energies above 0.3 mega-electronvolt. For photon energies greater than 1 mega-electronvolt, pair production becomes significant: the photon is annihilated, with the production of a positive and a negative electron.

X ray analysis. The study of the structure of crystalline substances by studying the X RAY DIFFRACTION patterns they produce. Either a collection of small crystals or a single large crystal may be used. In some equipment, the crystal may be rotated so that diffraction at successive sets of crystalline planes can be studied. Diffraction may be investigated by either transmission or reflection. *See* LAUE DIAGRAM.

X ray astronomy. The study of sources of X RAYS in space. It is necessary to mount the instrumentation in space probes since X rays are absorbed by the Earth's atmosphere. The observations show a continuous *X ray background* spectrum with numbers of superimposed discrete X ray lines. *See also* PULSAR; BLACK HOLE; X RAY SPECTRUM.

X ray background. *See* X RAY ASTRONOMY.

X ray camera. A camera used in X RAY DIFFRACTION studies to reproduce a photographic record of the diffraction patterns.

X ray crystallography. The studies of crystals by X RAY DIFFRACTION.

X ray diagnosis. A process performed using an X RAY TUBE operating at less than 150 kilovolt, since it is only for low energy

X rays that there is a marked increase of their absorption with atomic number. In order to enhance the absorption of the organ under investigation relative to that of surrounding tissues. highly absorbent material is introduced into the organ, for example by taking a barium meal.

X ray diffraction.   The diffraction of X rays, usually the diffraction produced by crystals. It occurs because the wavelengths of X rays and the interatomic distances in crystals are of the same order of magnitude. Moreover, the periodic crystal lattice acts like a diffraction grating. *See also* BRAGG'S LAW; ELECTRON DENSITY MAP.

X ray microscope.   *See* MICROSCOPE.

X ray reflection.   (1) The true reflection of X rays, which occurs for glancing angles of incidence.

(2) The apparent reflection of X rays at a crystal surface, or by parallel planes of atoms in a crystal, which arises from X RAY DIFFRACTION.

X rays.   Electromagnetic waves of wavelength in the range 10^{-7} metre to 10^{-10} metre. The rays blacken photographic plates and will produce some ionization in matter. *Hard X rays* are the most penetrating X rays, i.e. those of shorter wavelength. *soft X rays* are the least penetrating X rays, i.e. those of longer wavelength.

X ray spectrometer.   An instrument for measuring the variation of energy distribution of X RAYS with wavelength. A crystal or fine diffraction grating disperses the X rays and the energies of the resulting beams are measured by IONIZATION CHAMBER, COUNTER or PHOTOMULTIPLIER or by PHOTOGRAPHY.

X ray spectrum.   A combination of a continuous and a superimposed line spectrum. The sharp lines occur in groups characteristic of the transfer of outer electrons to each of the various inner shells, i.e. K. L, M etc. (*see* ATOMIC ORBITAL; MOSELEY'S LAW). The transfer may be instigated by electron, X ray or gamma ray bombardment. The continuum arises from the retardation of charged particles, such as electrons stopped by the target of an X RAY TUBE; there is therefore a sharp cut off at the shortest wavelength present, corresponding to the energy of the bombarding particles. A typical X ray tube spectrum is shown in fig. X1.

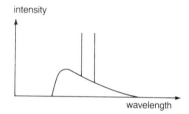

X1 X ray spectrum

X ray therapy.   The bombardment of tumours with X rays. For tumours on or near the skin an X RAY TUBE operating at not more than 250 kilovolt is used; for deep-seated tumours the required operating voltage is around 4 megavolt. Since all cells along the path of the beam are irradiated, the beam direction relative to the body is altered for successive treatments. but always in such a way that the tumour lies in the beam path; hence no cells except the tumour ones receive large dosages. Higher energy X rays penetrate soft tissue better and are less absorbed in bone than lower energy X rays.

X ray tube.   A thermionic tube, illustrated in fig. X2, which is used as a source of X rays. The glass envelope is evacuated as

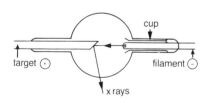

X2 X ray tube

fully as possible. Electrons emitted by the heated filament are accelerated towards the anode by a suitable potential difference, and are focussed towards the anode by a cup round the filament which is kept at a negative potential relative to the filament. The current through the tube is of the order of 20 milliampere. Less than 1% of the electron energy results in X ray production; the remainder appears as heat at the anode, which therefore requires cooling, usually by circulating water through it. The anode may be rotated so that the electron beam falls on different parts of it and thus spreads the heating.

X ray unit.   Symbol XU. A unit of length equal to $1.002\,02 \times 10^{-13}$ metre. It is used for expressing wavelengths of X rays and gamma rays.

x-y **recorder**.   A type of pen recorder with two inputs: one displaces the pen in the x direction and the other in the y direction, permitting a graph to be plotted of one varying electrical signal against another.

Y

Yagi aerial. An aerial used in RADIO ASTRONOMY and TELEVISION. It contains one radiating element and a number of parasitic ones.

yard. *See* Table 6A.

year. The TIME measure which is the basis of the calender. It is the time taken by the Earth to orbit the Sun and equals 365.25 mean solar day. (*See* SOLAR TIME). The time between successive arrivals of the Sun at the vernal EQUINOX is 356.242 mean solar day and is known as the *solar year* or *tropical year. See also* SIDEREAL TIME.

yellow spot. A small depression in the retina of the EYE where the sensitivity of the eye is greatest. The yellow spot contains only close-packed cones and so has the best colour discrimination and VISUAL ACUITY. *See* COLOUR VISION.

yield point. The point on the STRESS against STRAIN graph for a substance corresponding to a marked change in the internal structure of the substance due to slipping of crystal planes. As shown in fig. Y1, it is characterized by a large change in the slope of the graph from its constant value at low stresses. *See also* HOOKE'S LAW.

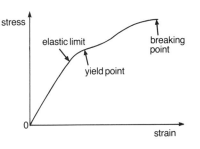

Y1 Stress v. strain plot

yield strain. The strain at the YIELD POINT.

yield stress. The stress at the YIELD POINT.

ylem. A hypothetical substance of density about 10^{16} kilogramme per cubic metre, consisting chiefly of neutrons. It may have been the ancestor of all nuclei.

yoke. A fixed piece of unwound ferromagnetic material which completes a magnetic circuit.

Young's fringes. FRINGES obtained using the apparatus illustrated in fig. Y2. Slit S is illuminated by monochromatic light of wavelength λ and so slits A and B act as coherent equal-intensity sources. Interference between the beams from these sources results in the formation of interference fringes as indicated. A bright fringe is observed at a point P if the pathlength difference of light rays from A and B is an integral number of wavelengths; if it is an odd number of half wavelengths destructive INTERFERENCE of the progressive waves occurs and the fringe is dark. It can easily be shown that the separation of like fringes is $\lambda l/(2d)$ where l and d are as indicated. If the fringe separation is measured then λ can be found.

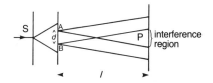

Y2 Production of Young's fringes

Young's modulus. The ratio of the tensile STRESS on a material to the resulting tensile STRAIN for stresses smaller than the ELASTIC LIMIT.

Yukawa particle. Another name for PION.

Yukawa potential. A potential used to explain the short-range STRONG INTERACTION occurring between nucleons in the atomic nucleus. The potential energy is assumed proportional to

$$(\exp(-\mu r))/r$$

where r is the nucleon separation and μ is a constant.

Z

Z boson. Another name for Z PARTICLE.

Zeeman effect. The splitting of spectral lines into components of slightly different wavelength when the light source is placed in a strong magnetic field. The observed state of polarization of the components depends on the direction, relative to the magnetic field, from which the source is viewed. Treating the electron as a classical oscillator leads to the prediction that there should be two lines, circulary polarized in opposite directions, for each original line when viewing the source in the direction of the magnetic field. For viewing at right angles to the magnetic field three plane polarized lines are predicted: two symmetrically placed on either side of the third, which is expected to occupy the undisplaced position. Moreover, the magnitude of the splitting is predicted to be proportional to the magnetic field strength and could therefore be used to estimate the magnetic fields associated with stars and sunspots. Sodium light is found to behave as predicted.

The anomalous Zeeman effect is in fact the effect exhibited by most radiation. The number of components and the shifts are different from the ones predicted by the Zeeman effect but are explicable if electron SPIN is taken into account.

Zener breakdown. A type of breakdown found in a reverse biased P-N JUNCTION, particularly if the DOPING LEVEL is very high. Such a junction has a high built-in potential across it and the DEPLETION LAYER is narrow, so that electrons may be excited directly from the valence band to the conduction band. Breakdown is accompanied by a very sharp current increase.

Zener diode. A semiconductor diode for which a sudden increase in reverse current occurs when the reverse voltage reaches a certain value, usually less than 6 volt. Such diodes are used for regulating the voltage in a circuit.

zenith. The point on the CELESTRIAL SPHERE which lies vertically above an observer.

zero point energy. The energy of vibration of an atom at the temperature of ABSOLUTE ZERO. Its value is $h\nu/2$ where h is the PLANCK CONSTANT and ν the frequency of vibration.

zero potential energy. In theoretical calculations this is taken to occur at infinity. In contrast for most practical work the surface of the Earth is considered to be at zero potential energy.

ZETA. An experimental FUSION REACTOR. The name is an acronym for zero energy thermonuclear apparatus.

Zeta potential. Symbol ζ. The gradual part of the potential change at a liquid-solid boundary. Its value is

$$4\pi\eta\nu/(\varepsilon_r E)$$

where ν is the liquid speed, η and ε_r respectively the liquid viscosity and relative permittivity and E is the electric field.

z modulation. The variation of the brilliance of the spot on the screen of a CATHODE RAY TUBE in accordance with the magnitude of a signal.

zodiac.   A band on the CELESTIAL SPHERE containing the apparent annual path of the Sun and the orbits of the Moon and planets. It extends about 9° on either side of the ECLIPTIC and is divided into 12 equal parts named after constellations.

zodiacal light.   Sunlight which reaches Earth after being scattered by small meteoric particles around the Sun. It is seen in the west just after sunset and in the east just prior to sunrise.

zone.   Crystal faces which intersect in parallel edges; the edge direction is the zone axis.

zone of silence.   A localized region where sound or radio waves from a given source do not penetrate. The waves are either reflected or refracted away from the region and may be received at some distance from the source, beyond the zone.

zone plate.   A plate, illustrated in fig. Z1, with annular concentric zones which are alternately transparent and opaque and have radii such that the zone areas are approximately equal. For some particular observation point, the distance of the periphery of the zone from the point increases by half a wavelength from zone to zone; the contributions from each transparent zone are hence in phase at the point and so the intensity of illumination there is much greater than in the absence of the zone plate. Thus in a sense the plate acts like a lens.

Z1 Zone plate

zone refining.   A technique for purifying materials. Its main use is in semiconductor manufacture. A sample in bar form is slowly moved past a heater so that a molten zone passes along the length of the bar. Impurities concentrate in the melt and so are transferred to one end of the bar. By using several passes, impurity levels as low as one part in 10^{10} can be achieved.

zoom lens.   An adjustable compound lens, mainly used in cinematic and television cameras, whose magnification is continuously variable without loss of image sharpness.

Z particle.   A massive electrically neutral particle which takes part in WEAK INTERACTION.

Further reading suggestions:

Undergraduate level:

University Physics, 6th edition. F. W. Sears, M. W. Zemansky and H. D. Young, 1982 London: Addison-Wesley.

To Acknowledge the Wonder: The Story of Fundamental Physics. E. J. Squires, 1985 Bristol: Adam Hilger.

Pre-undergraduate level:

Advanced Level Physics, 5th edition. M. Nelkon and P. Parker, 1982 London: Heinemann Educational.

Ordinary Level Physics, 4th edition. A. F. Abbott, 1984 London: Heinemann Educational.